Advances in Solar Research at Eclipses from Ground and from Space

NATO Science Series

A Series presenting the results of activities sponsored by the NATO Science Committee. The Series is published by IOS Press and Kluwer Academic Publishers, in conjunction with the NATO Scientific Affairs Division.

A. Life Sciences	IOS Press
B. Physics	Kluwer Academic Publishers
C. Mathematical and Physical Sciences	Kluwer Academic Publishers
D. Behavioural and Social Sciences	Kluwer Academic Publishers
E. Applied Sciences	Kluwer Academic Publishers
F. Computer and Systems Sciences	IOS Press
1. Disarmament Technologies	Kluwer Academic Publishers
2. Environmental Security	Kluwer Academic Publishers
3. High Technology	Kluwer Academic Publishers
4. Science and Technology Policy	IOS Press
5. Computer Networking	IOS Press

NATO-PCO-DATA BASE

The NATO Science Series continues the series of books published formerly in the NATO ASI Series. An electronic index to the NATO ASI Series provides full bibliographical references (with keywords and/or abstracts) to more than 50000 contributions from international scientists published in all sections of the NATO ASI Series.
Access to the NATO-PCO-DATA BASE is possible via CD-ROM "NATO-PCO-DATA BASE" with user-friendly retrieval software in English, French and German (WTV GmbH and DATAWARE Technologies Inc. 1989).

The CD-ROM of the NATO ASI Series can be ordered from: PCO, Overijse, Belgium

Series C: Mathematical and Physical Sciences – Vol. 558

Advances in Solar Research at Eclipses from Ground and from Space

edited by

Jean-Paul Zahn
Observatoire de Paris,
Meudon, France

and

Magda Stavinschi
Astronomical Institute of the Romanian Academy,
Bucharest, Romania

Kluwer Academic Publishers

Dordrecht / Boston / London

Published in cooperation with NATO Scientific Affairs Division

Proceedings of the NATO Advanced Study Institute on
Advances in Solar Research at Eclipses from Ground and from Space
Bucharest, Romania
9–20 August, 1999

A C.I.P. Catalogue record for this book is available from the Library of Congress.

ISBN 0-7923-6623-9

Published by Kluwer Academic Publishers,
P.O. Box 17, 3300 AA Dordrecht, The Netherlands.

Sold and distributed in North, Central and South America
by Kluwer Academic Publishers,
101 Philip Drive, Norwell, MA 02061, U.S.A.

In all other countries, sold and distributed
by Kluwer Academic Publishers,
P.O. Box 322, 3300 AH Dordrecht, The Netherlands.

Printed on acid-free paper

Printed in the Netherlands.

TABLE OF CONTENTS

PREFACE

The aim of this Advanced Study Institute was to give an account on the most recent results obtained in solar research. Bucharest was chosen to host it, because the capital city of Romania was located right in the middle of the totality path of the last eclipse of the millennium, on 11th August 1999; furthermore the phenomenon was close to reach there its longest duration: 2m 23s. Such a total eclipse is not only a very spectacular event which draws the crowds: to astronomers, solar eclipses still offer the best conditions for observing the lower part of the corona.

The Sun plays a crucial role in our very existence. It was responsible for the formation of the Earth, and rendered this planet fit to host living beings, providing the right amount of heat, and this for a long enough span of time. Quite understandably, it has always been a prime target of human curiosity, and more recently one of scientific investigation. During the last century, it was realized that the Sun is a star like billions of others; we learned since that it draws its energy from the nuclear fusion of hydrogen, and we are now able to estimate its age and life expectancy.

Progress has been extremely fast in the last years, especially thanks to the observations carried out in space, which have brought a wealth of new data. Two missions have been particularly successful: the Yohkoh spacecraft launched by Japan and the Solar and Heliospheric Observatory (SOHO) which was built in collaboration between Europe and the US. Whereas only the surface layers could be scrutinized in the past, we are now able to probe the interior of the Sun with the acoustic waves which are observed at the surface, and we can explore the corona in the X-rays and in the far ultraviolet. Ground based research also is gaining from the increased sensitivity of modern detectors, and global observation networks monitor the Sun in permanence.

The results are spreading through journals and specialized meetings, but we felt the need of presenting them in a more organized and pedagogical way, for the benefit of the younger scientists. This ASI proved quite attractive; it was attended by 76 participants from 18 different countries: 61 students and 15 lecturers.

During the first days of the ASI the participants prepared for observing the eclipse, and learned what can be gleaned from such events. They watched the eclipse from the Bucharest Observatory, where the Sun kept playing hide and seek behind some light clouds. Many distinguished guests shared their enthusiasm, such as the President of Romania Emil Constantinescu and NASA Administrator Daniel S. Goldin.

Thereafter the series of lectures covered the domains of solar physics where recent progress has been achieved, with emphasis put on the physical interpretation of the phenomena, rather than on their detailed description. The posters presented by the students are published in the supplement of *Romanian Astronomical Journal*, vol. 9, 1999. Highlights of the social programme were an excursion to Sinaia, a visit to the Village Museum in Bucharest, a barbecue in the gardens of the Bucharest Observatory, and a memorable football game among participants, which ended in the dark, long after sunset. Friendly links were established between participants of different countries. Even better: Nilay and Diaa met there, and are now married.

This meeting and these lectures were made possible by NATO, which sponsored them in the frame of its outstanding programme of Advanced Study Institutes. The organisers express their thanks to the directors of the Scientific Affairs Division, Drs. L. Veiga da Cunha and F. Predrazzini, for their kind assistance. They are grateful also to their home institutions, the Astronomical Institute of the Romanian Academy and the Observatoire de Paris, for generously granting both technical and financial support. Finally they wish to renew their warm thanks to the colleagues who contributed to the success of this ASI by their exciting lectures, and in particular to Petr Heinzel, Zadig Mouradian and Jay Pasachoff, who assisted them in the Organising Committee.

Magda Stavinschi and Jean-Paul Zahn

1	Victor PECHSHEROV	16	Pablo PEREZ	31	Gennady MARCHENKO	46	Ulyana LEIKO
2	Igor VESELOVSKY	17	Anatoly ARSENTIEV	32	Alexander RODIN	47	Lev KUROCHKA
3	Ural BAYAZITOV	18	Ines PEREZ	33	Susanna TOKHCHUKOVA	48	Vasyl DANYLEVSKY
4	Rafik SALAKHUTDINOV	19	Maxim BELOUS	34	Vladislav-V. PUSTINSKI	49	Shahinaz YOUSEF
5	Diaa GAD EL-MAWLA	20	Faruk SOYDUGAN	35	Artur BORODIN	50	Esha BEBARS
6	Paul BLAGA	21	Ahmet ERDEM	36	Petr BIRZA	51	Zadig MOURADIAN
7	Nilay YILMAZ	22	Andrei ZHUKOV	37	Irina PANAMARCHUK	52	Yutaka UCHIDA
8	Cristina Olivia BLAGA	23	Caner CICEK	38	Cosmina DUTAN	53	Klaus GALSGAARD
9	Syuzo ISOBE	24	Kostadinka KOLEVA	39	Dana KOVALEVA	54	Magda STAVINSCHI
10	Alexey ROSAEV	25	Ibrahim BULUT	40	Barbara CADER-SROKA	55	Jean-Paul ZAHN
11	Momtchil DETCHEV	26	Vladimir GARAIMOV	41	Jana KASPAROVA	56	Suzy COLLIN-ZAHN
12	Alicja HEINZEL	27	Guillaume BLANC	42	Almaz GALEEV	57	Pierre MEIN
13	Ana DZIEMBOWSKI	28	Lionel BIGOT	43	Subhon IBADOV	58	Pierre CUGNON
14	Sergey YAZEV	29	Michal VARADY	44	Elena MALANUSHENKO	59	Wojciech DZIEMBOWSKI
15	Mykola PISHKALO	30	Mohamed A. SEMEIDA	45	Edward BARANOVSKY	60	Petr HEINZEL

LIST OF PARTICIPANTS

Lecturers

Suzy COLLIN
Observatoire de Paris, Section de Meudon, F-92195 Meudon, France

Pierre CUGNON
Observatoire Royal de Belgique, Ringlaan 3, B-1180 Brussels, Belgium

Wojciech DZIEMBOWSKI
Copernicus Astronomical Centre, Polish Academy of Sciences,
Ul Bartycka18, PL-00716 Warsaw, Poland

Jean-René GABRYL
Observatoire Royal de Belgique, Ringlaan 3, B-1180 Brussels, Belgium

Klaus GALSGAARD
Mathematical and Computational Sciences, University of Saint Andrew,
Saint Andrew KY16 9SS, UK

Petr HEINZEL
Astronomical Institute, Ondrejov Observatory, CZ-25165 Ondrejov,
Czech Republic

Iraida KIM
Sternberg Institute, Moscow State University, Universitetskii prospect 13,
RU-119899 Moscow, Russia

Andre MANGENEY
Observatoire de Paris, Section de Meudon, F-92195 Meudon, France

Georgeta MARIŞ
Astronomical Institute of the Romanian Academy, Str. Cuţitul de Argint 5,
RO-75212 Bucharest, Romania

Pierre MEIN
Observatoire de Paris, Section de Meudon, F-92195 Meudon, France

Zadig MOURADIAN
Observatoire de Paris, Section de Meudon, F-92195 Meudon, France

Jay PASACHOFF
Hopkins Observatory, Williams College, Williamstown, MA 01267, USA

Henk SPRUIT
Max-Planck Institut für Astrophysik, Karl Schwarzschildstrasse 1,
D-85740 Garching, Germany

Magda STAVINSCHI
Astronomical Institute of the Romanian Academy, Str. Cuţitul de Argint 5,
RO-75212 Bucharest, Romania

Yutaka UCHIDA
Astronomy Depart., University of Tokyo, Bunkyo Ku, Tokyo 113, Japan

Jean-Paul ZAHN
Observatoire de Paris, Section de Meudon, F-92195 Meudon, France

Students

Anatoly ARSENTIEV
Institute of Solar-Terrestrial Physics, Siberian Division, Russian Academy of Sciences, RU-664033 Irkutsk, Russia

Edward BARANOVSKY
Crimean Astronomical Observatory, UKR-334413 Nauchny, Ukraine

Ural BAYAZITOV
Bashkir State University, New Information Technologies Department,
RU-450076, Bashkortostan, Ufa-76, a/b 7507, Russia

Esha BEBARS
National Research Institute of Astronomy and Geophysics, Solar Research Department, Helwan, Cairo, Egypt

Alexandre BELINSKI
Sternberg Astronomical Institute, Moscow State University, Universitetskii prospect 13, RU-119899 Moscow, Russia

Maxim BELOUS
Institute of Astronomy of the Russian Academy of Sciences, 48 Pyatnitskaya st., RU-109017 Moscow, Russia

Lionel BIGOT
Observatoire de la Côte d'Azur, BP 4229, F-06304 Nice Cedex 4, France

Petr BÎRZA
Astronomical Institute of the Romanian Academy, Str. Cuţitul de Argint 5,
RO-75212 Bucharest, Romania

Cristina Olivia BLAGA
"Babeş-Bolyai" University, Faculty of Mathem. and Computer Sciences,
1, Kogalniceanu str., RO-3400 Cluj-Napoca, Romania

Paul BLAGA
"Babeş-Bolyai" University, Faculty of Mathem. and Computer Sciences, 1, Kogalniceanu str., RO-3400 Cluj-Napoca, Romania

Guillaume BLANC
Service de Physique des Particules, DAPNIA/CEA Saclay, F-91191 Gif-sur-Yvette, France

Artur BORODIN
Institute of Solar-Terrestrial Physics, Siberian Division, Russian Academy of Sciences, RU-664033 Irkutsk, Russia

Ibrahim BULUT
Canakkale 18 Mart University, Art and Science Faculty, Physics Department, TR-17100 Canakkale, Turkey

Barbara CADER-SROKA
Astronomical Institute, Wrocklaw University, Ul Kopernika 11, PL-51140 Wroclaw, Poland

Caner CICEK
Astrophysics Section, Department of Physics, Canakkale 18 Mart University, TR-17100 Canakkale, Turkey

Mihail CODRESCU
Cooperative Institute for Research in Environmental Sciences, University of Colorado, 325 Broadway, Boulder CO 80303, USA

Vasyl DANYLEVSKY
Astronomical Observatory of Kiev University, Observatorna str. 3, UKR-254053 Kiev, Ukraine

Giulio DEL ZANNA
Centre for Astrophysics, University of Central Lancashire, 69 Selborne St., Preston PR1 4LA, UK

Momtchil DETCHEV
Institute of Astronomy, 72, Trakia Blvd, BG-1784 Sofia, Bulgaria

Alina DONEA
Astronomical Institute of the Romanian Academy, Str. Cuţitul de Argint 5, RO-75212 Bucharest, Romania

Cristiana DUMITRACHE
Astronomical Institute of the Romanian Academy, Str. Cuţitul de Argint 5, RO-75212 Bucharest, Romania

Cosmina DUTAN
Observatoire de Paris, Section de Meudon, F-92195 Meudon, France

Ahmet ERDEM
Canakkale 18 Mart University, Art and Science Faculty, Physics Department TR-17100 Canakkale, Turkey

Ladislau FARKAS
Astronomical Institute of the Romanian Academy, Piaţa Axente Sever 1, RO-1900 Timişoara, Romania

Diaa GAD EL-MAWLA
Institut für Theoretische Astrophysik , Universität Heidelberg, Tiergartenstrasse 15, D-69121 Heidelberg, Germany

Almaz GALEEV
Depart. of Astronomy, Kazan State University, RU-420008 Kazan, Russia

Vladimir GARAIMOV
Special Astrophysical Observatory, St.Petersburg branch of SAO, Pulkovskoe Shosse 65, RU-196140 St. Petersburg, Russia

Dan GRECU
Institute of Physics, Str. Atomiştilor 111, RO-76900 Platforma Magurele, Bucharest, Romania

Subhon IBADOV
Institute of Astrophyics, Tajik Academy of Sciences, 734042 Dushabe, Tajikistan

Jana KASPAROVA
Astronomical Institute, Ondrejov Observatory, CZ-25165 Ondrejov, Czech Republic

Kostadinka KOLEVA
Institute of Astronomy, 72, Trakia Blvd, BG-1784 Sofia, Bulgaria

Dana KOVALEVA
Institute of Astronomy of the Russian Academy of Sciences, 48 Pyatnitskaya st., RU-109017 Moscow, Russia

Lev KUROCHKA
Astronomical Observatory of Kiev University, Observatorna str. 3, UKR-254053 Kiev, Ukraine

Ulyana LEIKO
Astronomical Observatory of Kiev University, Observatorna str. 3, UKR-254053 Kiev, Ukraine

Elena MALANUSHENKO
Crimean Astronomical Observatory, UKR-334413 Nauchny, Ukraine

Gennady MARCHENKO

Astronomical Observatory, Kharkov State University, str. Sumskaya 35, UKR-310022 Kharkov, Ukraine

Elena MOISE
National Optical Astronomy Observatories, 950 N. Cherry Avenue, PO Box 26732, Tucson AZ 85726-6732, USA

Andrea MUNTEANU
Faculty of Physics, University of Bucharest, 5 Pârvan str., RO-75212 Bucharest, Romania

Irina PANAMARCHUK
Crimean Astronomical Observatory, UKR-334413 Nauchny, Ukraine

Victor PECHSHEROV
Institute of Solar-Terrestrial Physics, Siberian Division, Russian Academy of Sciences, RU-664033 Irkutsk, Russia

Marcos PENALOZA-MURILLO
University of Essex, Department of Biological and Chemical Sciences Central Campus, Wivenhoe Park, Colchester, Essex CO4 3SQ, UK

Ines PEREZ
Avenida Lucas de Vega 5, sobreatico, SP-38208 La Laguna, s/c de Tenerife, Spain

Pablo PEREZ
Departamento de Astrofisica y Fisica del Aire, Universidad Complutense de Madrid, Av.Complutense s/n, SP-28040 Madrid, Spain

Mykola PISHKALO
Astronomical Observatory of Kiev University, Observatorna str. 3, UKR-254053 Kiev, Ukraine

Miruna POPESCU
Astronomical Institute of the Romanian Academy, Str. Cuţitul de Argint 5, RO-75212 Bucharest, Romania

Vladislav-Veniamin PUSTINSKI
Tallinna Observatory, Tallinn, 13513, Estonia

Alexander RODIN
Pushkino Radio Astronomy Observatory, Lebedev Physics Institute of the Russian Academy of Sciences, RU-142292 Pushkino Moscow, Russia

Alexey ROSAEV
PB Drilling, Svoboda str.8/38, RU-150000 Yaroslavl, Russia

Ilia ROUSSEV
Armagh Observatory, College Hill, Armagh BT61 9DG, UK

Rafik SALAKHUTDINOV
Institute of Solar-Terrestrial Physics, Siberian Division, Russian Academy of Sciences, RU-664033 Irkutsk, Russia

Mohamed Ahmed SEMEIDA
National Research Institute of Astronomy and Geophysics, Solar Research Department, Helwan, Cairo, Egypt

Faruk SOYDUGAN
Canakkale 18 Mart University, Art and Science Faculty, Physics Department, TR-17100 Canakkale, Turkey

Susanna TOKHCHUKOVA
Special Astrophysical Observatory, Russian Academy of Sciences, RATAN-600, RU-357140 Zelenchukskaja, Karachai-Cherkessia, Russia

Vlad TURCU
Astronomical Institute of the Romanian Academy, Str. Cireşilor 19, RO-3400 Cluj-Napoca, Romania

Michal VARADY
Astronomical Institute, Ondrejov Observatory, CZ-25165 Ondrejov, Czech Republic

Igor VESELOVSKY
Institute of Nuclear Physics, Moscow State University, RU-119899 Moscow, Russia

Sergey YAZEV
Institute of Solar-Terrestrial Physics, Siberian Division, Russian Academy of Sciences, RU-664033 Irkutsk, Russia

Nilay YILMAZ
Ege University Solar Energy Institute, Bornov, TR-35100 Izmir, Turkey

Shahinaz YOUSEF
National Research Institute of Astronomy and Geophysics, Solar Research Department, Helwan, Cairo, Egypt

Luca ZANGRILLI
Department of Astronomy, University of Florence, Largo Enrico Fermi 5, I-30125 Florence, Italy

Andrei ZHUKOV
Department of Radiation and Computational Methods, Institute of Nuclear Physics, Moscow State University, RU-119899, Russia

ECLIPSES FROM THE STANDPOINT OF CELESTIAL MECHANICS

MAGDA STAVINSCHI
Astronomical Institute of the Romanian Academy
Str. Cuţitul de Argint 5, Bucharest, RO-75212, Romania

1. The 11 August 1999 TSE Had its Maximum in Romania

On August 11, 1999, the last total solar eclipse of this millennium occurred. Its maximum was in Romania. Here were:

- the maximum duration: 2 m 23 s (at Râmnicu Vâlcea);
- the maximum height of the Sun: 59°;
- the maximum coverage of the Sun: 103%;
- the only European Capital situated exactly on the central line of the totality band – Bucharest;
- the only professional astronomical observatory lying on this line: Bucharest Observatory of the Astronomical Institute of the Romanian Academy;
- a second one, in Timişoara, also situated in the totality band, belongs to the same Institute;
- maybe the best sky for the eclipse observation, especially at its maximum, at Râmnicu Vâlcea. Obviously, there were some clouds but not everywhere.

The '99 total solar eclipse was an exceptional event, as:

- it was the last TSE visible from Europe for the next decades;
- it was the last TSE of this millenium;
- the year when it took place (the last one before the "end of the world") had became an inexhaustible source of the most absurd interpretations. We are speaking about the year 2000, and this one recalls that the Christian Era was established in the 6th century by Dionysius Exiguus, born at Tomis (now Constanţa), in Romania;
- the totality band practically bisected Europe, from NW to SE (the most populated zones);
- it crossed nine European countries, namely England, France, Benelux, Germany, Austria, Hungary, Yugoslavia, Romania and Bulgaria (unfortunately, most of them were under clouds);
- it took place in midsummer, at suitable hours;
- it also took place in full tourist season, so a number without precedent moved to the band of totality;
- it was close to the maximum of solar activity, so it offered the possibility to observe very interesting solar phenomena;
- it was the most spread eclipse by mass-media, including the transmission live of the images of the lunar shadow on the Earth, obtained by the space mission MIR.

J.-P. Zahn and M. Stavinschi (eds.),
Advances in Solar Research at Eclipses from Ground and from Space, 1–22.

2. Where Are from the Differences between the Predicted Durations of the TSE?

The maximum duration of the totality phase was appreciated very differently:

- 2 m 23.0 s (by NASA);
- 2 m 26.8 s (by Bureau des Longitudes, Paris);
- 2 m 22.8 s (by the Belgian astronomer J. Meeus).

Is there an error in these computations (especially today, when the powerful computers are able to fix the characteristic instants of an eclipse with a very high accuracy)? In which way the scientists of the antiquity have succeeded in forecasting the future eclipses, or even the periods of their come back? There are some questions of small importance for those who deal with solar physics, but very important for celestial mechanics.

It is an obvious fact that, to answer these questions, the motions of the three bodies essentially implicated into an eclipse – the Sun, the Earth, and the Moon – had to be very well known.

The man first knew these motions only as a consequence of observation: the day-night alternation, the succession of seasons, the repetition of eclipses. Then, the 19th century brought, as a natural outcome of the advances of celestial mechanics, a much better knowledge of the mechanism that generates the eclipses. In fact, this mechanism is nothing but a particular case of the three-body problem. Yes, this is just a particular case, but is the most spectacular from the standpoint of the ground-based astronomy.

Let us try to have an overview of the motions of the three bodies that take part in this spectacular phenomenon which is the eclipse, especially that of the Moon (the most complicated) and Earth (the most interesting for us).

3. Moon's Motion

The Moon is too close to us to have a simple motion. The perturbation from the Earth makes a correct knowledge of its motion very difficult, even now, when it was conquered by the man.

3.1. MOON'S PERIODS

To help those who try to understand the mechanism of the lunar motion, implicitly that of the eclipses, we must recall some notions.

The mean interval between two successive Sun-Moon conjunctions (New Moon) is called Moon's synodic period, or lunation. Its duration can be determined knowing the epochs of eclipses separated by long time intervals. The value accepted today is:

29 d 12 h 44 m 02.8 s.

The necessary time for the Moon's mean longitude to perform an increase of 360° with respect to a fixed equinox is called Moon's sidereal period. It is:

27 d 07 h 43 m 11.5 s.

The anomalistic period is the interval elapsed between two consecutive transits of the Moon at perigee; it has the value:

27 d 13 h 18 m 33.1 s.

The nodal period is the time interval between two successive transits of the Moon through the ascending (or descending) node of its orbit; it is:

27 d 05 h 05 m 35.8 s.

The Moon's orbit can be regarded, in a second approximation (the first being the Keplerian one), as an ellipse whose focal axis rotates slowly in the plane of motion. This rotation (apsidal motion) is mainly due to the perturbing attraction of the Sun. As a consequence, the lunar perigee advances slowly around the orbit, completing one circuit in 8.85 years. This makes the anomalistic month be about 5.5 hours longer than the sidereal month.

For a terrestrial observer, the Moon moves with respect to the stars, from West to East, with 13° 10' 35" per day. Such a motion is easily observable, especially when the Moon crosses a celestial zone rich in stars.

The Moon's right ascension increases with 24 hours during a tropical revolution:

27 d 07 h 43 m 04.7 s.

The Moon's orbit plane does not coincide with that of the terrestrial orbit (the ecliptic). The angle between the two planes is equal to 5° 08' 43". Their intersection (the line of nodes of the lunar orbit) determines two points (nodes) on the celestial sphere. The ascending node marks the point in which the Moon crosses the ecliptic plane from the Southern hemisphere to the Northern one. For the descending node, the situation is converse. The line of nodes undergoes a retrograde motion, completing a circuit of the ecliptic in 18 y 11 d 08 h, or 6798.5 d. This period, fundamental for eclipse prediction, is called Saros. The Saros is equivalent to 223 lunations, or 239 anomalistic revolutions, or 242 nodal periods.

The above motions may be studied in the mass point approximation. But the Moon also presents motions for which the finite body approximation is necessary.

One knows that the Moon always shows the same face to the terrestrial observer. But this is only a rough statement.

The lunar rotation axis is not normal to the ecliptic; it is inclined to this plane by an angle of 83° 11' – 83° 29'. This makes the apparent center of the lunar disk undergo a periodic shift, called the libration in latitude. It was first pointed out by Galilei, in 1636.

There also is a libration in longitude, discovered by Hevelius in 1654. Since the circumterrestrial trajectory of the Moon is not circular (even in the Keplerian approximation), its orbital velocity is not constant. As a consequence, the Moon exhibits some extra 7° 45' round its Western and Eastern limbs.

Lastly, the Moon experiences a diurnal libration, too. It occurs for we see the Moon from different sides of the Earth at moonrise and moonset, so we can again see slightly round the Eastern and Western limbs, although by less than 1°.

The consequence of the three librations is: we can see about 59% the lunar surface.

3.2 SHORT OVERVIEW OF LUNAR THEORY

Let us see now which are the main principles of the Moon's motion.

The first one who approached mathematically the Moon's motion, trying to explain it, was Isaac Newton himself. In his *Principia*, Newton had identified some causes of the perturbations of the Moon's elliptic motion. Let us review the facts that observations had revealed at that time upon the dynamics of our satellite.

It was known that the Moon could be considered to be moving in an elliptic orbit with the following features:

- the line of nodes has a quasi-uniform retrograde motion, by whose virtue it performs a complete tour along the ecliptic in 6793 days (about 18 years, 8 months);
- there exists a small periodic inequality that shifts the ascending node from its mean position by ± 1° 26';
- the inclination oscillates, in ~173 days, between 5° 00' and 5° 18', keeping a constant mean value.

The Moon's longitude undergoes three main periodic inequalities:

- the *evection*: 1° 16' 26" sin $[2(L_S - L_M) - \zeta_M]$;
- the *variation*: 39' 30" sin $[2(L_S - L_M)]$;
- the *annual equation*: - 11' 10" sin ζ_S,

where the following notation was used: L_S = Sun's mean longitude, ζ_S = Sun's mean anomaly, L_M = Moon's mean longitude, ζ_M = Moon's mean anomaly.

Newton has deeply studied the variation, and the method he resorted to was considered by Laplace to be one of the most remarkable issues of the *Principia*.

As to the shift of the Moon's apogee, which could not be explained within the framework of the inverse-square force model, Newton considered a slight change of his law by adding a radial inverse-cubic perturbing force. This model was used later by Clairaut, who eventually abandoned it in favor of the classical Newtonian one. Clairaut based his theory on the integration by series of the differential equations of the three-body problem. We have to mention that Euler and d'Alembert had also obtained these equations, independently and at the same time.

D'Alembert also proposed a Moon's theory, which presented many analogies with that of Clairaut. The main similitude was the use of the true longitude as independent variable.

Euler has proposed two different theories. The most important idea was the hierarchization of the various inequalities by orders. The first order inequalities had to be wholly calculated, then those of the second order had to be deduced from the first step, and so forth. This allowed the separation of the initial problem in several simpler problems.

Laplace resumed and developed the theories of Clairaut and d'Alembert, but his method is much more refined. He obtained all the second and third order inequalities, even some of the fourth order. Laplace's tables, which synthesized his results, gave the Moon's position with an accuracy better than 30".

Two memoirs have competed for the 1820 prize of the French Academy of Sciences: one by Damoiseau, the other by Plana and Carlini. The proposed subject had been the elaboration of lunar tables based only on the Newtonian theory of the universal attraction. The two memoirs have constituted the bases for two lunar theories. Although Damoiseau's theory seems to be more accurate (as compared to Hansen's famous model), the latter one was much appreciated by outstanding scientists as Laplace and Poisson.

Another interesting theory belongs to the Swedish astronomer Gyldén. He started from the equations of the perturbed motion, and truncated them in order to perform an exact integration. The solution, although approximate, is much more accurate than the elliptic motion; it represents an intermediate model between the Keplerian motion and the real one.

The theories of Lubbock and Pontécoulant are based on the same principles: to obtain directly the perturbations of the Moon's longitude, latitude and inverse of the geocentric radius vector. The authors had used the mean longitude instead of the true longitude. Lubbock's theory is limited to the first approximations, whereas Pontécoulant's work is much more extended.

Delaunay proposed a new method. He showed that, if the perturbing function is reduced to its nonperiodic part and only one periodic term, then the equations that provide the derivatives of the orbital elements are integrable. This leads to a separation of the work into a series of distinct operations, but all having the same nature.

A Romanian name, Constantin Gogu (1854-1897), must be added to the above scientists. His Ph.D. thesis, defended at Sorbonne, was entitled *Sur une inégalité lunaire à longue période due à l'attraction perturbatrice de Mars, et dépendant de l'argument* $\omega + l - 24\, l' + 20\, l''$ (*On a Long-Periodic Lunar Inequality Due to Mars' Perturbing Attraction and Depending on the Argument* $\omega + l - 24\, l' + 20\, l''$). This contribution corroborates Delaunay's results, while the second part of the thesis, *Sur une objection présentée par M. Stockwell contre la théorie du mouvement de la Lune de Delaunay* (*On an Objection Presented by Mr. Stockwell against Delaunay's Theory of the Moon's Motion*), pointed out an error in Stockwell's calculations. These results were published in *Mémoires de l'Observatoire de Paris*, tome XVIII. The same Gogu proved, after hard and lengthy calculations, that the 7.5" inequalities of the Moon's longitude (claimed by Neison in *Monthly Notices of the Royal Astronomical Society*, 1878) are insignificant.

In his thesis, Gogu took into account both the inclination of Mars' orbit and the Sun's perturbing influence; the departure point was Delaunay's *Théorie du mouvement de la Lune* (1860). Gogu's results are quoted in several textbooks of celestial mechanics and mathematical encyclopaedias, as well as – much later – in the works on which the Apollo lunar missions were based.

3.3. SECULAR ACCELERATION

The Moon is the only celestial body for which the observations pointed out a secular acceleration. This was discovered by Halley (1693), who used some ancient lunar eclipses mentioned in Ptolemy's *Almageste*, some other eclipses recorded by the Arabs at the end of the 9th century, and the observations of his time. The investigations performed by Dunthorne (1749), Tobie Mayer and Lalande corroborated Halley's results, and fixed the value of this acceleration to 6.7" – 10".

Lagrange and Laplace investigated theoretically this problem, proposing several hypotheses. Laplace gave the answer: a secular variation in the eccentricity of a planetary orbit causes an acceleration in the mean motion of the satellite. In this way, he proved the existence of a term in t^2 in the Moon's longitude. The main cause was the attraction of Venus and Mars.

A new and important check for both the lunar tables and the secular acceleration was provided by Baily (have you seen the marvelous "Bailey's beads"?) in 1811: the records of ancient eclipses. On this basis, Laplace's value for the secular acceleration was estimated as follows:

Airy: 10.72", 12.18";
Hansen: 11.93", 11.47", 12.18", 12.56".

At the same time, the theoretical computations of Adams and Delaunay provided the values 1.66" and 5.07", respectively. Lastly, Laplace's explanation has been proved to be realistic; it entails a secular acceleration of 6.11".

Among the researches that contributed to a better knowledge of the Moon's motion, we have to mention:
- Hill's researches, especially on the variation, but on the other inequalities, too;
- Adams' investigations on the motion of the nodes;
- Hansen's detailed analysis of various inequalities.

4. Earth's Rotation

Among many motions our planet is performing, the fundamental one for the humanity has two components: the circumsolar orbital motion and the rotation about its axis.

4.1 IRREGULARITIES OF THE EARTH'S ROTATION

More than three centuries ago, Edmund Halley (1695) suggested that the Earth's rotation is not uniform. In a work treating the longitudes, he compared ancient astronomical observations to the ones contemporary to him, and remarked the secular acceleration of the Moon's circumterrestrial orbital motion (we have already told about this phenomenon). More precisely, there was an advance of roughly 11" with respect to the uniform time used to elaborate lunar ephemerides (today's ephemeris time). Part of this value was explained, as we have seen, by Laplace. What about the remaining acceleration?

The observational results about the Moon's motion found a supplementary explanation in Kant's theory (1754), which associated the lunar dynamics to the variability of the terrestrial rotation via the phenomenon of tides. Baily's contribution in the 19th century brought new data that confirmed the existence of deviations in the Moon position. At that epoch, the astronomical time scale was based on the "clock" Earth; every nonuniformity of this clock had a correspondent in the apparent motion of heavenly bodies. The Moon was the closest celestial body; hence its apparent motion was the fastest; consequently, it was the first astronomical object to which an irregular dynamics could be detected. Since the Newtonian celestial mechanics was not able to fully explain the observed deviations, one began to doubt the uniformity of the clock Earth, therefore the terrestrial rotation uniformity.

To provide a satisfactory explanation to these issues, lots of hypotheses can be formulated:

- the gravitational potential could contain, besides the Newtonian $1/r$ term, a supplementary small term in $1/r^2$ (this hypothesis was first considered by Newton himself, then by Clairaut; in our century, the Bulgarian physicist Manev brought a physical support to this assumption);
- the gravitational constant could be time-variable or anisotropic (see the theories formulated by Dirac, Will, Vinti, etc.);
- the classical celestial mechanics considers the planets to be point masses or, at most, rigid bodies, but this is valid only in a first approximation. To obtain more realistic models, one has to take into consideration their rheological properties (the inner frictions, due to tides, affect their mean motions);
- the numerical calculations could be insufficiently accurate (this was proved by resuming them via modern techniques);
- the Earth's rotation angular velocity could be variable.

Comparing the positions of the Moon, Sun, Mercury and Venus, calculated in the uniform Ephemeris Time scale (Newtonian gravitational time scale), with the respective observed positions expressed in the Universal Time scale (given by the Earth's rotation), De Sitter pointed out the fact that the differences were proportional to the mean motions. Consequently, these differences had to be assigned to the diminution of the angular velocity of the Earth's rotation.

If the hypothesis according to which the Earth is not uniformly rotating were true, then this fact would be reflected in nonuniformities observed in the motion of other celestial bodies. This was confirmed by Fortheringham (1924) and Spencer Jones (1939). A more technical explanation will be given below.

During the deformation of the elastic Earth under tidal influences, the tidal energy turns to thermal energy, for instance, during the friction of the tidal flows with the ocean ground, or during the periodic deformations of the solid globe. Every friction dissipates energy, which must be taken over starting from the available kinetic energy that can be nothing but the terrestrial rotation energy. This is a loss of angular momentum, in other words, a diminution of the angular velocity of the Earth's rotation. But the angular momentum conservation law leads to the compensation of this one elsewhere within the dynamical system. An increase of the angular momentum, transferred from the rotating Earth, requires an increase of the mean distance of the Moon, that is, on the basis of Kepler's third law, a corresponding decrease of the mean motion (or, equivalently, a deceleration of the Moon's orbital mean longitude).

But, although the lunar motion theory was further and further improved along two centuries, a high residual acceleration still remains insufficiently explained. Indeed, we do not fully know the Moon itself yet, and all positional observations (except the radiointerferometric ones) are reported to the Moon's mass center; all ephemerides refer to it. But in general we observe events referring to the visible lunar disk edge, which is irregular and situated at an angular distance of 900" from the mass center.

4.2 ECLIPSES CONTRIBUTION TO THE EARTH'S ROTATION THEORY

However, a great number of observations are able to remove such errors. For instance, the observation of total solar eclipses (involving the whole lunar edge) is not affected by the above mentioned errors, although it entails lots of errors of interpretation.

There are eclipse records more than 5000 years old (the lunar eclipses of 3450 BC or 2283 BC). A great amount of eclipse records, performed by the Babylonian, Sumerian, Egyptian, Chinese, or Greek astronomers, allowed a satisfactory estimate of the eclipse time and visibility zone. The comparison of these observations with the computed data points out systematic differences that might be removed only admitting the fact that at Hipparchos' time (second century BC) the day duration was 40 ms shorter than the actual one.

About one million days were elapsed since the oldest solar eclipse recorded in China (709 BC). In average, every day was 0.021 s shorter that today. This would lead to an accumulated error (ΔT) of about... 7 hours. If the Earth always had such a rotation, in other words, if the day duration were constantly changing, then ΔT were a linear function. So, when Jesus Christ was born, this parameter had a value of 3.5 hours. The same parameter would be of only 1 hour around the year 1000. The analysis of the ancient observations shows that this affirmation is only partly true.

Let us exemplify with some ancient total solar eclipses observed by the Chinese astronomers. The best records refer to the eclipses of 181 BC, or AD 454, 761 and 1275. The first was seen as total in Ch'ang-an, the Chinese Capital at that time (today the town Xi'an). The value of ΔT lies between 3.28 and 3.53 hours. This is hence the error accumulated in time between the year 181 BC and today. For the eclipse of 25 June 1275, the value of ΔT is about 0.36 hours.

The analysis of all eclipses recorded during the last 2500 years indicates a lengthening of the day duration of 1.7 ms per century. This means 1.7 s in 100 000 years, whereas the observations of the Sun (declination measurements starting from 1760, and right ascension measurements starting from 1835) and Venus (right ascensions since 1835), as well as the observations of Mercury's transit across the solar disk (since 1677) give for the last two centuries a lengthening of the day duration of only 1 s in 100 000 years.

But the interpretation of these modern measurements is based only on two centuries of observations, and this could be irrelevant due to the accidental variations of the Earth's rotation. Even several millennia constitute an interval too short as compared to the Earth's age. To formulate precise conclusions about the nonuniformities of the angular velocity of the Earth's rotation, one searched for data originated in the primary era, namely 500 million years ago. To this end, one used the fossil clocks represented by the shells of invertebrates, whose growth depended on the daylight duration. In this way one concluded that the Devonian year had 400 days. Since the Earth's orbital motion around the Sun does not seem to have undergone changes along the geologic eras, we must admit that the day duration was modified. The Devonian day was only 22 hours long. Supposing that the deceleration rate has not changed in time, it results a lengthening of the day duration of at least 2 ms per century, or a relative variation $d\omega/\omega$ of $2\cdot 10^{-8}$ per century. This value corresponds to the antiquity, as the eclipse observations point out.

There is another issue deduced from the ancient eclipse records. Besides the secular dragging of the Earth's rotation, which acts on a time scale comparable with the age of the Earth-Moon system, there also is a continuous acceleration that could be associated to the variations of the geomagnetic field. The most recent determinations give the values $0.05\cdot 10^{-11}$ s/day for the secular acceleration due to tides, and $200\cdot 10^{-11}$ s/day for the maximum periodic accelerations produced by the bimonthly tides. The nonperiodic rotational accelerations, which persist along several years, lie between $-1.5\cdot 10^{-11}$ s/day and $+0.3\cdot 10^{-11}$ s/day.

The value admitted today for the secular variation of the terrestrial rotation velocity is 0.00164 s per century, therefore a continuous lengthening of the day duration of $4.5\cdot 10^{-8}$ s/day.

5. Occurrence of Eclipses

5.1. TYPES OF ECLIPSES

In the point mass approximation, when the Moon is exactly between the Sun and the Earth, a solar eclipse occurs. In the same approximation, when the Earth is just between the Sun and the Moon, a lunar eclipse takes place.

Therefore, the solar eclipses occur at New Moon, whereas the lunar ones – at Full Moon.

But the three bodies are not point masses, and this relax the conditions for the occurrence of eclipses.

At New Moon, our satellite is between us and the Sun, and projects a shadow cone in the opposite side of the latter one. The length of this cone is not constant; it depends on the Moon's position with respect to the Sun. Whether the cone vertex reaches the terrestrial surface or not, this depends on the Moon's position with respect to the Earth. The length of the cone varies between 55 and 63 terrestrial radii (59 in average); the calculations were made after subtracting one terrestrial radius. This means that in general the cone vertex remains above the Earth's surface.

The distance between the Moon and us varies between 363 600 km (at perigee) and 405 500 km (at apogee). This makes the Moon remain for a time at a distance smaller than the average from us. It happens that the shadow have a length of 57 terrestrial radii, while the distance that separates the Moon from us decreases at 55 terrestrial radii only. Then the shadow cone touches the Earth's surface.

The solar eclipses occur when the Moon is near a node of its orbit and has the same longitude as the Sun. At that instant, either the first sheet of the shadow cone, or the second sheet, or only the penumbra that surrounds them, touches the Earth's surface.

For an observer situated in the first sheet of the shadow cone, the lunar disk covers completely the Sun's disk; the eclipse is total. If the observer lies in the second sheet of the shadow cone, the Moon's disk is smaller than the Sun's one; the eclipse is annular. Finally, for a site situated in the penumbra, the lunar disk covers the solar one only partly; the observer sees a partial eclipse.

It results that there exist three types of solar eclipses: total, annular, and partial, according to the intersection of the umbral cone with the terrestrial surface.

But, when we talk about the whole terrestrial surface, we may consider six types of solar eclipses:

P - partial eclipse, when only a part of the penumbral cone of the Moon touches the Earth.

T - total eclipse: a central eclipse at which the first sheet of the shadow cone intersects the Earth. At a central eclipse, the axis of the shadow cone cuts the terrestrial surface; the so-called central line does exist.

A - annular eclipse: a central eclipse at which the second sheet of the shadow cone intersects the Earth.

A-T - annular-total eclipse: a central eclipse, total for a part of the transit, and annular for the rest.

(T) - total noncentral eclipse: a noncentral eclipse at which only a part of the first sheet of the shadow cone intersects the Earth (in the polar regions); the axis of the shadow cone does not touch the Earth.

(A) - annular noncentral eclipse: a case similar to the previous one, but now only a part of the the second sheet of the shadow cone intersects the Earth.

5.2. RECURRENCE OF ECLIPSES

The eclipses repeat under quasi-identical conditions after time intervals that depend on the synodic and nodal revolutions of the Moon. More precisely, such an interval must be a linear combination of integer numbers of lunations and nodal periods (see Table 1 below).

TABLE 1. Eclipse's periods

Period	Combination Saros-Inex	Lunations	Days	Years
Semester	5 I – 8 S	6	177	0.49
Hepton	5 S – 3 I	41	1211	3.32
Octon	2 I – 3 S	47	1388	3.80
Tritos	I – S	135	3987	10.92
SAROS	**S**	**223**	**6585**	**18.03**
Meton cycle	10 I – 15 S	235	6940	19.00
INEX	**I**	**358**	**10 572**	**28.94**
Exeligmos	3 S	669	19 756	54.1
Tetradia	19 I + 2 S	7248	214 038	586.0
Heliotrope	58 I + 6 S	22 102	652 685	1787
Megalo-Saros	58 I + 7 S	22 325	659 270	1805
Accuratissima	58 I + 9 S	22 771	672 441	1841
Horologia	110 I + 7 S	40 941	1 209 012	3310

6. Calculation of a Solar Eclipse for a Given Site

6.1. THEORY

The circumstances of a solar eclipse for a given place can be determined very accurately. One needs for this only the equatorial ephemerides of the Sun and of the Moon, and the geographic coordinates of the observation site.

To this end, we could calculate – for different instants around the moment of conjunction (for instance, at every 10 minutes) – the topocentric coordinates of the Sun and of the Moon, as well as the apparent diameters of these bodies, corrected for parallax. With these data, we obtain the differential coordinates of the lunar center with respect to that of the Sun, and the distance between the two centers as a tabular function of time. To find the moments of the exterior contacts, we have to determine, by interpolation, the instants at which the distance between the centers is equal to the sum of the two apparent semi-diameters. As to the interior contacts, they correspond to the instants at which the distance between the centers is equal to the difference of the two apparent semi-diameters.

Similar tabular functions can be constructed for the magnitude of the eclipse, for the position angle of the line that joins the two horns of the visible Sun's croissant, for the angular distance that separates these horns, and so forth.

6.2 BESSEL ELEMENTS

A solar eclipse can be calculated if the observer is on a definite point of the Earth's surface, or if he is somewhere outside the Earth.

For us the first point of view is more interesting. The position of the Moon or of the Sun is affected by the parallax. We observe the Moon advancing eastward with respect to the stars till the lunar limb touches the solar limb externally or covers the star. In the case of total solar eclipse there are four contacts: two for the partiality, two for totality.

When the solar light is obscured by the Moon, a solar eclipse occurs. The eclipse can take place only when the Moon is on or near the ecliptic. On Figure 1, M and S are the Moon and the Sun at conjunction in longitude, M' and S' the same positions when their apparent distance is the least, N is the nodal point of the lunar orbit, i_M the inclination of the lunar orbit at conjunction, β_M the latitude of the Moon at conjunction, and Σ the distance M'S'. We could consider SMN as a plane triangle.

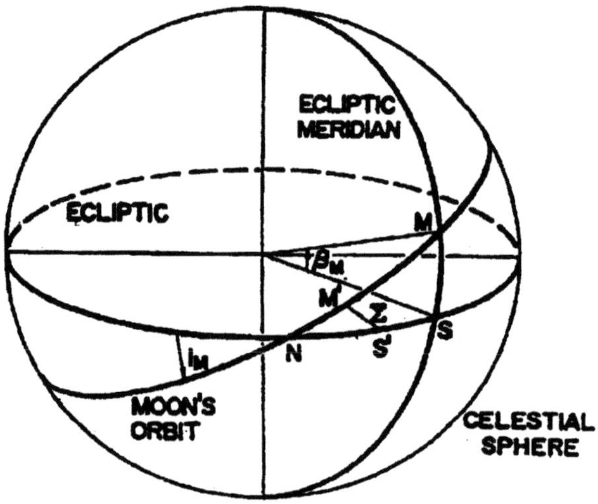

Figure 1

In this triangle we have:

$$SS' = \beta_M \tan \gamma.$$

If λ is the quotient of the motion of the Moon in longitude (SP) and that of the Sun (SS'), we have:

$$SP = \lambda\, SS' = \lambda\, \beta_M \tan \gamma,$$

$$S'P = SP - SS' = \beta_M (\lambda - 1) \tan \gamma,$$

and

$$M'P = MS - SP \tan i_M = \beta_M \tan \gamma \tan i_M.$$

From these follows that

$$\Sigma^2 = (\mathrm{M'P})^2 + (\mathrm{S'P})^2 = \beta_M{}^2\,[(\lambda - 1)^2 \tan^2\gamma + (1 - \lambda \tan i_M \tan\gamma)^2].$$

Differentiating this expression with respect to γ, setting it equal to 0, we finally obtain

$$\Sigma = \beta_M \cos I,$$

where

$$\tan I = \lambda \tan i_M / (\lambda - 1).$$

The apparent distances of the centers of the Sun and the Moon may be less than Σ by the difference of the horizontal parallaxes π_M and π_S of the Moon and of the Sun, respectively:

$$\text{minimum apparent distance} = \Sigma - (\pi_M - \pi_S).$$

The eclipse occurs when this distance is less than the sum of the semi-diameters of the two bodies, k_M and k_S:

$$\Sigma - (\pi_M - \pi_S) < k_M + k_S.$$

For every eclipse, the ephemerides also contain the so-called Bessel elements. Their use makes the calculation of an eclipse much easier. We shall see how such a calculation is performed.

For a given observation site, characterized by its geographic coordinates (φ = latitude, λ = longitude) and by its altitude h, we must determine:

- the instants of the exterior contacts (beginning and end of the partial eclipse);
- the instants of the interior contacts (beginning and end of the total or annular phase);
- the instant of the maximum of the eclipse;
- the co-latitude and the zenithal distance of every contact.

In the sequel we shall use the notation of *Connaissance des Temps.* As unit of length, the terrestrial equatorial radius will be chosen; it follows that the Moon's radius will be $k = 0.272274$.

The fundamental plane chosen by Bessel, $x\mathrm{T}y$, contains the Earth's center T and is normal to the axis of the shadow cone (Figure 2). It is an equatorial Cartesian coordinate system. The x-axis is the intersection of the fundamental plane and the terrestrial equator. The y-axis, normal to the x-axis, is directed northward. The z-axis, parallel to the shadow axis, has its positive direction towards the Sun. The reference frame is right-handed, determining in this way the positive direction on the x-axis. In addition, the Earth's rotation axis is contained in the Tyz-plane. In this frame, L(x, y, z) is the Moon's center, whereas S(x, y, $z + q$) is the Sun's center. Obviously, q stands for the LS distance.

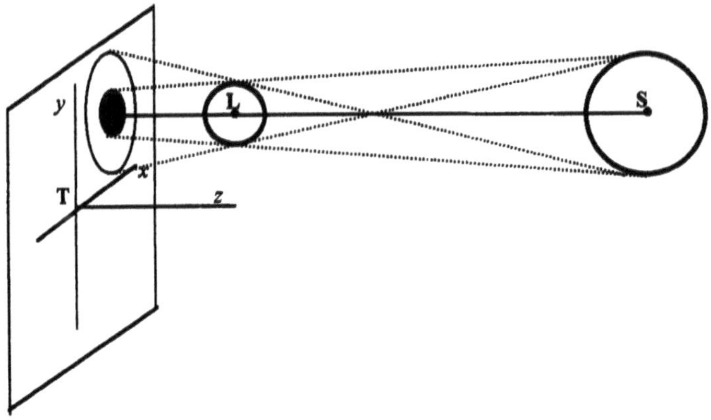

Figure 2

Let f_i and f_e respectively be the half-angles at the vertex of the umbral and penumbral cones; by convention, $f_i < 0$ and $f_e < 0$. The fundamental plane intersects the two cones following circles whose radii are u_i and u_e, respectively. These radii are easily calculated according to the formulae

$$u_i = z \tan f_i + k \sec f_i ,$$
$$u_e = z \tan f_e + k \sec f_e .$$

Let A be the observation site, and let (ξ, η, ζ) be its coordinates in the Txyz-frame. The circular sections obtained by cutting the two cones by a plane containing A, parallel to the fundamental plane, have P as common center and l_i, l_e as respective radii. These radii are obtained from the formulae

$$l_i = u_i - \zeta \tan f_i ,$$
$$l_e = u_e - \zeta \tan f_e .$$

The distance p of the point A to the center of these circles is given by

$$p = \sqrt{(x-\xi)^2 + (y-\eta)^2} .$$

At the instants of the exterior contacts, the point A belongs to the surface of the penumbral cone, hence

$$p = l_e .$$

At the interior contacts, A is on the shadow cone, and

$$p = |l_i|.$$

If l_i is positive, the point A is lying on the first sheet of the shadow cone, and the eclipse is total for the given site. If l_i is negative, the point A is lying on the second sheet of the shadow cone, and the eclipse will be annular. For all values of p such that

$$|l_i| < p < l_e,$$

the eclipse is partial.

Finally, the eclipse is central when $p = 0$.

The quantities x, y, z, u_i, u_e, f_i, f_e are given for each eclipse at every 10 minutes. To determine p, one has to construct the table of the coordinates of the point A for the same instants; then p, l_i, l_e are easy to find.

To calculate (ξ, η, ζ), we must know the geographic coordinates of the observation site, as well as the orientation of the reference frame Txyz. This orientation is fully determined by the hour coordinates of the Tz-axis. But the ephemeris provides, besides the above mentioned quantities, the declination d of this direction and the Greenwich hour angle H. If the longitude λ of A is positive westward, then the hour angle of Tz in the given place is $H - \lambda$.

Let us first report the Cartesian coordinates of the place A to a system of axes obtained by rotating Txyz around Tx such that Ty coincides with the Earth's rotation axis. Then the coordinates of A will be

$$\begin{aligned} x' &= \rho \cos\varphi' \sin(H - \lambda), \\ y' &= \rho \sin\varphi', \\ z' &= \rho \cos\varphi' \cos(H - \lambda), \end{aligned}$$

where

$$\begin{aligned} \rho\cos\varphi' &= \cos u + (h/6378140)\cos\varphi, \\ \rho\sin\varphi' &= 0.996647\sin u + (h/6378140)\sin\varphi, \end{aligned}$$

and

$$\tan u = 0.996647 \tan\varphi ,$$

h being the altitude of the observation site, expressed in meters.

To come back to the initial system, it is sufficient to rotate the system around Tx with the angle d:

$$\xi = \rho \cos \varphi' \sin (H - \lambda),$$
$$\eta = \rho \sin \varphi' \cos d - \rho \cos \varphi' \sin d \cos (H - \lambda),$$
$$\zeta = \rho \sin \varphi' \sin d + \rho \cos \varphi' \cos d \cos (H - \lambda).$$

6.3. CALCULATION

Every Bessel element b is represented on a time interval (t_0, t_1) by a power series of time, except $\tan f_e$ and $\tan f_i$, which are considered to be constant on this interval. A Bessel element is calculated for an instant t via the truncated series

$$b = b_0 + b_1 T + b_2 T^2 + b_3 T^3 .$$

with $T = t - t_0$ expressed in hours.

The hourly variation of a Bessel element is

$$\dot{b} = b_1 + 2b_2 T + 3b_3 T^2 .$$

Let us denote:

$$U = x - \xi , \quad \dot{U} = \dot{x} - \dot{\xi} ,$$
$$V = y - \eta , \quad \dot{V} = \dot{y} - \dot{\eta} .$$

With this, we can calculate:

(a) *The maximum magnitude*

The departure value t_d is considered to be the epoch of the eclipse maximum. The instant of the maximum is calculated by adding to t_d the quantity

$$\tau_m = -(U\dot{U} + V\dot{V}) / (\dot{U}^2 + \dot{V}^2) .$$

The next iteration considers $t_m = t_d + \tau_m$ as a new departure value.

The maximum magnitude is given by

$$g = \frac{l_e - l_m}{l_e - l_i} ,$$

for a total or annular eclipse in the considered place, or

$$g = \frac{l_e - l_m}{2l_e - 0.5465},$$

for a partial eclipse, where

$$l_m = \sqrt{U^2 + V^2}.$$

(b) *The exterior and interior contacts*

One considers approximate values extracted from the eclipse map as departure values (t_d) for the exterior contacts. As departure values for the interior contacts (if they exist), one considers the above determined value t_m.

For every departure value t_d, one calculates the following quantities:

$$\beta = \frac{U\dot{U} + V\dot{V}}{\dot{U}^2 + \dot{V}^2}, \quad \gamma = \frac{U^2 + V^2 - l^2}{\dot{U}^2 + \dot{V}^2}, \quad \theta = \pm\sqrt{\beta^2 - \gamma},$$

with $l = l_e$ or $l = l_i$, while θ has the sign of β.

The instants of the exterior contacts are calculated using

$$t = t_d - \beta + \theta,$$

while the interior contacts are deduced from

$$t = t_d - \beta - |\theta|$$

for the first contact, and

$$t = t_d - \beta + |\theta|$$

for the second contact.

As for the maximum, the calculation is performed by iterations, the new departure values being the previously determined values t.

(c) *The co-latitude and the zenithal distance*

The co-latitude P of a contact point is given by

$$\tan P = U / V,$$

where $\sin P$ has the same sign as U, except for the interior contacts of a total eclipse, where $\sin P$ and U have opposite signs.

The zenithal distance Z of a contact point is given by

$$Z = P - \Gamma,$$

where the parallactic angle Γ is approximately defined by

$$\tan \Gamma = \xi/\eta,$$

and $\sin \Gamma$ has the same sign as ξ.

So, the differences mentioned at the beginning are due to the difference between theories for the apparent positions of Sun and Moon. Generally, the calculations are based on the numerical integration DE200/LE200 of Jet Propulsion Laboratory.

Bureau des Longitudes, now Institut de Mécanique Céleste et Calcul des Ephémerides, uses the analytic theory of M. Chapront-Touzé and J. Chapront, and for the Sun the analytic theory VSO P82 of P. Bretagnon. These ones are sufficiently close to give similar results.

But all the national organisms, excepting NASA, apply an empirical correction in longitude and latitude to calculate the apparent positions of the Moon. It is +0.50" in longitude and -0.25" in latitude. Without these, NASA obtains other moments for conjunction and a small difference in determining the central line of the band of totality.

The second factor explaining the differences mentioned at the beginning of our presentation is the mean radius adopted for the Moon, k. It is expressed in terrestrial radius units. Before 1982 two different values of k were used: the first one (0.272 4880) in the general case and a second one (0.272 2810) only to calculate the quantities connected with the umbra for the total eclipses. To avoid the discontinuities introduced by the use of two different values for central eclipses, the International Astronomical Union chosen in 1982 a unique value $k = 0.272\ 5076$ for any calculus of the eclipses. Only NASA continues to use the IAU value, but the second one ($k = 0.272\ 2810$) for annular eclipses too, where from are the differences between NASA predictions and those provided by other organisms. Another difference is due to the difference between Terrestrial Time (uniform) and Universal Time (nonuniform, connected to the Earth rotation). Different values give different moments for conjunction and different values of the moments and longitudes in the phases of the eclipses.

7. Other Similar Phenomena

7.1 OCCULTATIONS

At a first sight, an occultation and an eclipse seem to be the same phenomenon: for the one as for the other, a body is hidden. But, during an occultation one celestial body passes in front of the other and the light of the second body is obscured by the first one.

The better known occultations happen when the Moon covers a star, sometimes a planet. Due to the inclination of the Moon's orbit with respect to the ecliptic and to the movement of the nodes of the Moon's orbit, all the stars in a belt of 10° around the ecliptic are occulted at some time during a period of about nine years.

The determination of the moments of the occultation of the stars by the Moon helps us to know the lunar orbit better.

The stars could be occulted by the asteroids or by the planetary satellites, too. So, we could better know the diameters of the bodies which are occultating (in angle) and ascertain whether they are binary systems (pairs of stars orbiting around their common center of gravity). The best such phenomenon we could observe was the occultation of 28 Sagittarii by Titan, Saturn's satellite, in 1989. The Galilean satellites of Jupiter are occulted regularly by the planet. The Jovian satellites could be occulted each other when the Earth is close to their orbital plane.

The disappearance or reappearance of a body are very different from a body to another. For Jupiter, the giant of the Solar System, it takes more than one minute; for an asteroid less than one second, as for a star is less still, by a factor of 10, sometimes nearly instantaneously, evidence of the absence of the lunar atmosphere.

It is not easy to determine the apparent diameter of stars. Many stars are close binaries (doubles) and the decline of the light of the occulted star suffers the diffraction. However, to determine the perturbations of the lunar orbit and the irregularities of the Earth's rotation, the exact times of the occultation of a star are very useful.

Don't forget: the astronomers discovered the rings of Uranus during a stellar occultation on March 10, 1977. The planet passed between the Earth and a bright star. The event was observed by many astronomers who were surprised that the light from the star was briefly obscured several times before and after the disk of Uranus occulted it. So was discovered a new planet with rings.

Another well known occultation is that of Crab Nebula by the solar corona. It takes place each year in June. This is a radio source (Taurus A). During the occultation its radio emission is much weaker.

Strictly speaking, a solar eclipse is a particular case of an occultation.

To predict an occultation we have to assume that the occultation is a simplified solar eclipse where the Sun is replaced by a star. Since the star may be considered at an infinite distance from us, we could replace the distance Moon-Sun by infinity and consider the diameter of the Sun k_S to be zero. It means that the two shadow cones become a shadow cylinder whose diameter is equal to the apparent diameter of the Moon, 2 k_M. The axis of this cylinder passes through the star. So, its direction is determined by the direction of the star.

All the ephemerides of the lunar occultations give the number of the star or its coordinates and some informations: variable or double star, magnitude, informations on the phenomenon (disparition or immersion, reappearance or emmersion), the elongation of the Moon (0° at New Moon, 180° at Full Moon) and the main coefficients a and b (minute/grade), used to calculate the occultation for a site of longitude λ and latitude φ, situated within 200-300 km from a central station. The local station time T of approximate contact is

$$T = T_0' - a\,(\lambda - \lambda') + b\,(\varphi - \varphi'),$$

where T_0', λ' and φ' are referring to the central station.

The coefficients a and b have the expressions:

$$a = [p \sin (N + \psi) + r \cos (N + \psi)] \cdot 1.0472/(n \cos \psi),$$
$$b = [q \sin (N + \psi) + s \cos (N + \psi)] \cdot 1.0472/(n \cos \psi)$$

in which the constant 1.0472 was chosen such that, if $(\lambda - \lambda')$ and $(\varphi - \varphi')$ are expressed in arcmin, T-T' results in seconds of time.

The coefficients p, q, r, s are yielded by the formulae:

$$p = \rho \cos \varphi' \cos h,$$
$$q = - \rho \sin \varphi' \sin h,$$
$$r = \rho \cos \varphi' \sin \delta \sin h,$$
$$s = \rho \cos \varphi' \cos \delta + \rho \sin \varphi' \sin \delta \cos h,$$

where ρ = the geocentric distance of the observer, h = hour angle of the star ($h = h_G + \lambda$; h_G = Greenwich hour angle of the Moon's shadow axis), δ = declination of the star.

The quantities n, N and ψ characterize the relative motion of shadow and observer. We have to mention that $N+\psi$ is the position angle of the contact measured from the North point of the lunar limb.

The ephemeris give also some informations of the resulting occultation: G – grazing occultation, A – below or too near the horizon, N – no occultation and S – sunlight interferes.

7.2. SATELLITE ECLIPSES

The eclipse is the passage of the shadow of a celestial body in front of another one. The planet satellites could be eclipsed by the shadow of their planets. The most easy observed are the eclipses of the Galilean satellites of Jupiter. The three innermost moons (Io, Europa and Ganymede) disappear into the shadow of the planet at each revolution, though the fourth one – Callisto – is not eclipsed every time. These bodies are not very small, so some minutes elapse between the first contact with the shadow and the totality. These satellites move nearly in the same plane as Jupiter moves around the Sun and, at practically every revolution, we could observe:

- the eclipse of the satellite when it passes through the planet's shadow;
- the occultation of the satellite when it disappears behind Jupiter;
- the transit of the satellite across the disk of the planet;
- the transit of the shadow of the satellite across the planet's disk.

Using the eclipses of Jupiter's satellites, the Danish astronomer Ole Roemer noticed discrepancies between the observed and calculated time of such eclipses; he deduced that it is due to the difference in the travel time of light when the Earth is nearest to planet or farther away from it. So, these eclipses offered the first proof of the finite speed of light.

Not only the natural satellites could be occulted. The same happens for a space probe. During the beginning and the end of these occultations, signals received on the Earth from the space probe have penetrated the planet's atmosphere and inform us about its density and composition.

7.3. ECLIPSES OF BINARY STARS

A binary system has the components moving on orbits around the common mass center. One component comes from time to time between the other one and the Earth: we observe an eclipse. In this moment we receive a smaller amount of light. This reached its minimum during a total eclipse of the star. We could observe two eclipses during each revolution. We know more than 3000 eclipsing binaries.

The first who observed an eclipse of a binary star was Geminiano Montanari. He observed in 1667 Beta Persei (Algol), discovering its variability. But only a century later the phenomenon was explained by John Goodricke.

If the orbits of the two stars are very elliptic, the motion is not uniform on the orbit, and the eclipses are not equally spaced. The space between two eclipses as the duration of light minima helps us to determine the eccentricity and the orientation of the orbit.

We could observe eclipses more or less dark, function on their sizes. If the star of higher surface brightness is behind, the eclipse is darker than in the opposite case.

7.4. TRANSITS OF MERCURY AND VENUS

When the Sun is eclipsed by the Moon, we have a solar eclipse; when it is eclipsed by one of the inner planets we observe a transit of the planet.

The inner planets, Mercury or Venus, are situated sometimes exactly between us and the Sun. Of course, they have to be close to one of the nodes of their orbits, at the inferior conjunction (i.e. when moving between the Earth and the Sun). In these moments it passes north or south of the Sun due to the inclination of its orbit. If the conjunction occurs when Mercury is near one of the nodes of its orbit, the planet crosses the solar disk as a small black circular spot, visible with a telescope. For Mercury it happens in November (ascending node) or early in May (descending node), as for Venus it happens early in December (ascending node) or early in June (descending node). Mercury passes more often than Venus. The shortest interval between two successive transits is seven years.

The transit of Venus could be seen even by the naked eye, of course protected for the solar light. When the transit is central, it takes about eight hours.

The transit of Venus has a historical significance: In 1679 Edmund Halley used it to find the parallax and from it the distance of the Sun. The transits of June 1761 and 1769, respectively of December 1874 and 1882 (the transits of Venus occur in pairs distanced by more than a century) were very careful observed. The next pair of Venus' transits will be on June 8, 2004 and June 6, 2012.

Much later than Halley, Innes used the transit of Mercury for some special investigations about the Earth's rotation.

For the parallax determination you have only to fix the moments of the contacts of the disks of the planet and of the Sun from different terrestrial points. Of course, you could see other interesting phenomena: when Venus is partially overlapping the solar disk, the part of the limb of the planet that extended beyond the Sun is surrounded by a radiant aureole, due to the Venusian atmosphere.

The observations of the transits help us to determine the diameters of the planets. The amount of light intercepted during the transit is function of the planetary diameter. So, measuring the diameter is measuring the amount of sunlight.

7.5. ECLIPSE PHENOMENA IN RADIOASTRONOMY

The phenomenon is similar to the optical one. The difference is the optical refraction which affects the eclipse observation is replaced by the electronic refraction whose existence is conditioned by the ionized medium. The high sensitivity of electronic refraction to the presence of electrical particles should be noted, so this refraction is a very sensitive criterion of presence of a highly rarefied atmosphere which happens to be ionized precisely.

Many radiosources can be occulted by the Moon. During this phenomenon we could determine the exact position and size of the source as the upper limit for the density of the lunar atmosphere.

The radio sources could be occulted by Solar corona too. We know that the light of the solar corona is mainly due to the sunlight scattered on the free electrons. The decrease of the brightness with the distance from the solar border helps us to derive the structure of the corona, i.e. the function giving the electron concentration at any height above the photosphere.

References

1. Danjon, A. (1959) *Astronomie generale*, J. & R. Sennac, Paris.
2. Espenak, F., Anderson, J. (1997) *Total Solar Eclipse of 1999 August 11*, NASA Reference Publication No. 1398.
3. Gogu, C. (1882) *Sur une inégalité lunaire à longue période due à l'attraction perturbatrice de Mars, et dépendant de l'argument* $\omega + l - 24\,l' + 20\,l''$, Mélanges Mathématiques, Vol.1 (10), Gauthier-Villars, Paris.
4. Mueller, I. I. (1969) *Spherical and Practical Astronomy as Applied to Geodesy*, Frederick Ungar Publ. Co., New York.
5. Oppolzer, T. von (1887) *Canon der Finsternisse*, Denkschriften der Kaiserlichen Akademie der Wissenschaften, Mathematisch-Naturwissenschaftliche Klasse, 52, Wien.
6. Tisserand, F. (1894) *Traité de mécanique céleste*, tome III, Gauthier-Villars, Paris.

THE WILLIAMS COLLEGE EXPEDITION TO RAMNICU VALCEA

JAY M. PASACHOFF
Williams College–Hopkins Observatory
Williamstown, MA 01267, U.S.A.

Abstract. The Williams College expedition to Râmnicu Vâlcea, Romania, made observations of the solar corona (a) to compare with the Extreme ultraviolet Imaging Telescope (EIT) on the Solar and Heliospheric Observatory; (b) to search for the source of coronal heating through a study of the oscillatory power spectrum of coronal loops in an emission line typical of high temperature; (c) mapped the coronal temperature using ultraviolet continuum observations; (d) made videos and still photographs of coronal motions and details; and (e) collaborated with the scientists at the Astronomical Institute in Bucharest.

1. Introduction

The August 11 total solar eclipse, visible from a narrow path extending across Europe and as far as India, provided wonderful overlapping observations from the ground and from space. Our team observed from Râmnicu Vâlcea, Romania, and had three major experiments. The sky was very clear throughout totality.

The path was close to that of the 1715 eclipse viewed by Halley (Pasachoff 1999a, 1999b).

2. The SOHO/ground-based eclipse experiment

In one of our experiments, features in the Sun's corona were traced from lower to greater heights using joint observations from space by scientists at NASA's Goddard Space Flight Center and from the ground at the August 11 total solar eclipse by our team. Space observations from the Extreme ultraviolet Imaging Telescope (EIT) on the Solar and Heliospheric Observa-

J. P. Zahn and M. Stavinschi (eds.),
Advances in Solar Research at Eclipses from Ground and from Space, 23–30.

tory (SOHO) show the Sun's hot outer atmosphere, the corona, silhouetted in front of the Sun and extending slightly outward. Eclipse observations, which are possible to make only every eighteen months or so when a total solar eclipse crosses the Earth's surface, are able to extend those observations to greater heights above the Sun's everyday surface. NASA Goddard and Williams College scientists took advantage of just such an occasion at this 1999 total eclipse. The result can improve scientists' understanding about how the Sun's magnetic field shapes and guides gas that leaves the Sun, some of which eventually envelops the Earth.

U.S. participation in the EIT experiment on SOHO is headed by Joseph Gurman of NASA's Goddard Space Flight Center, Greenbelt Maryland. SOHO is a mission of international cooperation between ESA and NASA. The SOHO EIT instrument was developed by an international consortium led by Principal Investigator Dr. Jean-Pierre Delaboudinière of the Institute for Space Astrophysics at the University of Paris - Sud in Orsay, France. EIT obtained an image of the Sun just four minutes after the eclipse ended in Râmnicu Vâlcea. The Williams College team for this experiment was headed by Jay Pasachoff, Stephan Martin, and Bryce Babcock, supported in part by a SOHO-EIT Guest Investigator grant from NASA. On site, the group included included Rebecca Cover and Misa Cowee. After two weeks of setting up our telescopes, electronics, and computers, the eclipse team observed two minutes and 23 seconds of totality in completely clear skies. Preliminary lineups of the EIT observations and the ground-based observations were immediately posted on the World Wide Web, and Williams students and faculty will continue working further on these data in the months to come. The eclipse data overlap not only the EIT observations on their inside but also, at their outside limit, other SOHO observations of the outer corona from the Large Angle Spectrographic Coronagraph (LASCO), and eventually a composite image will cover the whole range of the corona as it appeared on the day of the eclipse.

The experiment uses a lens that gives an image at the same scale and with a green filter that matches a filter in the C1 coronagraph, one of the telescopes in the coronagraph system on SOHO; this coronagraph was operable during the 1998 eclipse though is no longer working, so the eclipse observation in the coronal green line covers a gap between the lower solar levels and the lowest height covered by LASCO's C2 coronagraph. This observation was conceived in collaboration with the late Dr. Guenter Brueckner of the Naval Research Laboratory in Washington, D.C., principal investigator of that experiment, and is now in collaboration with Russ Howard, Dennis Socker, and other scientists at NRL.

3. The Oscillation Experiment

Another of the major experiments is a search for rapid oscillations in the corona, with periods of about 1 second. Pasachoff and his colleagues have developed techniques over the last two decades to observe in the so-called "coronal green line," the forbidden line of thirteen-times ionized iron (written [Fe XIV]), with time resolution so fast that such short periods can be detected. Oscillations with periods in that short range are predicted by some theories that hold that the extreme coronal heating is caused by vibrations of magnetic loops. The loops of gas, held in place by the Sun's magnetic field, have been observed, and the question is whether their vibrations bring enough energy into the corona to heat it sufficiently. The experiment is supported by a grant from the Atmospheric Sciences Division of the National Science Foundation. The sites at which they looked on the edge of the Sun was the one that NASA's Transition Region and Coronal Explorer spacecraft (TRACE) imaged at extremely high spatial resolution at the time of the eclipse, after consultation between the Williams team and the TRACE team at the Lockheed Martin Scientific Laboratory and the Smithsonian Astrophysical Observatory. The Williams group on site was headed by Babcock, Kevin Russell, and Timothy McConnochie. Alexander Ene and Misa Cowee were in charge of an Ocean Optics spectrometer to monitor the central wavelength of the DayStar [Fe XIV] filter and Mark Kirby helped at the eclipse time.

Previous work on this topic has been described by Pasachoff (1993) and Pasachoff et al. (1995a, 1995b). Phillips et al. (2000), after discussions with us, carried out a similar experiment. They used a higher cadence and a correspondingly lower statistical level of accuracy in the digitization, and the data should be complementary.

4. The Temperature Experiment

The remaining major experiment is a map of the temperature of the corona, using a technique of comparing electronic images of the corona taken at special ultraviolet wavelengths. Following theoretical work by Lawrence Cram of the University of Sydney, and subsequent recalculation by Kevin Reardon of our group, these wavelengths are chosen to include two such at which the difference between the shape of the everyday Sun's spectrum and the corona's spectrum is especially striking. The experiment is supported by grants from the Committee for Research and Exploration of the National Geographic Society. The Williams group on site included Lee Hawkins, Sara Kate May, and Alex Ene.

Figure 1. Dual composite image made from a composite of 12 photographic images of the August 11, 1999, total solar eclipse, with a composite SOHO EIT image of the solar disk superimposed in place of the dark lunar disk. The EIT image is a combination of the three coronal wavelengths (Fe IX/X, XII, and XV) combined to bring out coronal structure. (Eclipse photo (c) 2000 Wendy Carlos and Jay M. Pasachoff; SOHO/EIT image courtesy of NASA Goddard Space Flight Laboratory, EIT Group, courtesy of Joe Gurman)

5. Photography

We imaged the corona at a variety of scales and with a variety of cameras. Two videos were made with Canon camcorders. The XL1 was operated by Paul Rosenthal and Joey Shapiro on its own tracking mount; the Elura was operated by Darik Vélez on the temperature experiment's tracking mount. Vélez has edited various of the eclipse and expedition images onto a composite tape, and Feng Zhao has worked at Williams with digital video and put various images onto the World Wide Web at http://www.williams.edu/astronomy/eclipse99.

Pasachoff, Ridgeley, Seaton, and others operated Nikons with 500 mm, 600 mm, and 750 mm lenses. Seaton used a Hasselblad and Deborah Pasachoff used a Nikon to take wide-field images including scientists in the foreground and the eclipse in the background sky. Samples of Pasachoff's photographs of the partial and total eclipse have appeared in the February 2000 issue of National Geographic Magazine in their Geographica section.

6. Radial Filter Image

Jonathan Kern made a full-color radial filter image, using a coelostat and telescope of his own design (www.ligo-la.caltech.edu/ jkern/eclipse99).

Pasachoff and Wendy Carlos have made a composite image showing radial-filter information, similar to the one they made of the 1998 eclipse that is posted at www.williams.edu/astronomy/eclipse98. The new photo, posted at www.williams.edu/astronomy/eclipse99, combines 12 photographic images from the eclipse and also 3 SOHO EIT eclipse-day images of the corona silhouetted against the solar disk.

7. Eclipses and Space Research Together

The complementary sets of data compiled from the ground during the eclipse and from satellites in space show the value of continued ground-based eclipse studies. For example, the Extreme-ultraviolet Imaging Telescope (EIT) on SOHO gives wonderful images of the corona on the disk and peering over the moon's limb, but doesn't go far into the coronal images made at eclipses. Farther out, the Large-Angle Spectroscopic Coronagraph (LASCO) on SOHO makes a type of artificial eclipse, but has to hide the inner 1.5 solar radii because of internal scattering. So the region of the lower corona is still best studied at eclipses. And comparing the eclipse images of the corona with eclipse-day images made from various space instruments on SOHO, Yohkoh, and the Transition Region and Coronal Explorer (TRACE) extends the range of observation both spatially and through the spectrum.

TRACE makes very high-resolution observations of coronal loops, using ultraviolet radiation, but can take images only every few seconds (or even every 20 seconds). Our oscillation experiment on the ground made coronal-loop observations 10 times per second. So though we don't match TRACE in spatial resolution, we have better temporal resolution, and perhaps together the analysis of all the data will show waves moving through or along the coronal loops. Even the next solar spacecraft planned won't match the temporal observation we had.

One of our experiments was originally conceived to match the field of view and filter band of the innermost coronagraph on LASCO, and we made joint observations with them at the 1998 eclipse. When we compare our finally reduced data with theirs, we hope to be able to use our superior coronal images to measure the scattered light in this "C1" coronagraph.

Unfortunately, the C1 coronagraph is now defunct, a result of the mishap with SOHO from which the other instruments recovered.

In some decades, perhaps space instruments will be able to match a natural eclipse, but that time has not yet come. And another point is the relative prices of space vs. eclipse experiments. SOHO cost perhaps $300 million. For less than 1/10 of 1% of that cost, a major set of eclipse experiments can be mounted. It is surely worth that small percentage even if you consider the eclipse experiments merely ground-based support for the solar satellites aloft.

Eclipse science retains a valuable place in the astronomical set of investigations, just as ground-based astronomy remains important and vital even after the launch of the Hubble Space Telescope.

8. Liaison with the Astronomical Institute

Following lengthy discussions with Magda Stavinschi and her staff, Christina Reynolds and Rossen Djagalov were at the observatory in Bucharest for the eclipse, bearing a large-format Hasselblad camera to make extremely high-resolution coronal images during totality at the 6-m focal length astrograph. Unfortunately, a cloud covered the Sun just at the wrong moment, and totality was obscured from this particular Bucharest site, though the eclipse was visible from various locations in Bucharest.

9. The Eclipse Team

An unusual aspect of the Williams College experimental team is that it includes so many undergraduate students. They include Williams College students Kevin Russell, Sara Kate May, Rebecca Cover, Daniel Seaton, Joey Shapiro, Misa Cowee, Darik Vélez, and Rossen Djagalov; Keck Northeast Astronomy Consortium Summer Fellow Alexandru Ene, a student from Romania who studied at Middlebury College and who is transferring to Harvard; and Mark Kirby, from Deep Springs College in Dyer, Nevada. Recent Williams alumni Timothy McConnochie '98 of JPL, Pasadena, and Christina Reynolds '97 of the University of North Carolina were also on site. The equipment team was headed by Bryce Babcock, staff physicist at Williams College. Scientific staff also included R. Lee Hawkins of Wellesley College, now at Appalachian State University; Stephan Martin of Williams College; and Jonathan Kern, Optics Scientist at the Laser Interferometer Gravitational-Wave Observatory of Caltech. Allan Ridgeley of the Rutherford Appleton Laboratory, England; Prof. Marek Demianski of the Copernicus Astronomical Center in Warsaw; Mitzi Adams of the NASA Marshall Space Flight Center, and Paul Rosenthal, MD, of Williamstown, joined the group in Romania.

We are grateful to George Walther for his work on the equipment in the shop at Williams College. Student participation in the expedition was funded by the NSF, National Geographic, and NASA grants; by the Keck Northeast Astronomy Consortium; by the Massachusetts Space Grant, which is funded by NASA; and by the Safford Fund and the Brandi Fund at Williams College. A collaboration among Williams College, the Rutherford Appleton Laboratory, and the Astronomical Institute of the Romanian Academy of Sciences is funded by NATO.

10. Future Eclipses

The next opportunity to obtain ground-based coronal eclipse observations will come on June 21, 2001, when a total eclipse of the Sun crosses southern Africa. See Pasachoff (2000) and Espenak and Anderson (1999), http://umbra.nascom.nasa.gov/eclipse/010621/rp.html. The following total eclipse will occur on December 4, 2002, and will also cross southern Africa, ending at Sunset in Australia.

Links about experiments can be found at the Web site of the International Astronomical Union's Working Group on Eclipses:

www.williams.edu/astronomy/IAU_eclipses.

References

1. Espenak, F. and Anderson, J. (1997) *NASA Reference Publ. 1383 for the 1998 eclipse*
2. Espenak, F. and Anderson, J. (1998) *NASA Reference Publ. 1398 for the 1999 eclipse*
3. Espenak, F. and Anderson, J. (1999) *NASA Techn. Rep. 1999-209484 for the 2001 eclipse*
4. Golub, L. and Pasachoff, J.M. (1997) *The Solar Corona* (Cambridge Univ. Press)
5. Menzel, D.H. and Pasachoff, J.M. (1970) Solar Eclipse *National Geographic* **138**, #2, 222-233
6. Pasachoff, J.M. (1992) The Darkness That Enlightens *National Geographic* **181**, #5, 36-37
7. Pasachoff, J.M. (1999a) Halley and his maps of the total eclipses of 1715 and 1724 *Astronomy & Geophysics* (Royal Astronomical Society) **40**, 18-22
8. Pasachoff, J.M. (1999b) Halley as an eclipse pioneer: his maps and observations of the total solar eclipses of 1715 and 1724 *J. Astron. History and Heritage* **2(1)**, 39-54
9. Pasachoff, J.M. (1973) The Solar Corona *Scientific American* **229** #4, 68-79.
10. Pasachoff, J.M. (1993) Coronal Morphology and Heating-Mechanism Observations at Total Eclipses Through 1992 *IAU Colloq. 144 Solar Coronal Structures, Tatranska Lomnica, Slovakia*, 523-527
11. Pasachoff, J.M. (2000) *Peterson Field Guide to the Stars and Planets* (Houghton Mifflin).
12. Pasachoff, J.M. and Covington, M. (1993) *Cambridge Guide to Eclipse Photography* (Cambridge University Press).
13. Pasachoff, J.M., Babcock, B.A. and Reardon, K.P. (1995a), Coronal Heating Studies at the 1994 Total Eclipse, *Proceedings of the International Symposium on the Total Solar Eclipse of November 3, 1994* (M.S. Raljevic, F. Zaratti and J. M. Pasachoff, eds.) *Revista Academia Nacional de Ciencias de Bolivia* **69**, 18-21.

14. Pasachoff, J.M., Babcock B.A., Diaz, J.S., Reardon, K.P. and Nichols-Kiley, R. (1995b) Study of the High-Frequency Coronal-Loop Oscillation Spectrum at the 1994 Total Solar Eclipse, *Bull. Am. Astron. Soc.* **27**, 1427.
15. Phillips, K.J.H., Read, P.D., Gallagher, P.T., Keenan, F.P., Rudawy, P., Rompolt, B., Berlicki, A., Buczylko, A, Diego, F., Barnsley, R., Smartt, R.N., Pasachoff, J.M. and Babcock, B.A. (2000) SECIS: The Solar Eclipse Coronal Eclipse Imaging System, *Solar Physics*, in press.

HELIOSPHERIC EXTENT OF THE CORONAL PHENOMENA

G. MARIŞ
Astronomical Institute of the Romanian Academy
75212 Bucharest 28
ROMANIA

1. Introduction

The solar photosphere has been watched regularly since 1750; the patrol of the chromosphere has been made daily for the last 50 years. The solar corona, observed at the ground only during total solar eclipses a long time ago, has been observed with special coronographs since the '30s. However, the last three decades, beginning with OSO 7 and Skylab space missions, contributed enormously to our present knowledge of the solar corona. Today the corona is watched both from ground and from space. The solar space missions Yohkoh and SOHO are registering coronal phenomena in the X and UV radiations in the ecliptic plane, while Ulysses is crossing over the polar solar region (at an angle of about 87° with the ecliptic plan).

The heliosphere is the spatial cavity "designed" by the solar wind flowing into solar system and pushing the local interstellar plasma and magnetic field out of its way. It is the place of the Sun and its planetary system: the planets and their satellites, the asteroids and the comets, each of them surrounded by particular gas and plasma. This cavity is filled with the electromagnetic radiation (from gamma radiation to radio emissions) of solar origin and with the corpuscular radiation consisting of solar wind (electrons and ions) and more energetic particles sent out by the solar corona or entering from outside (galactic cosmic rays). Dust particles and rare micrometeorites populate the inner heliosphere producing the zodiacal light by scattering of white sunlight.

The solar wind and other more energetic solar particles "drive" the solar coronal field into the interplanetary space. This magnetic field, the "frozen in" wind plasma, is structuring the whole interplanetary magnetic field (IMF) in two or four sectors of opposite polarity. In this way, the coronal phenomena, sources of energetic particles, play a major role in the morphology and the dynamics of the IMF and influence the planetary magnetospheres and atmospheres.

The state of the interplanetary medium inside the earth's orbit is mainly determined by the solar initial (boundary) conditions, therefore, generally, the heliosphere is made up of two great regions, the inner and the outer heliosphere, with the earth's orbit as a border [16]. Beyond 1AU, there is a stronger interaction between the solar particle streams owing to the strongly bent spiral magnetic field; the coronal prime imprints on the interplanetary plasma get increasingly lost as the solar wind expands further and the dynamic phenomena of interplanetary origins prevail.

Nevertheless the prominent role is played by the Sun in shaping and structuring the heliosphere.

J.-P. Zahn and M. Stavinschi (eds.),
Advances in Solar Research at Eclipses from Ground and from Space, 31–47.

Heliosphere knowledge was the primordial aim for many spacecrafts beginning with Helios 1 and 2 missions [16]. The principal progress made in the last years in heliospheric studies come from the data obtained by Pioneer and Voyager spacecraft moving away from the Sun at about 3 AU/year. The Ulysses and SOHO missions (ESA–NASA) also have a significant contribution in the remote sensing of the inner heliosphere. Studies of the heliosperic plasma could help to answer some questions about solar wind, such as: How do the properties of the SW evolve with distance from the Sun? How does the SW interact with the interstellar medium? What mechanisms are responsible for the SW accelerations?

The large-scale coronal structure and phenomenology are very complicated by the solar magnetic field dynamics and its interaction with coronal plasma. There are coronal streamers, coronal mass ejections, coronal holes, and solar flares as proven contributors in the solar wind. There are also polar plumes, the small-scale coronal phenomena, which could play a major role in the solar wind particle acceleration. The first observations of the solar corona and its phenomena were obtained at total solar eclipses. Further eclipse observations could still be a good and necessary complement to the sophisticated space coronal observations.

2. Heliosphere

Leverett Davis made the first suggestion about the existence and nature of the heliosphere in 1955 in connection with the cosmic rays dynamics.

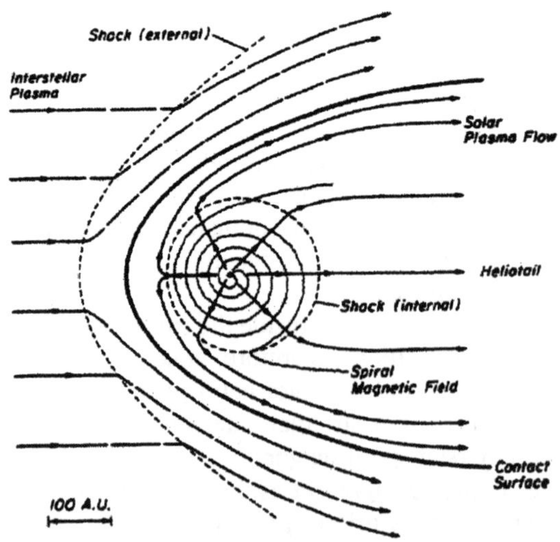

Figure 1. Schematic sketch of a plausible heliosphere generated by the flow of the solar wind and its interaction with the interstellar gas [7].

The simplest signification of the concept is that the solar wind, propagating at a great distance from the Sun, beyond 1 AU, presses the local interplanetary plasma and the magnetic field forming a cavity – the heliosphere. The pressure of the solar wind and the pressure of the interstellar medium, of opposite directions, reach equal values along a surface called the heliopause (the contact surface). Inside this surface there are almost only plasma and magnetic field of solar origin (figure 1).

The solar wind flows supersonically out to a spherical "termination shock", supposed to be at 60 160 AU from the Sun, where the wind speed becomes subsonic [7]. Beyond the shock, the solar wind is "forced" to continue its propagation in a direction parallel to the heliopause and to head for an open tail, which continues, far into the interstellar medium. So far, neither the internal shock of the heliosphere nor the heliopause have been detected by the Pioneer or Voyager missions up to a distance of 50 AU from the Sun.

Studies of the interaction between the solar wind and the interstellar medium were made based upon the available observational inferences concerning the relevant properties of the solar wind and the nature of the very local interstellar medium. The principal solar wind parameter controlling this interaction is the ram pressure of the supersonic flow, ρu^2, where ρ is the mass density, and u the flow speed of the solar wind [4].

Certain constituents of the interstellar medium can penetrate the heliopause and enter the heliosphere. They are:

(a) neutral gases (mainly H and He);
(b) interplanetary dust supposed to consist of 0.05 μm typical metallic grains, and the dust-to-gas ratio by mass of 1.01, leading to a mass density for dust of about 2×10^{-27}g cm^{-1};
(c) galactic cosmic rays (relativistic ions and electrons).

The penetration of interstellar hydrogen atoms and their effects on the solar wind in the region of supersonic and subsonic flow are controlled by the solar wind proton density $n_p u$ (n_p is the proton density), the solar wind speed u in the supersonic region and the solar wind temperature T in the subsonic region. Considering the heliosphere's evolution, induced by the changes of the coronal magnetic structure and transient phenomena in the corona with the solar cycle, some changes would be expected in the structure and dynamics of the solar wind. According to the observations made so far no major changes were found. However, there are several effects of the solar cycle in the solar wind evolution that should be pointed out. Thus:

(a) the number, duration and stability of the high speed streams in the solar wind are significantly higher before the solar minimum;
(b) the number of shocks varies 10 times from the minimum to the maximum;
(c) the average intensities of the energetic particles vary considerably in keeping with the evolution of the solar activity in the 11 years cycle;
(d) the intensity of the galactic cosmic rays which reach the inner heliosphere vary in inverse proportion to the solar activity (the Forbush effect).

The first systematic studies on the variations in the solar wind structure and parameters during the eleven years solar cycle is made based on the data of the Helios 1 and 2 missions [16]. For the period 1974 – 1982, which includes the minimum of 1976 and the maximum of 1979, such analysis were made.

The fact that the cosmic rays' particles have relativistic energies, great gyroradii and a behavior of individual particles makes them capable of crossing the heliopause.

After they enter the heliosphere, its collisions with the ambient plasma are negligible, so that the magnetic field determines the motion.

To be able to propagate up to the inner heliosphere these particles must have sufficient energy to overcome the effects of:

(a) the magnetic pressure of the interplanetary magnetic field;
(b) the diffusion due to the irregularities in the interplanetary magnetic field;
(c) magnetic drift on the current sheet in the heliosphere.

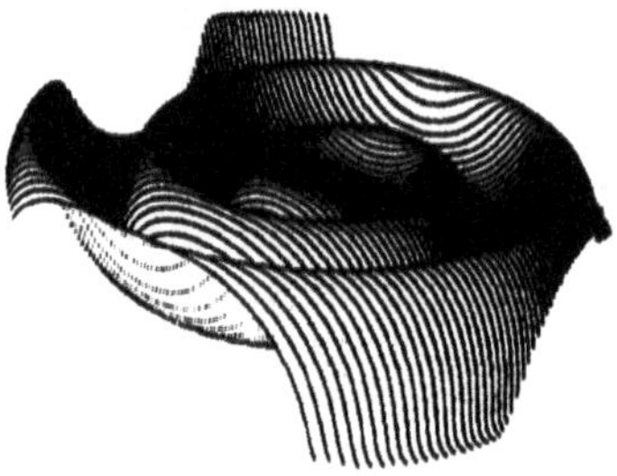

Figure 2. The heliospheric current sheet when solar magnetic dipole is inclined substantially relative to the solar rotation axis [10]

The magnetic field has an average given by the Archimedean spiral of Parker. Superposed on this average field are magnetic irregularities that scatter the cosmic ray particle and cause a random walk. It is important to note that collisions of cosmic rays with the ambient plasma are completely negligible, so that the magnetic field determines the motion. The cosmic rays are subject to magnetic drift on the current sheet in the heliosphere. A good image of these heliospheric current sheets is the well-known "ballerina" picture (figure 2).

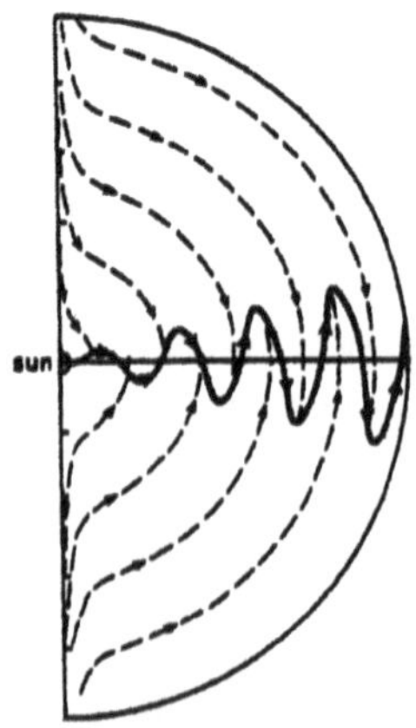

Figure 3. Drift trajectories of cosmic rays in the large-scale heliospheric magnetic field [15]

In such cases, when the current sheet is highly inclined near the Sun, the peak and troughs are extending to higher latitude and the drifting cosmic rays then encounter the current sheet sooner and are expelled more readily from the heliosphere (figura 3).

In the figure, the solid line is the heliospheric sheet with its peaks and troughs at the maximum and minimum latitude. The dashed lines show the trajectories of 2 GeV protons that enter over the Sun's magnetic poles. The trajectory represent the average motion on which would be superposed random walk and outward convection.

There are short-term increases and decreases in the cosmic-ray intensity, presumably in association with the oscillatory motion of the heliospheric current sheet. An alternative explanation for the cosmic rays modulation involves the coronal mass ejections. A coronal mass ejection (figure 4, left side), as a "blob" of plasma with its proper magnetic field flowing away from the corona, could oppose the entry of cosmic rays into the inner heliosphere. At large distances (> 10 $R_\odot$), the individual "blobs" merge together to form "barriers" that close off the inner heliosphere (right side of figure 4). In paragraph 3.2 the coronal mass ejections are shortly presented.

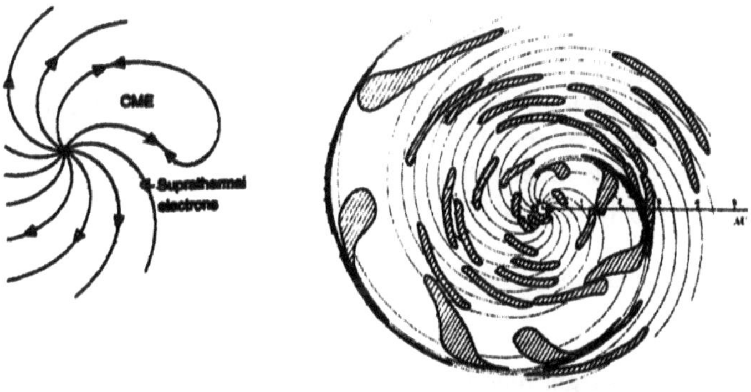

Figure 4. The initial "blob" of a coronal mass ejection (the left side) and the shaded "blobs" of many coronal mass ejections, pushed away by the rotating Sun (right side) [10, 15]

The heliosphere might be represented as a "stagnation" point flow occuring between the solar wind and an interstellar wind. The result is similar to the Earth's magnetosphere. Inside the heliopause, the solar wind passes through a termination shock that is elongated in the downstream direction and which is moving back and forth at speeds up to 100 km/sec.

3. Coronal Phenomena

The coronal formations observed at the total solar eclipses, beginning with the end of the past century, are proofs of the richness of the small scale and large-scale morphological structures existing in the solar corona. The morphology and dynamics of these structures are without exception determined by the coronal magnetic field and by its variation in time. The coronal observations from space, starting from the seventh decade of the 20th century (the OSO and Skylab space missions) in the UV radiation and in white light, and later in the X radiation, have confirmed and enriched the data on the coronal structures.

The external corona, without rigid and well-set boundaries, is a "connecting bridge" between the Sun and the Earth. The expansion tendency of the solar coronal hot and magnetized plasma cannot be "fenced in" and thus a flow of matter appears – the and magnetized plasma cannot be "fenced in" and thus a flow of matter appears – the solar wind. It leaves the solar corona continuously and propagates far into the interplanetary space, interacting with the planetary magnetospheres and determining the entire structure of the heliosphere.

At distances of several solar radii, in the upper corona, the open structures occupy extended regions and the solar wind flux becomes radial predominant. These regions of open magnetic field, named coronal holes are the best-established sources of the stationary solar wind, but also sources of other great speed streams, which can reach even 700 km/sec. The coronal mass ejections, generally associated with other eruptive phenomena such as the solar flares or the active prominences, are themselves sure sources of the solar wind, being responsible for approximately 5% of the coronal mass loss as solar wind. The coronal streamers and the interfaces between these streamers and the neighbouring coronal regions are, hypothetically, sources of small-scale, slow streams (of approximately 300 km/sec) in the solar wind. There are also small-scale dynamical phenomena observed at the base of the corona, such as the spicules, the macrospicules and the high speed jets. The polar plumes, other small-scale structures, observed at the total solar eclipses in the polar zones of the corona could play a significant role in the acceleration of the coronal plasma particles and in the mass moment and energy transfer in the solar wind.

3.1. CORONAL STREAMERS

The classic shape of a coronal streamer, as seen at a total solar eclipse, is that of a "helmet" filled with bright matter, situated in the lower corona ($r \leq R_\odot$) surrounded by a less bright "veil" which extends almost radially outward from the Sun. The bright region at the basis is made up of plasma captured in the magnetic loops (here $P_{gas}/P_{magn}<1$) deeply anchored in the photosphere. The more diffuse neighborhood consists of rarefied plasma that follows the structure of magnetic lines, open in the upper corona, extending far away into the solar wind (here $P_{gas}/P_{magn} > 1$).

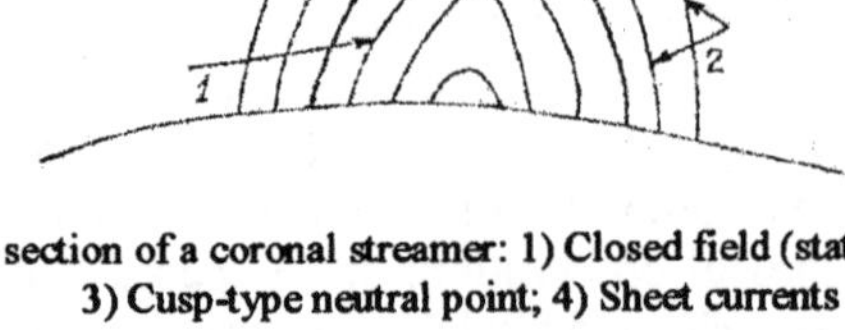

Figure 5. Schematic cross section of a coronal streamer: 1) Closed field (static); 2) Open field (expanding); 3) Cusp-type neutral point; 4) Sheet currents [8]

The regions of open and close field lines are separated by a streamer layer which converges in a magnetically neutral point situated on top of the "helmet", which corresponds, in three dimensions, to a neutral field line (figure 5).

The observations made at the total solar eclipses, which lasted for a couple of minutes, gave the impression that these streamers are long lasting, stable structures. Nevertheless, the observations with space instruments showed their daily variation due, firstly, to the differential rotation of the Sun and, secondly, as a result of the continuous evolution of the photospheric magnetic regions. The neutral layers, as interfaces between the magnetic fields of different polarities, which go down into the photosphere and merge in the boundary zones between active regions or between an active region and the weak polar magnetic field, justify the continuous evolution induced by the photospheric magnetic fields' dynamics. However, some coronal streamers can also undergo sudden morphological variations, sometimes even dramatically, associated with certain explosive phenomena of the solar activity in the near vicinity, namely: flares, surges, eruptive prominences. In certain conditions these phenomena are accompanied by coronal mass ejections which bring about a large scale restructuring, the coronal magnetic topology and the plasma itself. Thus, even a solar streamer can evolve rapidly transforming itself into a coronal mass ejection.

On the basis of all observation data and theoretical knowledge about the plasma behaviour in the magnetic fields, the magnetic nature of the coronal streamers was established. Two scenarios can be drawn up for the way in which a coronal stream forms, certain characteristics of the scenarios having been observed aboard the space missions. We present shortly these scenarios.

(a) A magnetic structure with closed field lines, containing very little plasma or no plasma at all, can gradually fill with plasma, coming from the inferior layers, through certain unspecific processes. When the thermal pressure of this plasma becomes of the same order as the magnetic pressure of the preexisting structure, an outward distension of the closed structure will begin. So, at the uppermost part and at the lateral sides of the inner closed magnetic structures, the lines will get a special extension which should allow us to consider them "open" at a first glance. Observation examples of this type of "filling" mechanism were registered with the coronagraph/polarmeter of the SMM mission (18 August 1980) and more recently, in radiation X, on Yohkoh. In these examples the bright central ray, which extends radially, far away in space, could not be seen. This could be due to the limitations of the instruments or, in such a mechanism in which successive layers of the closed field lines are somewhat long, extended, up to their "opening", the plasma content of the initial structure could "flow" outside.

(b) The other mechanism is via the reconnection of previously opened magnetic field lines. It requires, in order to form a coronal streamer, the presence in a region of sufficiently dense plasma of a bipolar magnetic field with an excess of "open" magnetic flux. Any energetic phenomenon transient in the upper atmosphere can produce such excess. At a gas pressure sufficient to balance the magnetic "stress" of the open field, the unbalanced $J \times B$ force makes the field lines of opposite polarities to come close. If the electric conductivity of the coronal plasma were infinite, a current sheet could form between regions of opposite polarities so big that it could keep the lines open until far, all throughout their length. In this current sheet the magnetic field gradient should be infinite and the field should change its direction discontinuously at the passage through the sheet. Nevertheless, in the more actual case of a conductivity sufficiently big but not infinite, the current sheet is replaced by a "diffuse region", of finite thickness across which the gradient of B is large. In this region the reconnection of the field lines will take place, while towards the bottom of the structure at the base of corona appears a zone of closed field lines, in a continuous development.

During the total solar eclipses, when the photosphere and even the solar chromosphere are obturated, the coronal streams can be registered photographically, and with photos of good resolution, inhomogenities of density and temperature, speeds and the magnetic field can be later determined. The observations at the total eclipses are essential because a very good photometric precision permits an absolute calibration for several space observations.

The observations of future eclipses should focus on experiments that are difficult to make on a space platform, such as: the analysis of the lines profiles for the emission lines in the intermediary corona, polarimetry in infrared, etc.

Some perturbations of the coronal streamers' morphology might be due to some transitory phenomena produced at great distances (such as, in another solar sector or even in the other hemisphere).

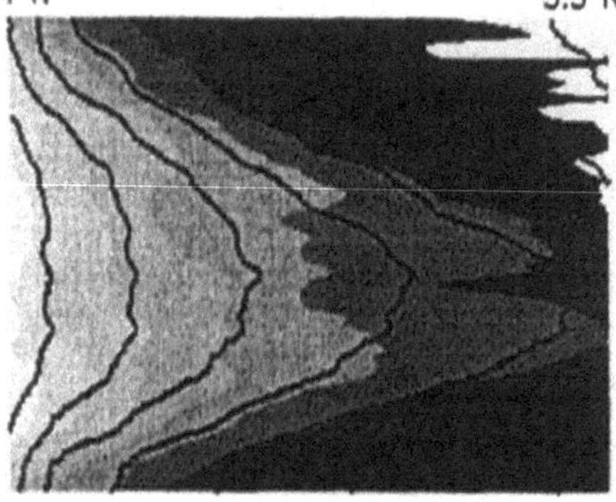

Figure 6. Comparison between HI and OVI images For the May 4, 1996 West limb streamer. Ly-α isophotes (solid lines) are superimposed on the OVI λ1032 image, given with a scale of grays [12].

The UltraViolet Coronograph Spectrometer (UVCS) on board the SOHO spacecrft, which aims to observe the coronal streamers, began the observations at the end of January 1996. Actually, the observations with this instrument are made in the wavelength intervals around line HI Ly-α (1145 Å – 1287 Å) and around the OVI resonance lines (984 Å – 1080 Å) at the wavelengths of 1032 Å and 1037 Å, respectively. With UVCS, on the third channel, one can also obtain records of the polarized radiance in the visible continuum. The instrument can observe the solar corona from 0 to 12 $R_\odot$ heliographic distance.

At the beginning of the observation period (January, 1996) the coronal streamers axis was placed in the equatorial plane of the Sun so that the shape of the solar corona was almost circular, a shape specific to the minimum activity. Later on, the structure of the corona complicated; the dynamics of this structure is very complex and cannot fit into a rigid model. The most unexpected thing observed is the structure difference between a streamer observed in Ly-α and OVI (figure 6). Thus, the coronal streamer observed on 4 May, 1996, appears in Ly-α as a singular structure, with a brightness maximum approximately along the axis, (the continuous line in figure 6), while in OVI, the streamer has a complex structure, being made up of 2 – 3 micro-streamers [13]. The cause of the structure difference between the observations in Ly-α and those in OVI could be due to:

(a) A *temperature effect*: an electronic temperature higher in the emission region of OVI can affect the electrons population more than that involved in the emission of the neutral hydrogen;

(b) A *speed effect*: the OVI lines are more sensitive to the Doppler effect than the HI Ly-α line;

(c) An *abundance effect*: if the oxygen is less abundant on the streamer axis than in the neighboring regions, its emission along the axis is diminished. It is also worth mentioning that HI, unlike OVI and other ions, has a particular behavior towards the local magnetic field, namely it can cross the field lines, thus modifying the concentration equilibrium in the presence of inhomogeneities.

From the measurements made with UVCS certain physical parameters of the coronal streamers could be deduced, studying the widths of the spectral lines and the ration of the OVI doublets. Thus, for the *kinetic temperature*, the component along the line of sight in several spectral lines could be determined. Here you are some of the values:

(a) For OVI: $T=3.1\times10^6$ K in the central part of the helmet region and $T=29.0\times10^6$K at 5 $R_\odot$;

(b) For HI: $T=(1.6-2.4)\times10^6$ K in the central part of the helmet region and $T=1.2\times10^6$K at 5 $R_\odot$.

There are no observation data on the velocity distribution (from which the kinetic temperature can be obtained) perpendicular to the line of sight. Indirect information can be nevertheless obtained for the OVI ion from the study of the ratio between the λ1037/λ1032 lines. The fact that this ratio shows a strong Doppler pumping of the λ1037 line to great heights in the solar corona permits the determination of the upper limit of the kinetic temperature of the OVI ions relative to the motions along the radial direction. Thus, for the coronal streamer observed on February 5, 1997, west of the solar disk, a kinetic temperature of $T=3\times10^6$ K was registered.

A lot of parameters are important for understanding the physical processes which take place in the coronal streamers. The OVI λ1037/λ1032 ratio also permits the determination of the *outflow velocity* in the solar corona [13]. For the June 4 east streamer, the value of the OVI λ1037/λ1032 ratio is between 0.3 and 0.4, slightly decreasing with the heliocentric distance from 1.5 $R_\odot$ to 3 $R_\odot$. This indicates a very small plasma velocity in the core of the streamer, which remains small (≤ 50 km/sec) even outside the core. UVCS can make a direct measurement of the *electron temperature* by determining the width of the Ly-α line scattered by the coronal free electrons.

3.2 . CORONAL MASS EJECTIONS

The coronal mass ejections are the strangest manifestations of the solar activity, because they involve the spitting out from the solar corona of huge masses of matter, of different speed, that are difficult to explain in the light of the present theories. In the most common case, the majority of the CMEs involve the eruption of the coronal plasma (sometimes even of the chromospheric one) from on region "stuffed" with magnetic field lines, initially closed (figure 7).

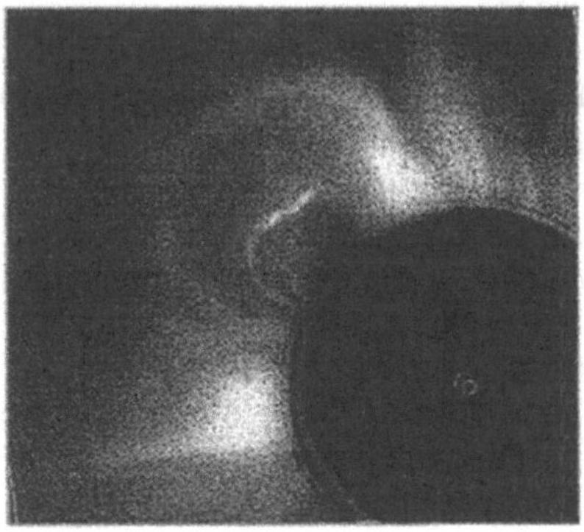

Figure 7. Coronal mass ejection seen by Solar Maximum Mission

Because the magnetic field is frozen in the high conductivity coronal plasma, the upper parts of these closed field lines are transported at the same time with the plasma ejected towards the exterior of the corona. The field lines are thus transformed into a huge "magnetic tongue", stretched far away into the interplanetary space. The region affected by the CME will have after the ejection a magnetic structure of open field lines. In many instances the ejections under the form of a closed loop have been observed to form through the disruption of a coronal streamer with the shape of a helmet, with the structure of a closed field in the lower corona. At other times these CMEs involve the eruption of a whole or only of a part of the loop type solar prominence.

Several parameters characteristic to these ejections can prove how interesting this phenomenon is. The movement of a CME matter in the solar corona takes place at speeds ranging between 10 and 2,000 km/sec. This involves the displacement of huge masses, sometimes even 2×10^{16} g of plasma being determined. In order to "lift" into the corona such a quantity of matter against the gravitation, at a speed of almost 2,000 km/sec, energies of 4×10^{32} ergs are necessary. We point out that the phenomenon takes place many times over a length greater than the 100 measured on the edge of the solar disk. All these characteristics prove that a CME can be considered as the most significant manifestation of the solar activity registered in the corona.

The CMEs were identified in the '70s, in the registerings of the chronographs aboard the OSO-7 missions and then on the Skylab. The most complete observations were made with the instruments of the SOLWIND mission between 1979 and 1985, and then aboard SMM in 1980 and between 1984 and 1989. Coronal mass ejections were identified also on the photos taken at the total solar eclipses of 1860 and 1893. At present, current observations are made from the ground with a chronometer at the Mauna Loa Observatory in Hawaii and also from space. The best spatial observations are supplied at present by the SOHO mission, in which experiments such as LASCO, EIT, CDS have registered CMEs both in the visible range and in the UV one.

A CME is most often quite well defined to allow relatively precise measurements of its location in the heliographic coordinates and its width (the angular dimension) expressed in degrees. Most of the CMEs appear right at the solar limb, and in this case the measurement of the angular dimension is quite precise.

A short analysis of a CME in the phase of minimum solar activity permits to point out several essential aspects, sometimes even strange, of this phenomenon, namely:

(a) The angular width of a CME is many times bigger than that of the associated solar activity phenomenon. The values of the CMEs expansion speeds, as well as the presence of an almost continuous acceleration, require the existence of some effective "driving" mechanisms over huge distances, of several solar rays.
The appearance of the CMEs in closed magnetic structures, those not connected with active regions included. CMEs appear also at high latitudes, far from the active regions, but in coronal structures of closed magnetic field lines. Consequently, any CME model has to be able explain the possibility of "breaking" these closed structures without resorting to strong magnetic fields and, therefore to greater densities of magnetic energy.

(b) The multitude of morphological forms of the CMEs often raises serious problems to a theoretical model.

(c) The expansion speeds of the phenomenon range between 10 km/sec and 2,000 km/sec, most of the CMEs often overpass the escape velocity from the solar gravitational field.

(d) A CME can trigger along with it masses of 2×10^{14} – 2×10^{16} g of solar plasma, as well as kinetic energies of 10^{29} – 10^{31} ergs (according to the observations on the SOLWIND, between 1979 and 1981).

Because the CMEs bring the eruption of the hot coronal plasma (of millions degrees) and the cold plasma of the prominence (of thousands degrees) the observation of such a phenomenon requires a multidisciplinary approach, instruments with a large view field and a sufficiently large temperature field.

Besides these large scale CMEs, recent observations from Yohkoh board (with the Soft X-ray Telescope) revealed frequent or even the continuous expansion of small coronal loops into space [3]. The reported velocity of such expanding loops above active regions were of the order of a few to several dozens km/s. The LASCO coronograph on board SOHO also revealed the frequent ejections of small loops within the streamer belts. It was suggested that these "blobs" might be the source of the slow speed solar wind. It is hard to know if the loops observed by Yohkoh and SOHO concern the same phenomenon.

A first objective is the study of the onset of a CME. Here we have to make the distinction between the observations of an event at the center of the disk, which is very probably directed towards the Earth, and an event observed at the limb, which is unlikely to be headed towards the Earth.

3.3. CORONAL HOLES

The coronal holes are large scale regions, of an open magnetic field structure, which extend from the surface of the Sun up to the interplanetary space. These structures were identified for the first time in the images obtained in the X radiation, on board the Skylab space mission, (1973 and 1974) as dark regions of great width, edged by bright points and covering huge zones from the poles of the Sun till the equator (figure 8). Later studies have proven that the coronal holes are short duration formations; they are observed as having almost the same shape throughout several solar rotations (figure 9).

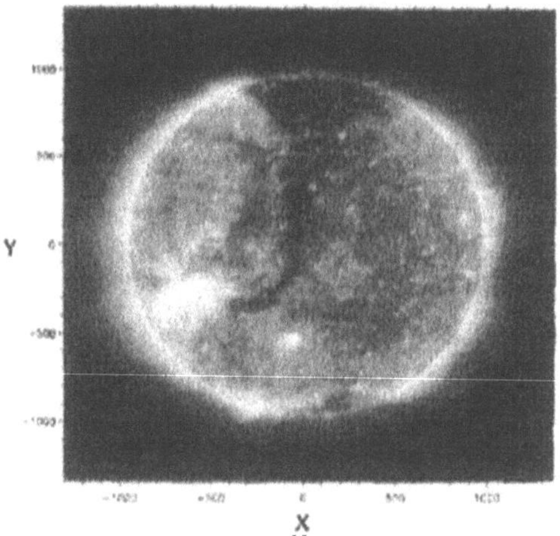

Figure 8. Coronal hole observed on board Skylab

According to their heliographic location, there are polar and equatorial coronal holes. In the phases of minimum solar activity coronal holes of great extension appear, that extend from the poles till the opposite hemisphere, including polar and equatorial regions into one single structure. It has been statistically observed that the passing of such regions over the central meridian of the Sun is associated with an increase of the geomagnetic activity, which has led to the identification of the coronal holes with the M regions, hypothetical causes of the recurrent geomagnetic storms. Krieger (1973) has traced the approximate trajectory of a high speed stream in the solar wind and has deduced that stream had its origin in an equatorial CH. Radio and optical observations from the ground and from space (OSO-7, Skylab) have later proven the truth of this conclusion. Later determinations of the physical parameters in the solar wind and in the coronal regions, as well as the modelling of the solar and interplanetary sectorial magnetic structure supported this association. Thus, the main arguments concerning the origin of the solar wind streams in a coronal hole were in chronological order:

(a) the almost perfect association between the passing at the central solar meridian of an equatorial CH and the registering of a rapid stream in the solar wind;
(b) the coincidence between the global maps of the high speed streams in the solar wind (obtained using observations of radio scintillation of some

compact radio sources) and those of the polar coronal hole (the dimension and shape of a coronal hole);

(c) the identification of the unipolar magnetic field sectors, from the high speed individual streams, with the unipolar sectors that fill the coronal holes;

(d) the electronic temperature in the upper coronal region deduced from the observations of the solar wind (beyond 0.3 $R_\odot$) is almost identical with the electronic temperature observed in a coronal hole, in the inner corona.

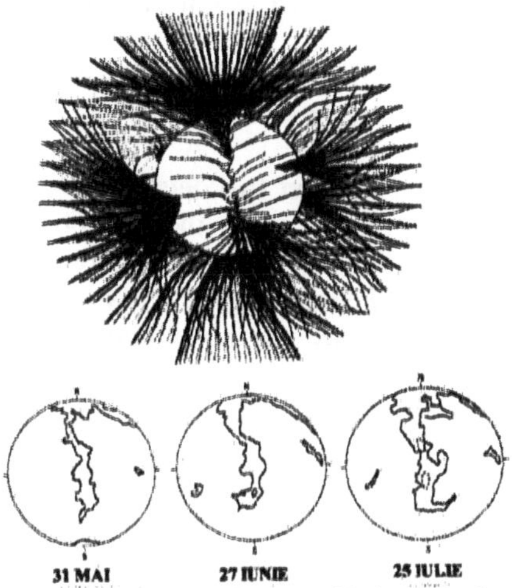

Figure 9. Coronal magnetic field (calculated, up) and a coronal hole during three solar rotations (down).

3.4. SMALL SCALE CORONAL STRUCTURES

Besides the large scale structures presented in the preceding paragraphs, the complex topology of the coronal magnetic field determines a series of small scale structures in the lower and intermediate regions of the corona, namely polar plumes or rays, plasmoids, jets.

Polar plumes are quasiradial structures, present in the zone of the polar corona. These somewhat long structures which follow the magnetic field lines open in the poles zone, have a higher density and a lower temperature than the plasma present in the coronal holes, in the space between the plumes.

The atmospheric part, the base of the polar plumes is at the intersection of the photospheric supergranular network, where the magnetic field concentrations can reach several hundreds gauss. It is supposed that the flux tubes, which represent the magnetic support of the polar plumes, are deeply rooted in the convective zone.

The main dynamic phenomenon associated with the polar plumes is the appearance in the solar wind of high-speed streams that are supposed to have originated inside the polar plumes. A possible acceleration mechanism of the plasma inside the polar plumes can have the following scenario:

The intense magnetic flux tubes which prolong the polar plumes deeply into the interior of the convective zone, undergo convective movements which become manifest at very different scales in that zone. Thus, MHD waves can be excited, andthen propagate at the surface and further into the atmosphere, through the interior of the plumes, which can act as wave-guides. The dissipation of these waves somewhere in the region of the interface between the chromosphere and the corona could then offer the quantity of energy to accelerate the plasma in the solar wind up to speeds of several hundreds km/s.

The launch of the SOHO space mission by NASA and ESA on 2 December, 1995, opened a new perspective for a systematic research of some characteristic phenomena for the solar activity, studied only sporadically until then, at a rather descriptive level. The possibility of simultaneously observing the Sun in several wavelengths brought forth so far unknown manifestations of some structures present in the solar atmosphere.

Before the SOHO space mission, the studies dedicated to the polar plumes were almost entirely based on images obtained on the occasions of the total solar eclipses.

Saito [14, 15] published the first two detailed studies on the morphology and physical properties of the polar plumes using the data obtained at the eclipses of 25 February, 1952 (Khartoum, Sudan), 30 June, 1954 (Holmhallar, Germany), 20 June, 1955 (the South of Vietnam), of 5 February, 1962 (New Guinea).

According to his conclusions, the polar plumes (polar rays) are distributed under the form of a large ring, of approximate 10°, at a distance of $10^\circ - 15^\circ$ from the pole. Taking into consideration also the three-dimensional disposal of the rays, Saito comes to the conclusion that their number is situated in the interval 30 – 40 in the minimum period of the solar activity. Regarding the lifetime of the plumes, Saito concludes that very few of them disappear in less than 6 hours or survive more than 24 hours. The plumes' width at a distance of 1.2 $R_\odot$ ranges between 0.02 $R_\odot$ and 0.07 $R_\odot$. On the basis of photometric measurements the electronic density is estimated to be approximate 5.4 times greater than that of the surrounding environment. By measuring the density gradient a kinetic temperature was obtained, of 1.2×10^6 K inside the polar plumes and of 1.0×10^6 K for the surrounding corona.

In their study on the polar plumes, Newkirk and Harvey (1967) focussed especially on the relation between the polar coronal plumes and the photospheric structures that correspond to them. The data used were images of the corona in white light, obtained at the eclipses of February 5, 1962, July 20, 1963 and May 30, 1965. For the data analysis were used statistical methods based on the Fourier transform. Assuming the existence of a cylindrical symmetry the authors established that a typical plume has at the center a density of about 10^8 cm^{-3} and a variation profile with the ray given by the relation:

$$N \approx 10^8\left(1-\frac{r}{3.6\times10^4}\right)^{1.6}. \qquad (1)$$

At 1.1 $R_\odot$, the radius of a typical polar plume is of 3.3×10^4 km and the average distance between plumes is of 7×10^4 km. On the basis of the analysis of the images from 33 eclipses ranging between 1889 and 1963, the two authors discovered that there is a direct relationship between the polar plumes and faculae. They also presented a qualitative model of plumes supposed to have the origin in the intense magnetic flux in the chromospheric network nodes.

An analysis of the polar plumes in EUV was made by Ahmad and Withbroe (1977) starting from the data obtained on board the Skylab in the lines MgX λ625 and OVI λ1032. The two authors extracted quantitative information on the plasma temperature inside the plume studied, using the ratio of the two spectral lines' intensity. Thus, the average temperature of the three polar plumes analyzed was of $T=1.1\times10^6$ K, a value very close to that of 1.2×10^6 K obtained by Saito under the assumptions of isotherms and the hydrostatic equilibrium. As for the electron density, Ahmad and Withbroe supposed an exponential decrease by distance *r* from the cylinder axis:

$$N_e(R,r)=N_0(R)\exp\left(-\left(\frac{r}{\beta}\right)^2\right), \qquad (2)$$

where $N_0(R)$ is the density on the axis at distance R from the center of the Sun, and parameter β is a measure of the plume's actual width. From the study made it became clear that the logarithmic decrease of the density at the plume center has a gradient slightly more abrupt than that observed at a higher altitude in white light. Ahmad and Withbroe advanced for the first time a hypothesis according to which the polar plumes contain approximately 15% of the plasma in the coronal hole and occupy about only 10% of its volume, but they could be a credible source for the streams in the solar wind.

Wang [17] issued the hypothesis according to which, the source of the high speed streams in the solar wind is placed in the space between the plumes. This space takes the greatest part of the volume of a coronal hole, although the density of the mass flux is greater inside the polar plumes. The model used by Wang is a stationary one, with the variables depend only on the heliocentric distance. It is described by the conservation equations of the mass, impulse and energy throughout a flux tube:

$$\frac{\rho u}{B}=\xi=const. \qquad (3)$$

$$\rho u\frac{du}{dr}=-\rho\frac{GM}{r^2}-\frac{dp}{dr}-\frac{d}{dr}\frac{\langle\delta B^2\rangle}{8\pi} \qquad (4)$$

$$B\frac{d}{dr}\left(\frac{(F_c+F_b+F_m+F_A)}{B}+\xi\left(\frac{5kT}{m_p}+\frac{u^2}{2}-\frac{GM}{r}\right)\right)=-n^2\Phi \qquad (5)$$

In the above equations, $p=2nkT$ is the thermal pressure, $\langle\delta B^2\rangle$ is the r. m. s. amplitude of the magnetic fluctuations, F_c and F_A are the densities of the conductive energetic flux and of the Alfven waves, respectively, and $\Phi(T)$ is a function for the radiation loss.

According to this model, energy has to accumulate at a scale comparable with *R* in order to produce mass fluxes and temperatures characteristic to the solar wind streams [11]. A supplementary energy has to be dissipated close to the coronal base of the plume, which will lead to the appearance of a very abrupt temperature gradient in that region to obtain the densities characteristic of the polar plumes. The model also shows that the base of the plume is very close to the hydrostatic equilibrium.

Wang and Sheeley [18] studied the polar plumes using the EUV spectroheliograms obtained on Skylab. The analysis of some images obtained in the emission line MgIX λ368 indicates the presence of two types of polar plumes. Some, lance shaped, show a well defined core in line MgIX, while others, under the form of sheets, have a more diffuse base which can extend over several supergranules. They observed that at the basis of all bright plumes there are bright points in the lines NeVII λ465 and HeII λ304, while the reverse is not true: not all the bright points in NeVII and HeII are associated with plumes observable in line MgIX.

The most complex set of data obtained so far on polar plumes is that from the observation campaign of 7-8 March 1996, with several instruments on SOHO, as well as the coronagraph Mk 3 in Mauna Loa. There were obtained series of the polar regions images, in several wavelengths, making thus possible a detailed analysis of some well individualized plumes, from the level of the photosphere to approximate 15 $R_{\odot}$.

The plumes observed originated in small concentrations of unipolar magnetic flux, of diameter about 2-5 arc seconds, situated at the intersection of the chromospheric network cells. They are colder and denser than the region in the surrounding coronal hole and can be clearly observed in the emission lines FeIX and FeXII. These lines mean an ionization temperature in the interval $1.0\text{-}1.5\times10^{6}$ K. Initially the plumes are in rapid expansion at the same time with the altitude growth, up to diameters of 20 – 30 Mm at heights of 30 Mm from the surface. Above the level of 1.2 $R_{\odot}$, the plumes can be observed in white light, up to around 15 $R_{\odot}$. With spatial scales smaller than 10 arc seconds, the plumes structure in the lower corona ($R<1.3R_{\odot}$) seems to be stationary over periods of at least 24 hours.

One can also determine from the EIT images in a qualitative way which regions are hotter by comparison with others, by comparing the relative brightness of the structures present in two wavelengths of Fe. The hotter structures issue more radiation from the more strongly ionized states. The most remarkable characteristic of the plumes is their continuous variability at small scale and their static nature at large scale. The shape of each plume's core changes dramatically at a time scale smaller than 10 minutes, with small filament structures (section of ~5 arc seconds) glowing and fading apparently in an aleatory way. These intensity variations come up to 10% of the plume brightness comparing with the interplumes regions and propagate outside at speeds of 300-500 km/s. This value is close to the estimated Alfvén speed, suggesting the presence of compression type waves, generated by possible small eruptions at the plumes basis.

4. Future Space Missions in the Heliosphere

Instead of conclusions we like to underline some ideas:

(a) The heliosphere is part of our galaxy which is swept by the solar wind;

(b) In this "cavity" the Earth with its magnetosphere is travelling through the different regions of the "ballerina" model;

(c) The understanding and forecasting of the solar phenomena are very important goals for the space weather knowledge.

(d) A lot of spacecrafts are flowing into the heliosphere bringing the data on its physical parameters.

A lot of projects are been prepared for the next heliospheric probes. **The Interstellar Probe** is a mission that will be sent out of the solar system to take data on the nearby interstellar medium and heliosphere-interstellar medium interactions. For the inner heliosphere observations **the Mercury-Messenger mission** is proposed to approach the Sun at around 0.3 AU.

5. References

1. Ahmad, I.A., Withbroe, G.L., (1977), Solar Phys. 53, 397
2. Bowers, R.L., Wilson, J.R. (1991) Numerical modelling in applied physics, Jones and Barlett Publ., Boston.
3. Harrison, R. A. (1997), CME Onset Studies, ESA SP-404, September 1997, 85-92
4. Holzer, T.E. (1989) Interaction between the solar wind and the interstellar medium, Annu. Rev. Astron. Astrophys. 27, 199-234
5. Low, B.C. (1996) Solar activity and the corona, Solar Physics 167, 217-265
6. Smitth, E.J. (1991) The Sun and interplanetary magnetic field, in Sonett, C.P., Giampapa, M.S., and Matthews, M.S. (eds.), The Sun in time, The University of Arizona Press, Arizona, pp. 175-201.
7. Jokipii, J.R. (1991) Variations of the cosmic-ray flux with time, in Sonett, C.P., Giampapa, M.S. and Matthews, M.S. (eds.), The Sun in time, The University of Arizona Press, Arizona, pp. 205-220.
8. Kopp, R.A. (1992) Theory of coronal streamer formation, Proceedings of the First SOHO Workshop, Annapolis, Maryland, USA 25-28 August 1992, pp. 53-61.
9. Koutchmy, S. (1992) Streamer eclipse observations, Proceedings of the First SOHO Workshop, Annapolis, Maryland, USA 25-28 August 1992, pp. 73-82.
10. Gosling J.T. (1996) Corotating and transient solar wind flows in three dimensions, Annu. Rev. Astron. Astrophys. 34, pp. 35-73.
11. Ghizaru, M.S., Goossens, M. (1996), Rom. Astron. J. 6, 21
12. Ghizaru, M.S. (1997), in Mouradian, Z., and Stavinschi, M., (eds.), Theoretical and observational problems related to solar eclipses, NATO ASI Series, Kluwer Acad. Publ., 77
13. Noci, G. et al., 1997, Te Quiescent Corona and Slow Solar Wind, ESA SP-404, September 1997, 75-84
14. Saito, K. (1965) Publ. Astron. Soc. Japan 17, 1.
15. Saito, K. (1958) Publ. Astron. Soc. Japan 10, 49.
16. Schwenn, R. (1990) Large-scale structure of the interplanetary medium, in Schwenn, R. and Marsch E. (Eds.) Physics of the Inner Heliosphere, Springer-Verlag Berlin Heidelberg pp. 99-181
17. Wang, Y-M., Sheeley N.R. Jr. (1995) ApJ 452, 457.
18. Wang, Y-M. (1994) ApJ 435, L153.

THE SOLAR CORONA: WHITE LIGHT POLARIZATION AND MODELLING OF THE LARGE SCALE ELECTRON DENSITY DISTRIBUTION

P. CUGNON and J. -R. GABRYL
Royal Observatory of Belgium
Avenue Circulaire, 3 B-1180 BRUXELLES

1. Introduction

The corona is the outer part of the solar atmosphere, visible in white light during the total solar eclipses. As such, it appears constituted of giant loops and of streamers, extending up to several solar radii. The corona is separated from the chromosphere by a narrow transition region with very high temperature gradients. In this very thin layer, the temperature increases from about 10 000 K in the chromosphere up to millions K in the corona. Though considerable progresses have been recently made in this field, the heating of the corona remains a rather unsolved problem of the physics of the corona. This high temperature and the low density imply that the corona is constituted of highly ionized thin plasma. The temperature appears roughly stable between 10^6 and $2\ 10^6$ K (Withbroe [1]). Hydrostatic equilibrium may be assumed for some local investigations, but the existence of the solar wind and of frequent big coronal mass ejections (CME) contradicts this assumption at larger scales.

The solar corona not only produces visible light, but also emits in the X, EUV, UV, and in the radio wavelengths. Space and ground-based observatories operating in these domains allow now deep insight in the different layers constituting this highly variable part of the solar atmosphere. Eclipse observations remain nevertheless a very useful way to investigate the large-scale structure of the corona and its long-term variability. This lecture gives describes a method for large scale modeling of the corona using white light observations taken during total solar eclipses, with some emphasis on the program of the *Royal Observatory of Belgium*.

1.1. RADIATIVE STRUCTURE OF THE CORONA

The radiative corona consists in three components:

- *The E-corona*: the E-corona extends up to 1.6 solar radii, and is constituted of all the ions emission lines. These lines are formed in the denser part of the low corona, and extend from the EUV to the nearby IR. The spectroscopic observation of the E-

J.-P. Zahn and M. Stavinschi (eds.),
Advances in Solar Research at Eclipses from Ground and from Space, 49–66.

corona informs us about element abundances, temperatures, non-thermal (MHD) phenomena and velocities. The variation of the line intensities as a function of time during solar cycle shows density and temperature increasing at maximum activity. The empirical density-temperature relation $N_e = 10^{-10}\, T^3$ (Waldmeier [2]) may be used for rough estimates. The E-corona is on average 20 times weaker than the continuum. Its influence is negligible in large pass-band photometry.

- *The K-corona (or electronic corona)* extends up beyond more than 4 solar radii, and is constituted by photospheric light scattered by free electrons (*Thomson scattering*). Its spectrum is identical to the photospheric continuum, but the absorption lines are absent due to electron thermal Doppler broadening, except for the intense H and K CaII lines. Indeed, from $\Delta\lambda/\lambda = v/c$ we obtain a broadening of 120 Å at 5000 Å, using an electron thermal velocity of $7\ 10^{6}$ m/s. Thomson scattering produces also *partially polarized light*, with the electric field parallel to the limb. This important property will be used in the forthcoming analysis.

-*The F corona (or dust corona)* becomes dominant beyond 2.4 solar radii and is difficult to distinguish from the K-corona, because it also consists in scattered photospheric white light. The main differences with the K-corona are the reddening due to a greater scattering efficiency of the dust grains in the blue and the visibility of the Fraunhoffer absorption lines in the absence of Doppler broadening. The F-corona is not polarized, remains stable with time, and presents a spherical symmetry up to approximately 2 solar radii. Beyond this limit, its flattening becomes noticeable, due to the preferential concentration of dust in the ecliptic plane.

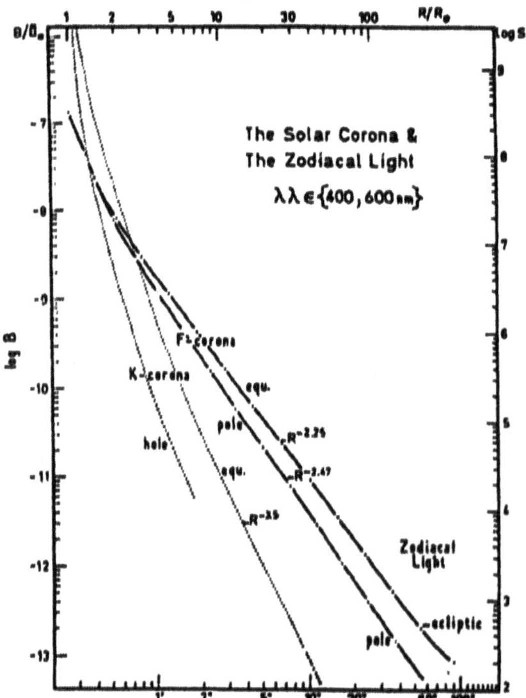

Figure 1: The F- corona (zodiacal light) and the K-corona compared (Koutchmy [3])

1.2. EVOLUTION OF THE GENERAL SHAPE OF THE CORONA WITH THE SOLAR CYCLE

In order to quantify the morphological changes in the shape of the solar corona during the 11-year cycle, Koutchmy & Loucif [4] proposed to characterize the eclipse corona by its phase in the cycle, its main deviation of big streamers with respect to latitude, and its flattening index. It appears clearly that the mean deviation as well as the flattening index follows the sunspot activity. The corona that appears flattened at minimum, exhibits a roughly spherical shape at maximum, as shown in Figure 2.

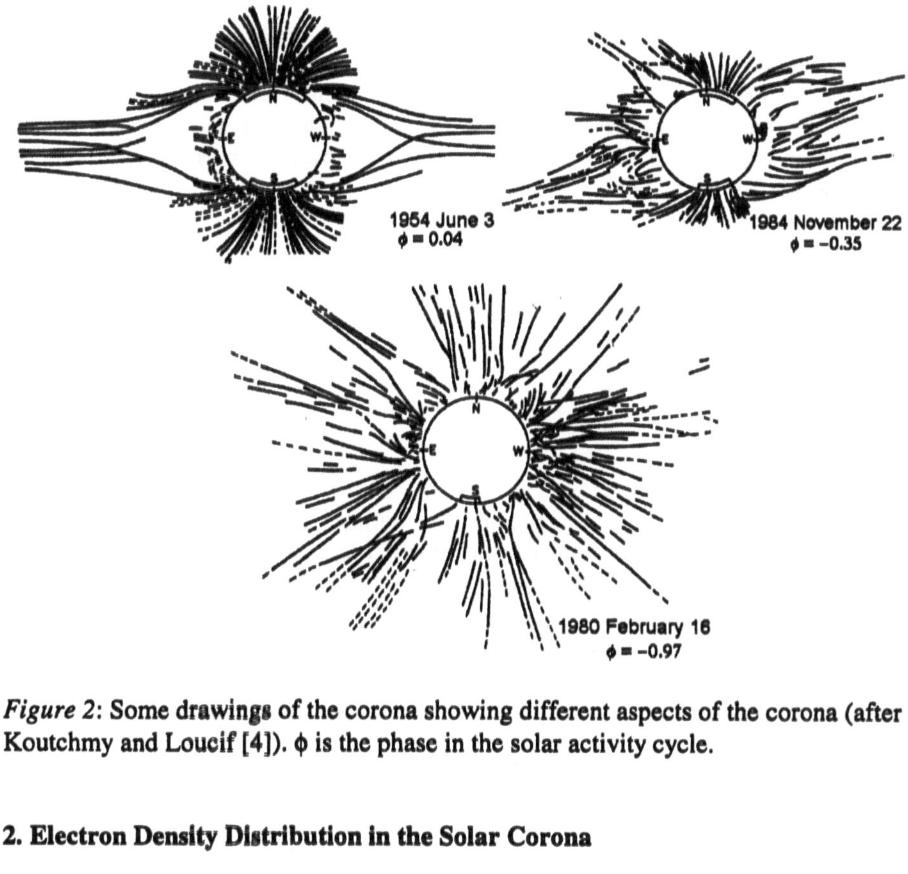

Figure 2: Some drawings of the corona showing different aspects of the corona (after Koutchmy and Loucif [4]). ϕ is the phase in the solar activity cycle.

2. Electron Density Distribution in the Solar Corona

2.1. SOME HISTORY

The first coherent approach of the problem is due to Schuster [5] who made the correct assumption that the light from the corona was due to scattering of the photospheric light. The acting particles were naturally unknown. With this simple assumption, however, Schuster could already theoretically explain that the polarization was *tangential* and related to the density of matter in the corona.

In 1930, Minnaert [6] identified the scattering particles to free electrons, and proposed the Thomson scattering mechanism. He also introduced in his developments the ***center-to-limb darkening***. The F-corona light was attributed to "Emissive unknown particles". Minnaert derived the first expression of the ***electron density distribution***, based on a simple radial decrease of the electron density, for which he made use of the polarized intensity, independent of the F-corona contribution. Baumbach [7] improved the description of the electron density decrease in the corona using a three terms power law of the inverse of the radius.

An important step forward was due to van de Hulst [8] who gave distinct descriptions of *minimum* and *maximum* types corona, by treating separately the *polar* and *equatorial* distributions. He also interpreted correctly the F-corona as scattered light by small solid particles, and calculated tables of electron densities.

The first latitude dependant axisymmetric models of the electron density distribution were due to Saito [9], who also used an inverse power law for the radial decrease. He also provided numerous tables, namely of the F-corona contribution.

2.2.THOMSON SCATTERING THEORY

A polarized electromagnetic wave encountering an electron communicates to this electron an acceleration given by (at the classical and non-relativistic approximation - terms in *v/c* neglected):

$$\mathbf{a} = \frac{e}{m_e}\mathbf{E}_0 \tag{1}$$

This excitation will result in an oscillatory « motion » of the electron giving rise to an oscillatory dipole emitting an electromagnetic radiation of which the electric field at a vector distance **l** is given by

$$\mathbf{E} = \frac{e^2}{4\pi\varepsilon_0 mc^2 l^3}\mathbf{l}\times(\mathbf{l}\times\mathbf{E}_0) \tag{2}$$

If ψ is the angle between the electric field direction of the incident wave and the direction of scattering, the square of the scattered field is

$$E^2 = \frac{e^4 E_0{}^2}{\left(4\pi\varepsilon_0 mc^2 l\right)^2}\sin^2\psi \tag{3}$$

In term of intensities, the angular distribution of the scattered radiation can then be written

$$I = \left(\frac{r_e}{l}\right)^2 I_0 \sin^2\psi = \sigma I_0 / l^2 \tag{4}$$

where r_e is the classical electron radius, which is equal to

$$r_e = \frac{e^2}{4\pi\varepsilon_0 mc^2} = 2.8\ 10^{-13}\text{cm} \tag{5}$$

2.3. THE GEOMETRY OF THE PROBLEM

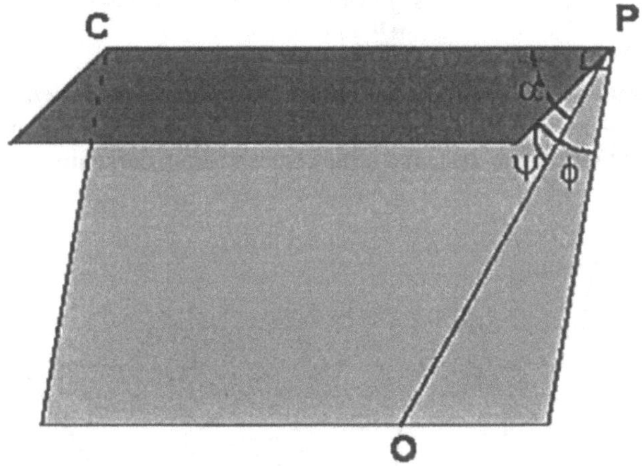

Figure 3: Thomson scattering for light emitted from a point source

We start from the simple case (Figure 3) of a point source C, an electron at P, and the line of sight PO making an angle α with the incidence direction CP. For a particular polarization direction making an angle ϕ with the plane CPO, the angle ψ as defined above is related to ϕ and α through

$$\cos\psi = \cos\phi\ \sin\ \alpha \tag{6}$$

so that

$$\sin^2\psi = 1 - \cos^2\phi\ \sin^2\alpha \tag{7}$$

All directions of polarization being equally represented in the incident wave, we have to average on the angle ϕ

$$\langle\sin^2\psi\rangle = (2 - \sin^2\alpha)/2 = 1/2 + (\cos^2\alpha)/2 \tag{8}$$

$\alpha = \pi/2$ corresponds to the case of a fully polarized scattered wave in the plane perpendicular to CP, so that the term 1/2 represents the proportion of tangential polarization. Alternatively, the term $(\cos^2\alpha)/2$ is the radial contribution to the polarization. The radiation is then always partially tangentially polarized.

The Sun being not a point source, we have to take into account all the surface element contributions included in the part of the photosphere intercepted by the cone tangent to the Sun originating in P (Figure 5). After some mathematics, the two following equations give the intensities scattered from P:

$$J_t = \frac{\sigma}{2}\pi N_e \sin^2\xi \int_{\cos\Omega}^{1} I(\varphi)(1+\cos^2\omega)\,d(\cos\omega) \tag{9}$$

$$J_t - J_r = \frac{\sigma}{2}\pi N_e \sin^2\xi \int_{\cos\Omega}^{1} I(\varphi)(-1+3\cos^2\omega)\,d(\cos\omega) \tag{10}$$

J_t and J_r are respectively the tangential and radial scattered intensity, $I(\varphi)$ is the intensity emitted by the Sun per surface unit, $I(\varphi) = B_{\odot}(1 - u + u\cos\varphi)$ where $B_{\odot}$ is the central brightness of the disc, and u the center-to-limb darkening parameter.

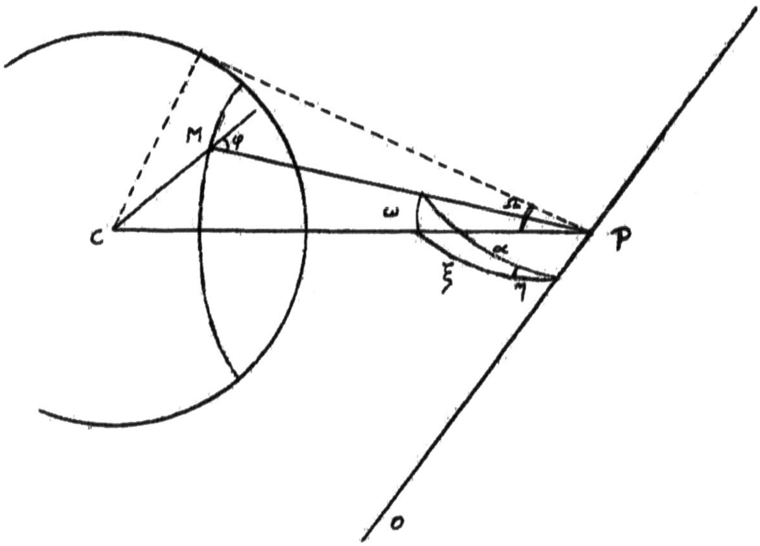

Figure 4 : geometry of the scattering of the photospheric light

After integration on the angle ω, the scattered intensities may be expressed as functions of the distance CP = r, of the electron density at P, and of the center-to-limb darkening parameter. Introducing now the functions $A(r)$ and $B(r)$ (Van de Hulst [8]),

$$J_t = \sigma N_e \pi B_{\odot} A(r) \;\; ; \;\; J_t - J_r = \sigma N_e \pi B_{\odot} \sin^2\xi\,[A(r) - B(r)] \tag{11}$$

These expressions still need to be integrated along the line of sight to obtain the correct general final forms:

$$K_t(x) = C\int_x^{\infty} N_e A(r)\frac{r\,dr}{\sqrt{r^2+x^2}} \;\; ; \;\; K_t(x) - K_r(x) = C\int_x^{\infty} N_e\,[A(r)-B(r)]\frac{x^2\,dr}{r\sqrt{r^2+x^2}} \tag{12}$$

Where $C = 2\pi\sigma \times 10^{-6} R_{\odot}$, if r is expressed in $R_{\odot}$ and K in units of $10^{-6} B_{\odot}$. The respective expressions of the total intensity and of the degree of polarization of the K-corona are then :

$$K(x)=2\,K_t(x)-(K_t-K_r)(x) \quad ; \quad p_k(x)=\frac{(K_t-K_r)(x)}{K(x)} \tag{13}$$

The expressions of A and $A - B$ are

$$A(\Omega)=\frac{1-u}{1+\frac{u}{3}}\left(\frac{2}{3}-\frac{\cos\Omega}{2}-\frac{\cos^3\Omega}{6}\right)$$
$$+\frac{1}{16}\frac{u}{1-\frac{u}{3}}\left[5+\sin^2\Omega-\frac{\cos^2\Omega}{\sin\Omega}(5-\sin^2\Omega)\ln\left|\frac{1+\sin\Omega}{\cos\Omega}\right|\right] \tag{14}$$

$$A(\Omega)-B(\Omega)=\frac{1-u}{1-\frac{u}{3}}\frac{\cos\Omega\,\sin^2\Omega}{2}$$
$$+\frac{1}{16}\frac{u}{1-\frac{u}{3}}\left[-1+3\sin^2\Omega+\frac{\cos^2\Omega}{\sin\Omega}(1+3\sin^2\Omega)\ln\left|\frac{1+\sin\Omega}{\cos\Omega}\right|\right] \tag{15}$$

here expressed as a function of the angle Ω, with $\sin\Omega = r^{-1}$

2.4. VAN DE HULST DENSITY DISTRIBUTION MODEL

Van de Hulst model gives a correct interpretation of the F-corona: photospheric light scattered by small dust particles (close-to-Sun part of the zodiacal light). The theory (Rayleigh or Mie scattering) predicts that this component is also tangentially polarized. However, this polarization is very small compared with that due to electron scattering. The distribution of dust is supposed to be spherical and to remain constant with time.

- Maximum corona : K_t and $K_t - K_r$, respectively the tangential and polarized intensities are expressed in powers of the inverse of the elongation to the Sun center. It is shown that this property is conserved for J_t and $J_t - J_r$, and the coefficients are then obtained by identification. Iterative processes are used to constrain both equations to produce similar results.

- Minimum corona and high latitude regions: the first step postulates isotropic scattering, yielding a first estimate of the electron density; the second step introduces a factor taking into account the anisotropy, with the integration limited to the solar disc.

The results are provided in tables, from 1 to 4 solar radii. The electron densities deduced close to the limb are of the order of 10^8 electrons/m^3.

2.5. SAITO NON-SPHERICAL MODEL

In order to account for the asphericity of the corona, Saito [9] proposed a latitude dependant form for the electron density:

$$N_e(r,\varphi) = \sum_{i=1}^{n} N_{0,i} \frac{1 - f_i \sin^{s_i} \varphi}{r^{b_i}} \quad (16)$$

Where the, b_i, f_i and s_i are quantities that are identified in three steps: the equatorial distribution ($\varphi = 0$) is first calculated using elliptical integrals. A minimization of the differences between the deduced theoretical and the observed radial equatorial behaviors of the K_t and K_t-K_r values allows then to determine the quantities $N_{0,i}$ and b_i. A similar method is used to identify the f_i from polar data (latitude ±90°). Finally, the s_i are determined using a few typical directions (e.g. 30°, 45° and 60°). For instance, here is given one the two analytical expressions obtained by Saito for the electronic density:

$$N_e(r,\varphi) = \frac{3.09\times10^8\ (1 - 0{,}5\ \sin\ \varphi)}{r^{16}} + \frac{1.58\times10^8\ (1 - 0{,}95\ \sin\ \varphi)}{r^{6}}$$

$$+ \frac{0.0251\times10^8\ (1 - \sin^{0.5}\varphi)}{r^{2,5}}\ \mathrm{e}^-/\mathrm{cm}^3 \quad (17)$$

3. The Scandariato-Gabryl-Clette-Cugnon (SGCC) Model

In order to treat the most general axisymmetric case with a minimum of constraints, namely without postulating a specific latitude dependence, our method [10] will make use of orthogonal functions of the angular variables, and of inverse powers of the elongation from the Sun center to describe the radial decrease. Some advantages of this method are, for instance:

- the flexibility of the model
- a not pre-imposed symmetry axis direction (not necessarily the Sun rotation axis)
- the natural separation of the angular and radial problems
- angular functions which are Legendre polynomials of the sine of the latitude, so that the problem may be solved at a chosen angular degree.

However, two still unsolved problems appear in the reconstruction process: numerical instability in the polar behavior of the model, and negative densities, which may be present in some limited regions.

We start again from the general form of van de Hulst integral equations (12). Any axisymmetric electron density distribution (Perry and Altschuler [11]) can always be represented by

$$N_e\,(r,\varphi)=\sum_{n}^{N_{max}} P_n(\sin\,\varphi)\; f_n\,(r) \tag{18}$$

where the f_n are radial functions for which we assume the following inverse power law structure:

$$f_n\,(r)=\sum_{i}^{I_{max}(n)} \frac{a_{i,n}}{r^{\,b_{i,n}}} \tag{19}$$

Introducing moreover the following change of variable,

$$cotg\,\xi=\frac{\sqrt{r^2-x^2}}{x}\quad for\,\chi\in\left[0,\frac{\pi}{2}\right]\;;cotg\,\xi=-\frac{\sqrt{r^2-x^2}}{x}\quad for\,\chi\in\left[\frac{\pi}{2},\pi\right] \tag{20}$$

the integral equations become

$$K_t(k,\varphi)=c\int_0^{\pi} A\,[\sum_n\sum_i P_n(\sin\,\varphi)\;a_{i,n}\;k^{\,b_{i,n}-1}\;\sin^{\,b_{i,n}-2}\chi\,]\,d\chi \tag{21}$$

$$(K_t\,-\,K_r)\,(k,\varphi)=c\int_0^{\pi}(A\,-\,B)\;[\sum_n\sum_i P_n(\sin\varphi)\,a_{i,n}\;k^{\,b_{i,n}-1}\sin^{\,b_{i,n}}\chi]\,d\chi \tag{22}$$

where $k=1/x$, and $c=\pi\sigma\times10^{-6}R_\odot$.

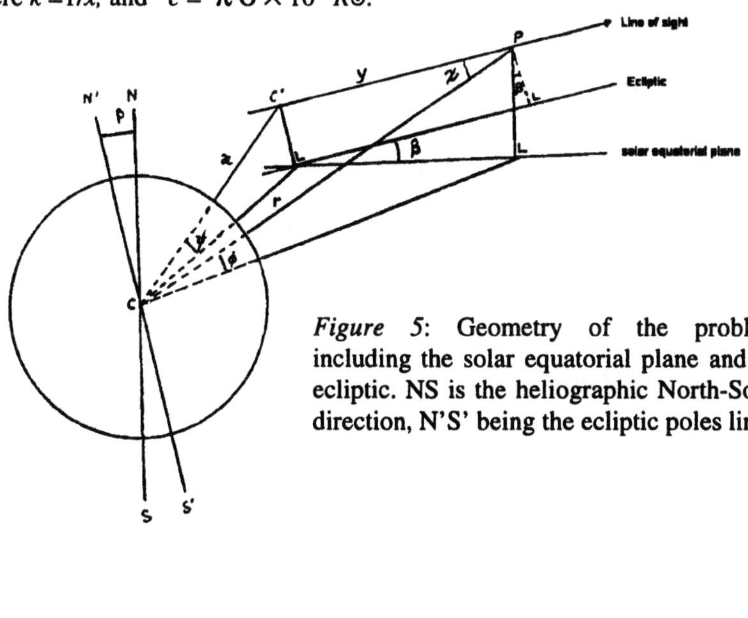

Figure 5: Geometry of the problem, including the solar equatorial plane and the ecliptic. NS is the heliographic North-South direction, N'S' being the ecliptic poles line.

With the help of Figure 5, the relationship between the angles ψ, β, φ and χ is straightforward :

$$\sin\varphi = \sin\psi \sin\chi \cos\beta + \sin\beta \cos\chi = \sin\psi \sin\chi \sin\beta_0 + \cos\beta_0 \cos\chi \qquad (23)$$

where $\beta_0 = \pi/2 - \beta$

Using the addition theorem of Legendre polynomials, we obtain expressions containing associated Legendre functions in χ. After some mathematics, including also a Taylor development of the functions A and B, equations (21) and (22) become

$$K_t(k, \psi_0) = \sum_n \sum_l \sum_m \sum_p E_{nlm} D_p k^{2p} \int_0^\pi P_n^m(\cos\chi) \sin^{b_{l,n}+2p-2}\chi \, d\chi \qquad (24)$$

$$(K_t - K_r)(k, \psi_0) = \sum_n \sum_l \sum_m \sum_p E_{nlm} C_p k^{2p} \int_0^\pi P_n^m(\cos\chi) \sin^{b_{l,n}+2p}\chi \, d\chi \qquad (25)$$

where

$$C_p = \frac{-p}{2p-3}\left[(1-u)\frac{p(2p-2)!}{(p!)\,2^{2p-1}} + \frac{u}{4p^2-1}\right]\frac{1}{1-\frac{u}{3}} \quad ; \quad D_p = \frac{p-2}{-p} C_p$$

$$E_{nlm} = c\, a_{l,n} k^{b_{l,n}-1} \frac{(n-m)!}{(n+m)!} P_n^m(\cos\beta_0) \cos(m\psi_0)(1 + \mathrm{sign}(m))$$

These integrals (24) and (25) may be performed immediately, yielding

$$\pi\, 2^m (-1)^n \frac{\Gamma\left(\frac{t+1+m}{2}\right)\Gamma\left(\frac{t+1-m}{2}\right)}{\Gamma\left(\frac{t+2+n}{2}\right)\Gamma\left(\frac{t+1-n}{2}\right)\Gamma\left(\frac{n-m+2}{2}\right)\Gamma\left(\frac{-n-m+1}{2}\right)} \qquad (26)$$

Where t is either $b_{l,n}+2p$, or $b_{l,n}+2p-2$, and is submitted to the following constraints

$$t + 1 > m; \quad (n + m)\,\text{even}; \quad n \geq m$$

Putting now for convenience

$$R(t,n,m) = (-1)^n \frac{\Gamma\left(\frac{t+m}{2}\right)\Gamma\left(\frac{t-m}{2}\right)}{\Gamma\left(\frac{t+1+n}{2}\right)\Gamma\left(\frac{t-n}{2}\right)}$$

$$KC_n^m = c\,\frac{(n-m)!}{(n+m)!} P_n^m(\cos\beta_0)(1+sign(m))\,2^{m+1}\sqrt{\pi}\prod_{j=0}^{\frac{n+m}{2}}\left(\frac{1}{2}-j\right)\left[\left(\frac{n-m}{2}\right)!\right]^{-1}$$

equation (22) and (23) take the following form

$$K_t(k,\psi_0)=\sum_{n\,i\,m\,p} a_{i,n}\, k^{\,b_{i,n}+2p-1}\, D_p \cos(m\psi_0)\, KC_n^m\; R(b_{i,n}+2p-1,n,m) \quad (27)$$

$$(K_t-K_r)(k,\psi_0)=\sum_{n\,i\,m\,p} a_{i,n}\, k^{b_{i,n}+2p-1}\, C_p \cos(m\psi_0)\, KC_n^m\; R(b_{i,n}+2p+1,n,m) \quad (28)$$

where the left-hand members contain the quantities deduced from the observation, while the right-hand members contains the coefficients and exponents to be identified.

3.1. SGCC MODEL : THE ANGULAR PROBLEM

Expressions (25) and (26) are sums of cos ($m\,\psi_0$) multiplied by radial functions :

$$K_t(k,\psi_0)=\sum_{m=0}^{N_{max}} B_m(k)\;\cos(m\,\psi_0)$$

$$(K_t-K_r)(k,\psi_0)=\sum_{m=0}^{N_{max}} A_m(k)\;\cos(m\psi_0)$$

The cosines constitute a set of orthogonal functions, so that the limitation to any maximum value of m appears natural, and corresponds to a certain level of complexity in the description of the corona. This property will be exploited to solve the problem in two steps. Considering the radial functions A_m and B_m as simple coefficients, it is now possible to solve the angular problem once fixed the maximum value of m, N_{max}. The K- tangential and polarized intensities supposed known on a pixels grid after reduction of the observations, a least square fit is applied on circular rings, leading to the determination of A_m and B_m.

$$S^2_{B,N_{max}}(k)=\frac{1}{N(k)-N_{max}}\sum_{j=0}^{N(k)}\left[K_t\,(k,\psi_0^{(j)})-\sum_{m=0}^{N_{max}} B_m(k)\cos(m\;\psi_0^{(j)})\right]^2 \quad (29)$$

$$S^2_{A,N_{max}}(k)=\frac{1}{N(k)-N_{max}}\sum_{j=0}^{N(k)}\left[(K_t-K_r)\;(k,\psi_0^{(j)})-\sum_{m=0}^{N_{max}} A_m(k)\cos(m\;\psi_0^{(j)})\right]^2 \quad (30)$$

3.2. SGCC MODEL : THE RADIAL PROBLEM

We define now two kinds of auxiliary functions F and G and regroup the terms in order to separate those terms related to the order m from those related to higher orders:

$$A_m(k) = KC_m^m \sum_{i=1}^{I_{max}(m)} a_{i,m} F_{i,m}(k) + \sum_{n=m+1}^{N_{max}} KC_n^m \sum_{i=1}^{I_{max}(n)} a_{i,n} F_{i,n}(k) \quad (31)$$

$$B_m(k) = KC_m^m \sum_{i=1}^{I_{max}(m)} a_{i,m} G_{i,m}(k) + \sum_{n=m+1}^{N_{max}} KC_n^m \sum_{i=1}^{I_{max}(n)} a_{i,n} G_{i,n}(k) \quad (32)$$

A method of inverse recurrence (Gabryl [12]) has been derived to solve the radial problem, starting with the determination of the coefficients $a_{i,m}$ and the exponents $b_{i,m}$ at the highest degree N_{max}, and using least squares fit. The number of radial terms is dynamically determined. In fact, a first couple of ($a_{i,m}$, $b_{i,m}$) is selected, which minimize the square function. The same method is applied to the residual, until the result is no more changed significantly. Gabryl has studied different approaches of the problem. One includes the careful examination of the behavior of the A_m and B_m radial functions, in order to localize the inflexion points. The above method is then applied to the different segments of constant curvature, with one couple ($a_{i,m}$ $b_{i,m}$) per segment.

4. Application to different eclipse observations campaigns

The electron density distribution being an important parameter in any physical modeling of the solar corona, we have applied this method to different eclipse results from campaigns organized by the Royal Observatory of Belgium.

Measuring the coronal brightness through a linear polarizer consists in projecting the *K-corona* components and filtering the *F-corona* intensity:

$$P(\alpha_k) = K_t \cos^2(\theta - \alpha_k) + K_r \sin^2(\theta - \alpha_k) + \frac{F}{2} \quad (33)$$

where θ and α_k are respectively the K_t and polarizer directions with respect to a given reference direction. Apparently, the measuring leads to a non-linear mixing of four quantities. However, trigonometric identities show that it is actually a linear mixing of three quantities X, Y and I known as the linear Stokes parameters :

$$P(\alpha_k) = \frac{1}{2}\left[X \cos(2\alpha_k) + Y \sin(2\alpha_k) + I\right] \quad (34)$$

with

$$X = (K_t - K_r)\cos(2\theta)$$
$$Y = (K_t - K_r)\sin(2\theta)$$
$$I = K_t + K_r + F$$

It is thus necessary to sample the polarization along at least three different directions to obtain the linear Stokes parameters. Furthermore, it may be shown that these parameters are deduced with the best accuracy when the sampling directions α_k are regularly spaced and constitute a closed system (n measurements spaced by $180°/n$). When $n > 3$ (oversampling), a least square method must be used to solve the problem.

Started in 1973 using a classical photographic device with a three-position polarizer, the observational program was improved in the early nineties through the use of a video CCD camera of which the output was "grabbed" in a computer and converted in digital images. This camera is also equipped with an automatically rotating filter. The data processing of the coronal images has been then highly fastened and the accuracy enhanced thanks to 24 directions sampling. The camera being however not able to cover the whole brightness dynamics of the corona, the sampling has to be repeated for different exposure times.

4.1. 1991, 1994 AND 1998 RESULTS

The major challenge to get the electron density distribution is to calibrate photometrically the total intensity I, because it is the key to the knowledge of the polarized brightness and of the tangential component of the K-corona., the polarization rate is indeed easily deduced from the linear Stokes parameters, and does not rely on the intensity scale.

$$p = \frac{\sqrt{X^2 + Y^2}}{I} = \frac{K_t - K_r}{K_t + K_r + F} \tag{35}$$

Thus, after the calibration of the total intensity, the absolute polarized brightness is directly derived, and using tabulated values of the F-corona, the absolute tangential brightness is obtained. Absolute calibration is theoretically possible using measurements of the photospheric brightness through a neutral density filter before and after totality. However, uncertainties in the filter transmission required to check the calibration using respectively the tabulated ratios of the total to the F-corona brightness for low intensity regions, and of the polarized to the K-corona brightness for high intensity regions. This provided a lower and an upper limit to the correction factor to be applied to the calibrated total intensity I.

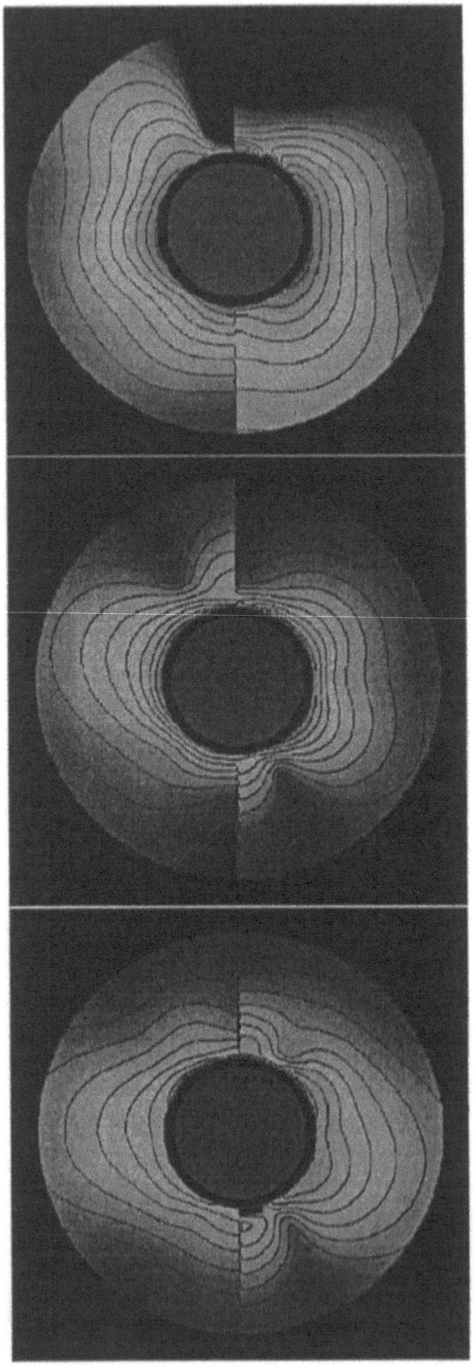

Figure 6: From top to bottom, electron density distribution plotted in logarithmic scale for 1991, 1994, 1999 total solar eclipses. The vertical direction corresponds to the solar rotation axis, North up. The step factor between two successive curves is 2 and the outermost curve respectively corresponds to 10^{11}, 10^{11} and 2 10^{11} e/m^3.

Up to now, three sets of observations have been fully processed to obtain electron densities: the 1991, 1994 and 1998 total solar eclipses. The analysis of the 1999 total eclipse observation is still in progress but preliminary results of polarization are already available. From our experience, it appears that the best modeling is reached when the maximum polynomial order is 6 or 7. We also noticed that two radial terms, sometimes three are generally sufficient to describe the decrease in density. Furthermore, the integral kernels introduce strong weighting, so that only the electrons close to the plane of the sky produce a strong contribution. Therefore we compute two distinct models with respect to the distribution axis. The reconstructions shown in figure 6 offer a good description of the main structures and of the radial gradients in high as well as in low-density regions.

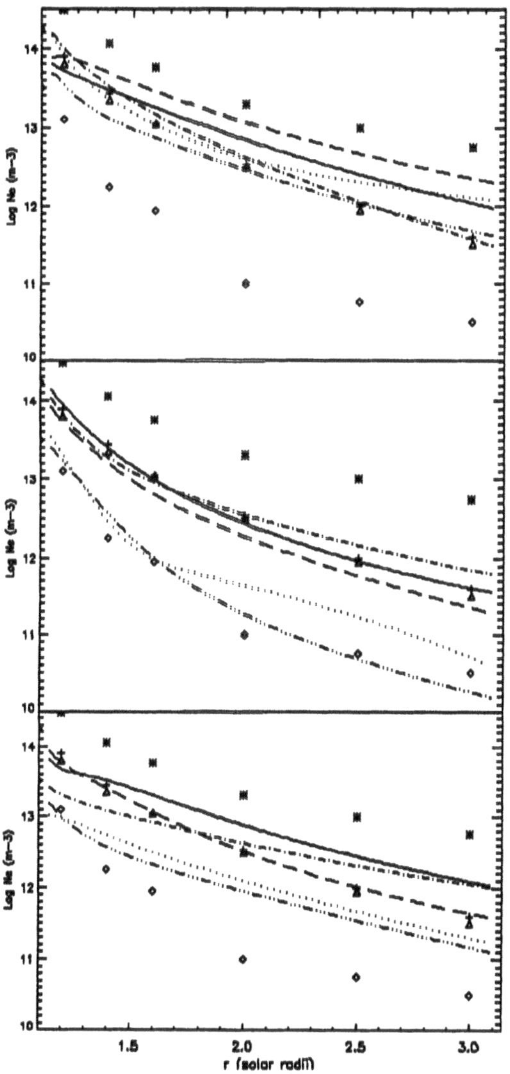

Figure 7: 1991 (top), 1994 (middle), and 1998 (bottom) electron density radial profiles. The dotted and dash-dot-dot-dotted profiles are relative to polar regions while the continuous and dashed profiles are relative to a typical streamer. Discrete symbols are Koutchmy's [12] tabulated values for a polar hole (◇), a streamer (*), the equatorial region during maximum activity (+) and minimum activity (△).

Up to now, only the 1991 corona model exhibits structures with high densities at high latitudes. In fact, against our own expectations, the 1998 density model is closer to that of 1994 (minimum type reconstruction), although the solar activity was in the ascent phase for two years. However, although the 1998 inner densities near the equator are just lower than or equal to those of 1994, the radial gradients are much smoother in the 1998 model. The 1998 equatorial values appear then of the same order of magnitude as the 1991 values in streamers at large distances. In that sense, the 1998 corona may be an intermediate type corona. Radial density profiles taken along typical structures from the different eclipse reconstructions lead to the same conclusion (Figure 7). Concerning the 1999 solar corona, it is still too early to derive conclusions. However, everyone agrees on the point that the observed radial intensity gradient was quite high. Although the shape is consistent with the maximum type corona, if the density gradient were that high, then the coronal temperatures would be rather low, as we shall see later.

These plots also suggest a kind of density regime. Considering that the measured extreme radial profiles (streamer and hole profiles) delimit a density area, the 1991 density area fills the band between the "equatorial" and "streamer" tabulated profiles, while the 1994 area fills the band between "polar" and "equatorial" profiles, and the 1998 area is approximately centered on the "equatorial" tabulated profiles. The transition between the minimum corona and the maximum corona seems thus to be first characterized by an increase of the electron density in the outer corona, so that the observed radial gradients are smoother. In the context of an isothermal hydrostatic corona, this means that the temperature is increasing (Badalyan, [13]).

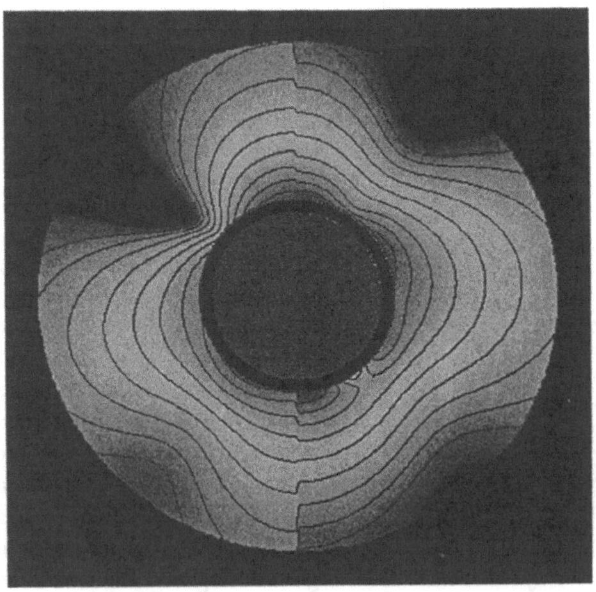

Figure 8: 1991 electron density distribution with the mean dipolar magnetic field as symmetry axis. The scale is logarithmic and the step between two successive curves is 2. The outermost curves corresponds to 2 10^{11} e/m³.

Another comparison has been attempted considering the mean dipolar magnetic field axis. Unfortunately, only the 1991 total eclipse provided an interesting configuration because of the strong differences between the solar rotation axis (p=+2°, B=+4°) and the mean dipolar magnetic field axis (p=-62°, B=-18.5°) (Gulyaev [14]). As shown in Figure 8, considering this physical axis as symmetry axis leads to a higher symmetry in the density distribution with high densities located in magnetic equatorial regions and above magnetic polar regions. For the other two observations, the differences in the axis directions were too small to give significantly different re-constructions. We hope that we will be able to perform such an analysis on the 1999 August 11th eclipse observations. However, as shown in figure 9, the 1999 corona may be symmetric enough to inhibit strong differences between models.

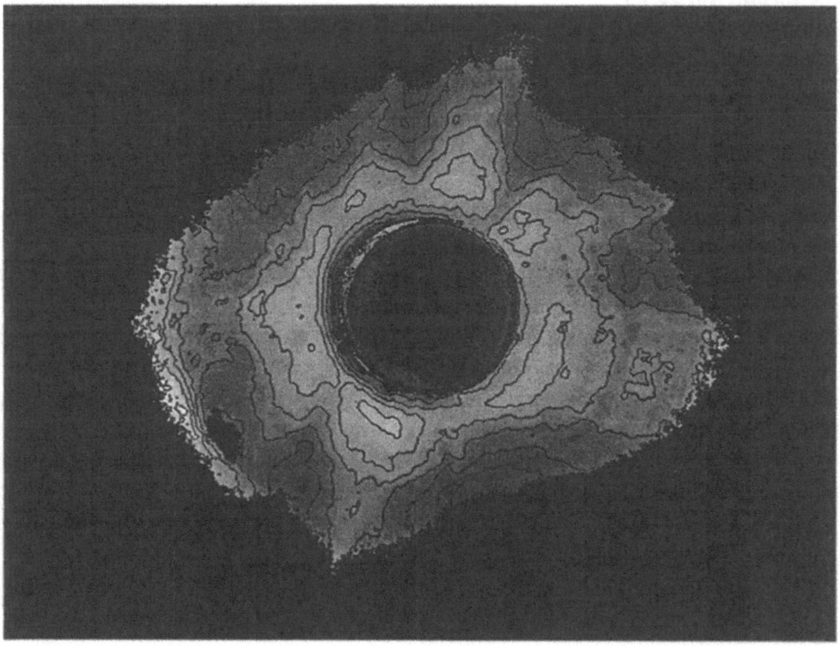

Figure 9: polarization map of the 1999 August 11th corona observed from Râmnicu-Vâlcea, Romania. The vertical direction is the projected solar rotation axis. The step between two successive curves is 5% and the outermost curve corresponds to 15%. The maximum of 43% is reached near the South, roughly at 1.5 $R_\odot$. The high polarization at the eastern side is an artifact caused by reflections in the optics.

References :

1. Withbroe G.L. (1988), The temperature, mass and energy flow in the corona and inner solar wind, *Astroph. Journ.*, **325**, 442-467
2. Waldmeier M. (1971), The solar corona in the eleven-year cycle, *Physics of the Solar Corona* (Proceedings of NATO ASI held at Athens, 6-17 September 1970), ed. C.J. Macris, Reidel, Dordrecht, 130-139
3. Koutchmy S. (1986), *Preprint 161*, partim. Structure de la couronne solaire, Inst. Astroph. Paris, 1-13
4. Koutchmy S. and Loucif M.L. (1989), Solar cycle varations of coronal structures, *Astron. Astroph. Suppl. Ser.*, **77**, 45-66
5. Schuster A. (1879), On the polarisation of the solar corona, *Monthly Notices of the R.A.S.*, **40**, 35-57
6. Minnaert M. (1930), On the continuous spectrum of the corona and its polarisation, *Zeitschrift für Astroph.*, **1**, 209-236
7. Baumbach S. (1937), Strahlung, Ergiebkeit und Elektronendicht der Sonnekorona, *Astronomische Nachrichten*, **263**, 121-134
8. van de Hulst H.C. (1950), The electron density of the solar corona, *Bull. Astron. Inst. Netherlands*, **11**, 135-150
9. Saito K. (1970), A non-spherical axisymmetric model of the solar K corona of the minimum type, *Ann. Tokyo Astr. Obs.*, **13**, 53-120
10. Gabryl J.-R. (1994), *Développement d'un modèle de densité électronique de la couronne solaire*, Mémoire de DEA, Université de Liège, Belgium
11. Perry R.M. and Altschuler M.D. (1973), Improved three-dimensional mapping of the electron density distribution of the solar corona, *Sol. Phys.*, **28**, 435-456
12. Koutchmy S. (1992), *Preprint 404*, partim. Allen – 3rd edition revised by Serge Koutchmy, Inst. Astroph. Paris, 1-3
13. Badalyan O.G. (1986), Polarization of the white-light corona under hydrostatic density distribution, *Astron. Astroph.*, **169**, 305-312
14. Gulyaev R.A. (1992), , Eclipse observations of the outer corona and the SOHO mission, *ESA-SP 348*, 133-136

OBSERVING THE SOLAR MAGNETIC FIELD

I.S. KIM
Sternberg State Astronomical Institute
Moscow State University
Universitetsky pr. 13, Moscow 119899, Russia

Abstract. Problems of "weak" magnetic fields diagnostics in the upper solar atmosphere are discussed, where the Zeeman splitting is 3-4 orders of magnitude less than the line width. This corresponds to magnetic observations of chromosphere, prominences and the corona requiring more advanced technique as compared with the photosphere. Direct and indirect methods of magnetic researches in the upper solar atmosphere are briefly reviewed. The "signal-to-noise" ratio of magnetic records is shown to strongly depend on the stray light in the feeding telescope. Concordance of performances of a feeding telescope and Stokes polarimeter are analysed to carry out reliable magnetic measurements. Eclipse coronagraphic polarimetric observations are concluded to be powerful for low-resolution magnetic diagnostics based on analysis of Hanle effect in prominences and the emission line corona. Employment of a new generation large-aperture reflecting coronagraph allows to expect advances in high-resolution magnetic observations of prominence and coronal structures.

1. Introduction

The nature of the Sun, our magnetic star, is based on solar magnetism. That's why numerous experimental and theoretical researches are devoted to observational and theoretical aspects of solar magnetism since the first successful measurements made by Babcock (1953). There are no doubts now that the observed solar structures (granules, filigree, spots, spicules, prominences, jets, arches and loops, coronal streamers and holes, etc.) manifest specific configurations of solar magnetic fields. However, the actual magnetic configurations of solar structures are still matter of debate. Existing theories require observational evidences for their verification. For instance,

J.-P. Zahn and M. Stavinschi (eds.),
Advances in Solar Research at Eclipses from Ground and from Space, 67–83.

advances in modeling of solar prominences were successfully presented at the IAU Colloquium N 167 (April, 1997, Aussois, France). But not a single contribution was devoted to recent magnetic observations made by direct methods.

"Solar Magnetic Fields", the excellent monograph written by J.O. Stenflo (1994) as well as his review (Stenflo 1999), can effectively introduce everyone to solar and stellar magnetism (Stenflo 1994). Since the first successful magnetic observations in prominences (Zirin and Severny 1961, Ioshpa 1962), summary of magnetic (polarimetric) researches in the upper solar atmosphere were presented by Rust (1966), Harvey (1969), Tandberg-Hanssen (1970), Smolkov and Bashkirtsev (1973), Leroy (1985), Kim (1989), Firstova et al. (1998), etc. Taking into account the topic of the current ASI I suggest to discuss observations of the "weak" magnetic fields which require advanced magnetographs (or polarimeters) as well as reducing parasitic signals caused by instrumental scattering and the sky brightness. What should be done to carry out reliable observations of "weak" magnetic fields? Total solar eclipse methods as well as employment of new generation large-aperture coronagraphic telescopes are suggested to be actively used for "weak" magnetic fields diagnostics.

The structure of my contribution is as follows. Comments on magnetic scales, "weak" magnetic fields, Stokes-vector presentation of the light, magnetographs and polarimeters, telescopes using as feeding optics are given in Section 1, "Introduction". Methods of magnetic observations are reviewed in Section 2, "Methods". Problems of recording "weak" magnetic fields by direct methods are discussed in Section 3, "Weak Magnetic Field Observations". Opportunities for future magnetic researches are presented in Section 4, "Prospects" (subsections "Eclipse Polarimetric Researches" and "Large-Aperture Coronagraphic Magnetic Observations"). The Reference list is not extensive as I cite publications closely related with the topic under discussion.

Let us first agree on the terminology used below.

1.1. SCALES

Observations of solar events allow to classify them in the frame of scale size:

- small-scale - $< 1''$ corresponding to about 725 km,
- medium-scale - $1 - 10''$,
- large-scale - $10 - 100''$,
- global-scale - comparable with the solar disk radius ($\approx 1000''$).

Space and ground-based imaging experiments reveal in the upper solar atmosphere (chromosphere, prominences and the corona) fine structures

TABLE 1. **Lande factor for several solar lines**

Element	λ, **Å**	**g**
Na I D_1	5889	1.33
Na I D_2	5896	1.17
Mg I b_1	5184	1.75
Fe I	5233	1.3
Fe I	5250	3
H I $H\alpha$	6563	1.05
H I $H\beta$	4861	1.06
He I D_3	5876	1.12
He I	4471	1.04
He I	10830	1.46
Fe X	6374	1.5
Fe XIII	10747	1.5
Fe XIV	5303	1.5

less than $1''$. However, typical "magnetic resolution" of magnetographs and polarimeters developed for observations of these solar structures corresponds to $3 - 10''$ (Baur et al. 1981, Den O.E. 1977, Stepanov A.I. 1989, Klepikov V.Yu. 1999). Subarcsec-scale and small-scale magnetic observations require employment of magnetographs and polarimeters. Researches of medium-scale and large-scale magnetic processes are often based on imaging experiments for magnetic tracers. For example, filaments evidently trace polarity reversals of longitudinal magnetic fields ($B_{//}$)at the level of the solar photosphere. The nature of solar structures seems to be a result of interrelation of different scale magnetic processes.

1.2. "WEAK" MAGNETIC FIELDS

The term "weak" magnetic fields is normally used when the Zeeman (magnetic) splitting is 2-4 orders of magnitude less as compared with the line width ($FWHM$ - full width at half maximum). Magnetic splitting $\delta\lambda_B$, is described by expression

$$\delta\lambda_B = 4.67 \times 10^{-13} g \lambda^2 B_{//},$$

where λ – the wavelength, $B_{//}$ – longitudinal component of magnetic field, g – Lande factor. On the average in the upper solar atmosphere (chromosphere, prominences and the corona) the line widths are $0.4 \div 1$ Å and for the field strength of 10-20 G the Zeeman splitting equals $2 \times 10^{-4} \div 3 \times 10^{-3}$ Å, or $2 \div 3$ orders of magnitude less as compared with the line widths. Lande factors for the most strong photospheric, chromospheric (prominence) and coronal lines are given in Table 1.

1.3. STOKES-PROFILES

Polarized light is described by the four Stokes parameters which are often defined in terms of the intensity measurements: I, Q, U, and V.

- I is intensity.
- The I, Q and U-parameters describe linearly polarized light.
- The I and V-parameters characterize circularly polarized light.

The degree of linear polarization is defined by the relation

$$p = \frac{I_{max} - I_{min}}{I_{max} + I_{min}} = \frac{(Q^2 + U^2)^{1/2}}{I},$$

where I_{max} I_{min} – maximum and minimum intensity passed through a polarizer. The direction of linear polarization angle α is determined by expressions

$$\cos 2\alpha = \frac{Q}{(Q^2 + U^2)^{1/2}}, \quad \sin 2\alpha = \frac{U}{(Q^2 + U^2)^{1/2}}.$$

The degree of circular polarization is given by

$$p_c = \frac{I_{right} - I_{left}}{I_{right} + I_{left}} = -\frac{V}{I}.$$

1.4. TELESCOPES

- The term "conventional solar telescope" indicates refractive or reflective solar telescope with traditional micro-roughness ($RMS \approx 50-100$ Å) of primary optics.
- Lyot type coronagraph invented by Lyot (1930) is a special refractive telescope with reduced level of scattered light. The coronagraphic method (the Lyot method) is to employ three main elements:
  - high quality single lens,
  - masking the solar disk light in the primary focal plane,

- masking the light diffracted at the edge of entrance aperture in the plane of Lyot stop.

Correct employment of Lyot method result in reducing the scattered light in 2 orders of magnitude (Lyot 1930, Sazanov 1968). It should be noted that the aperture of Lyot type coronagraphs does not exceed 53 cm (Gnevyshev et al. 1967) and is limited by technology problems.

- Reflective coronagraphs or coronagraphic telescopes indicate new generation telescopes with super-smooth reflective primary optics ($RMS \leq$ 10 Å) combined with masking both in the primary focal plane and in the plane of Lyot stop. Development of such kind of telescope is to come to large aperture coronagraphic researches.

1.5. MAGNETOGRAPHS AND POLARIMETERS

- As a rule, the term "magnetograph" is used when circular polarization is measured for the following Zeeman analysis.
- Polarimeters normally measure linear polarization for the following Hanle analysis.
- Stokes-polarimeters measure the whole Stokes-profiles to get vector of magnetic fields.

2. Methods

2.1. INDIRECT METHODS

Indirect methods require the knowledge of velocity and density, as well as the assumption on the dominance of the magnetic field energy over the kinetic one. The method allows to determine the lower limit of the magnetic field strength. The velocity is determined using prominence filtergrammes. Density is determined less accurately, but fortunately, the field strength is proportional to the density square root (Idlis et al. 1956). Further successful developments of this method (see, for instance, Ballester 1984) has provided data on magnetic fields of a number of active prominences. Evidently the method can be applied only to active features exhibiting rapid movements (loops, surges, eruptions, sprays, active region filaments)

We can expect much from solving the inverse problem of determining the magnetic field in the solar atmosphere using the known field at the photosphere level. The field normal component distribution is known from the magnetograms. The transverse field is identified by the Hα-fibriles orientation (Molodensky and Filippov 1988).

Comparative analysis of NLTE diagnostics and polarization measurements in terms of equilibrium MHD models of the K-S type seems to be a powerful method for deducing actual magnetic fields in prominences (Anzer and Heinzel 1998, Heinzel and Anzer 1999).

Significant progress in coronal magnetic diagnostics is expected from employment of radio methods. Spectral polarization researches of active region magnetospheres are actively carried out by RATAN-600 (Gelfreikh 1996, Ryabov et al. 1999).

Magnetic data can be deduced from analysis of Faraday effect.

2.2. DIRECT METHODS

As a rule, magnetic researches based on analysis of the Zeeman and the Hanle effects are considered as direct ones.

2.2.1. *Zeeman Analysis*

An outer magnetic field forces the atomic level to split. There are longitudinal and transverse Zeeman effects depending on the line-of-sight direction. Transverse observations correspond to linear polarization of all spectral lines: some of them are parallel to magnetic field (π–components), others are normal to magnetic fields (σ–components). Observations along magnetic fields result in σ–components with circular polarization.

The whole Stokes-profiles should be recorded to get reliable measurements of "weak" magnetic fields. Emission lines profiles (I–Stokes profiles) of prominences, chromosphere and the corona are well approximated by gaussians. If the Zeeman splitting is much more less as compared with the line width then V–Stokes profile can be fitted by the first derivative of the I-Stokes profile (Nikolsky et al. 1985). Then

$$V(\lambda) \approx \frac{dI}{d\lambda} = I_0 \frac{2(\lambda - \lambda_0)}{\Delta\lambda_D^2} \exp - \left(\frac{\lambda - \lambda_0}{\Delta\lambda_D}\right)^2,$$

where I_0 - peak of line intensity, λ_0 - wavelength of emission line, $\Delta\lambda_D$– Doppler width. Examples of typical prominence magnetic observations are presented below. The large aperture 50 cm Kislovodsk coronagraph (Gnevyshev et al. 1967) was used as feeding optics. The scanning Fabry-Perot magnetograph (Nikolsky et al. 1985) recorded the whole I and V Stokes profiles as well as the Doppler mark. Figure 1 shows high-resolution H_α-picture of an active prominence representing active region filament type. A black spot is the magnetograph pin-hole with diameter of 8″.

The peak of V is reached at

$$V_0 = I_0 \frac{2\delta\lambda_B}{\Delta\lambda_D^2} \exp - \left(\frac{\delta\lambda_B}{\Delta\lambda_D}\right)^2.$$

Figure 1. $H\alpha$-picture of an active prominence representing the active region filament type. The black spot is the magnetograph pin-hole with diameter of 8″.

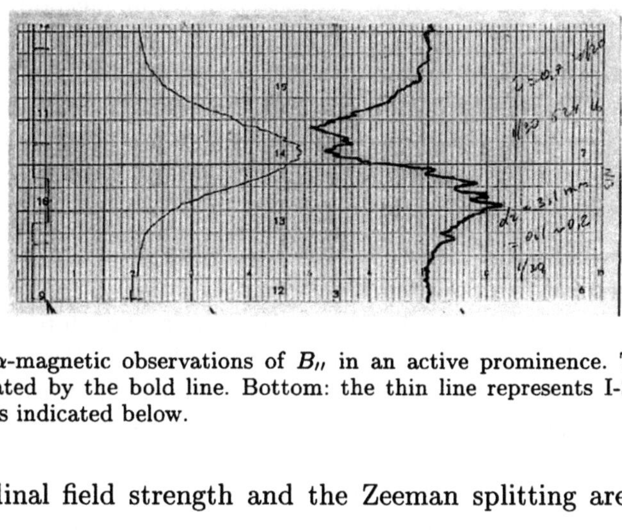

Figure 2. $H\alpha$-magnetic observations of $B_{//}$ in an active prominence. Top: V-Stokes profile is indicated by the bold line. Bottom: the thin line represents I-Stokes profile. Doppler mark is indicated below.

The longitudinal field strength and the Zeeman splitting are related by equation

$$B_{//} = \frac{\delta\lambda_B}{4.67 \times 10^{-13} g\lambda^2},$$

which could be rewritten as

$$B_{//} = \frac{1.07 \times 10^{12}}{g\lambda^2} \Delta\lambda_D^2 \frac{V_0}{I_0}.$$

It is seen that V_0/I_0 is a crucial parameter for magnetic measurements.

2.2.2. *Hanle Analysis*

It is known that polarization is produced when the spatial symmetry of the emission process is broken. Examples of symmetry breaking by anisotropic excitation are presented in prominences (resonance scattering), in the white-light corona (Thomson scattering), etc. That means that emission in prominences and the corona must be linearily polarized due to the directivity of the exciting field. For purely resonance scattering the polarization vector should be tangential to the solar limb, and the polarization degree increases with distance. A non-vertical "weak" magnetic field results in decreasing the polarization degree and rotating the polarization direction. That is Hanle effect. Strong magnetic fields results in disappearance of Hanle effect as the dipole axis becomes oriented along the field lines.

Attempts of coronagraphic observations of prominences and coronal emission line polarization were made since the 40ties. However, the usage of linear polarization measurements for magnetic researches has became successful thanks to the progress in the quantum theory of polarization radiation (House 1970, Stenflo 1976, Sahal-Bréchot et al. 1978, Landi Degl'Innocenti 1982). The main advantage of Hanle method is to determine the field vector and its high sensitivity for diagnostics of small field strength. The main disadvantage is ambiguous interpretation. As mentioned before 2 parameters (degree of polarization and polarization direction) are measured, and 3 parameters of the magnetic field should be determined.

The "broad-band" and the "narrow-band" modifications of Hanle method were successfully used in 70th-80th.

- The "broad-band" Hanle method was based on measurements of the whole emission line intensity, degree of polarization and polarization direction as for the small field strength polarization parameters are approximately constant within the line profile. The 4th Stokes parameter (circular polarization) equals to zero when the light from the whole line is used. Instrumental polarization should be less than 0.1% to make such measurements (Leroy et al. 1985).
- The "narrow-band" Hanle method is based on simultaneous measurements of Stokes-profiles of two emission lines (Athay et al. 1983).

Both "broad-band" and "narrow-band" methods effectively work for the field strengths $< 30\,G$. Magnetic researches of active structures have to be based on the Zeeman method.

3. "Weak" Magnetic Field Observations

Magnetic observations by direct methods require low level of instrumental polarization caused by parasitic signals. Input of instrumental polarization in conventional solar telescopes having two coelostat mirrors can be significant. Mutual orientation of the mirrors is changed with time. Similar situation would arise when large-aperture coronagraphs are used as feeding optics for magnetic observations because of 2 additional mirrors (Coude mirror combined with a mirror directed the light beam in horizontal direc-

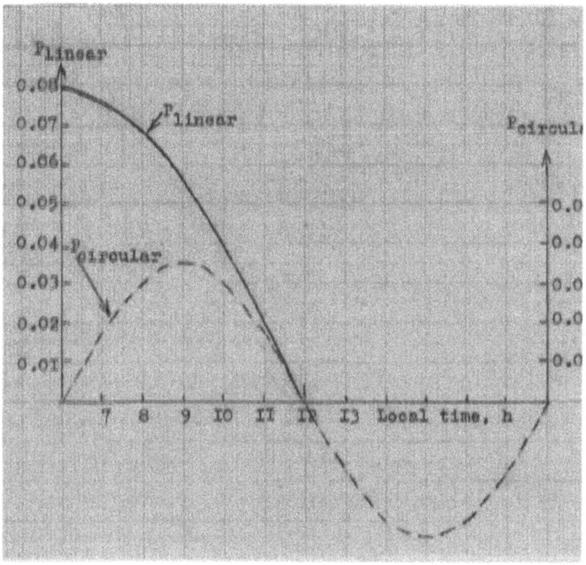

Figure 3. Behavior of circular and linear polarization of Kislovodck 50 cm coronagraph over the course of the day

tion). Figure 3 shows behavior of instrumental circular and linear polarization over the course of the day calculated for 50-cm coronagraph installed at Kislovodsk station (Russia).

It is seen that "weak" magnetic field observations would be very "noisy" about 9 am. Modification of optical sketch of the coronagraph was made to avoid rapid changes of instrumental polarization. The horizontal mirror was removed to attach the Fabry-Perot magnetograph to the polar axis tube. In case of one mirror (Coude one) instrumental polarization is approximately constant for the given Sun's altitude and can be compensated. Details and review of prominence magnetic data were presented by Kim (1989).

3.1. SIGNAL-TO-NOISE RATIO

Instrumental background is the main factor preventing magnetic recording in the corona. Scattered light in conventional telescopes equals $10^{-4} - 10^{-2}$ (Kim et al. 1995) Let us estimate the upper limit of the noise level for reliable magnetic diagnostics in the upper solar atmosphere. Let S/N is the signal-to-noise ratio. Let $(S/N) \geq 6$. That means

$$N \leq \frac{1}{6}S \text{ or } N \leq \frac{1}{6}V_0 .$$

In our case $S = V_0$ and N is the noise caused by the instrumental background, sky brightness, dark current, read-out noise, etc. Noise caused by instrumental background is dominant when high-altitude or eclipse measurements are made. $N = \sqrt{I_{sc}}$ when a photomultiplyer or a CCD are used as detectors. Using the expression for V_0 we obtain

$$N \leq I_0 \frac{2\delta\lambda_B}{6\Delta\lambda_D^2} e^{-\left(\frac{\delta\lambda_B}{\Delta\lambda_D}\right)^2}$$

Let I_0 is the peak intensity of the I-profile.

4. Prospects

4.1. ECLIPSE POLARIMETRIC OBSERVATIONS

As mentioned above instrumental polarization and the sky brightness are the main preventing factors. Significant improvement of "signal–to–noise" ratio is expected during total solar eclipses when the sky brightness reaches $(10^{-8} - 10^{-9})B_\odot$, where $B_\odot$ is the brightness of the solar disk.

Our first attempt of eclipse polarimetric researches of prominences was based on non-filter polarimetric observations of the white-light corona of 03 November 1994 in Criciuma (Brazil). Exposure times and negative colour film were chosen to get normal density images of the inner white-light corona. That's why the images of the E-limb obtained just after the second contact revealed "green" prominences imposed in the images of the white-light corona. Preliminary results of data reduction of the E-limb prominences were presented by Lebedev et al. (1997).

It's known that colour eclipse pictures of the white–light corona evidently reveal prominences which are purple features in slides or green ones in negatives. Prominence brightness in the H_α–line equals to $(10^{-2} \div 10^{-1}) \cdot B_\odot$. Then the contrast C of "prominence–to–corona" images recorded by the "red" sensitive layer may be described by the expression

$$C = \frac{B_{H\alpha}}{\int\limits_{550nm}^{710nm} B_{wc}(\lambda) S_\lambda \, d\lambda} \approx 13 \div 130.$$

Stokes-vectors are additive. Polarization degree of the white–light corona does not exceed 25% at $R < 1.2R_\odot$ while prominence brightness equals $13 \div 130$ of the corona one. That means that input of the white–light corona in polarization degree measurements at places of prominence location does not exceed $2 \div 0.2\%$ respectively.

Successful polarimetric observations were made on 03 November 1994 in Criciuma (Brazil) with a portable motion-picture coronagraph-polarimeter

(D = 55 mm, F_e = 339 mm) developed at Sternberg State Astronomical Institute of Moscow State University. A rotating polarizer was placed in the plane of Lyot stop. This results in diameter of 9 mm for the light beam at a polarizer. Absolute and relative calibrations were based on exposures of the solar disk and photometric wedges. Details of photometry, data reduction and the algorithm of the statistical processing were presented by Kim et al. (1996) and Belenko et al. (1997). Stokes–vector presentation and IDL software were used for analysis. After mutual centering the 24 frames the averaged density image $\mathbf{D} \equiv D(i,j)$ is created, where i, j are integer rectangular coordinates of pixel images when the point of reference is at the center of the Sun. The distribution of densities in this frame is used as a noise level indicator in statistical procedures of the polarization picture of the object under study. $\mathbf{S_k}$–intensity frames resulted from density ones contain information about the distribution of polarization parameters, which could be extracted as a result of a statistical processing.

24 equations for every (i, j) pixel can be written using the 24 consecutive frames

$$\mathbf{S_k} \equiv S_k(i,j) = \mathbf{I} + \mathbf{Q}' \cos\frac{\pi k}{12} + \mathbf{U}' \sin\frac{\pi k}{12}, \; k = 0, 1, \ldots, 23,$$

where the vector $(\mathbf{Q}', \mathbf{U}')$ is related with $(\mathbf{Q}, \mathbf{U})$ by $2\alpha_0$ rotation unknown beforehand with enough accuracy, and where α_0 is the position of the plane of polarization in the first frame.

Figure 4 shows a computer "polarization image" (distribution on polarization degree) of the E–limb prominences. It looks similar to a conventional prominence image revealing evidently the regions of prominences near the Moon's limb where the polarization varies from 2 to 5%. Polarization for prominences is found to be tangential to the solar limb within the limits of errors of about 2°.

It should be noted that the above–mentioned analysis was applied to non–filter, low–resolution, low-contrast and small–scale prominence images. During the eclipse of August 11, 1999 a modified filter polarimetric research of prominences was carried out. Narrow passband interference filters ($FWHM \approx 6$ Å and peak transmission $\tau_{max} \approx 40\%$ and a rotating polarizer were put near the plane of Lyot stop. Two identical portable coronagraph-polarimeters were installed at 2 observational sites along the totality pass.

- Motion-picture coronagraph-polarimeter observations were made in Shabla (Bulgaria). An ariel near infrared film was used as a detector.
- In Turhal (Turkey) a video version of a coronagraph-polarimeter was installed at mounting with tracking. A conventional SONY Hi-8 CCD TR-820 videocamera and TDK Hi-8 MP video tape was used as a

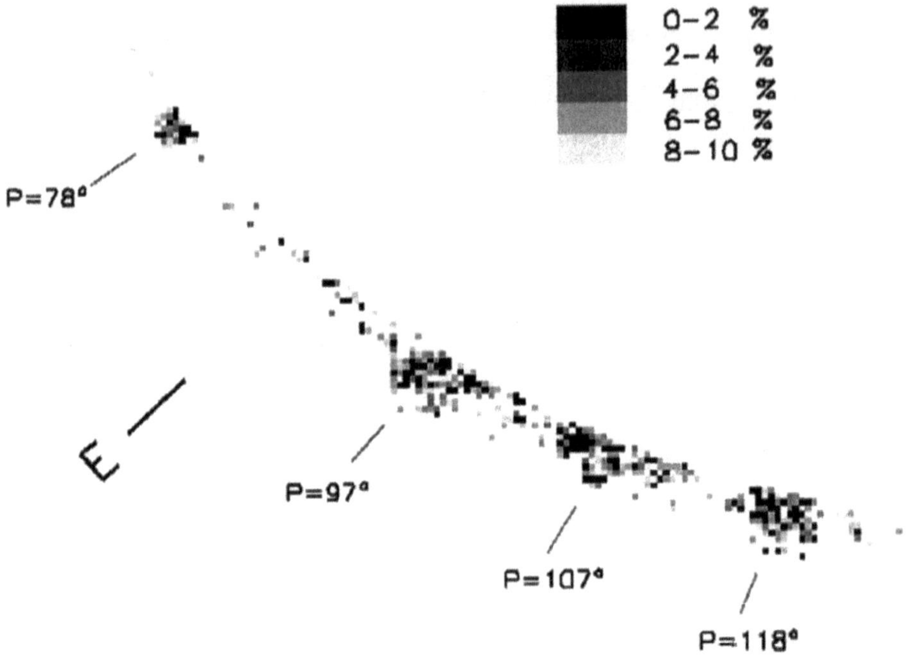

Figure 4. A computer "polarization image" (distribution on polarization degree) of the E–limb $H\alpha$-prominences of 03 November 1994.

detector for observations of the white-light corona, and prominences in the light of H_α, and D_3-lines.

Preliminary data reduction is being made now.

4.2. LARGE-APERTURE CORONAGRAPHIC MAGNETIC OBSERVATIONS

The main sources of the background in the final focal plane of a ground-based reflecting coronagraph are the sky brightness (at sea level - $(10^{-3} - 10^{-4})B_\odot$ and at altitude > 2.5 km - $(1-5)10^{-5}B_\odot$) and the stray light caused by diffraction at the entrance aperture and the microroughness of primary optics. Scattered light of the former mirror primary optics was caused by diffraction on the entrance aperture and microroughness of 100-200 Å. This prevented to record the faint objects near the bright ones. Figure 5 shows numerical simulation of scattered light in the primary focal plane of a 20 cm aperture mirror telescope (Kim 1998).

- Microrelief of mirror primary optics is modeled by superposition of phase gratings.
- The grating constant is accepted to equal to correlation length of microroughness inhomogeneity (5, 10, 20, 50 and 100 μ).

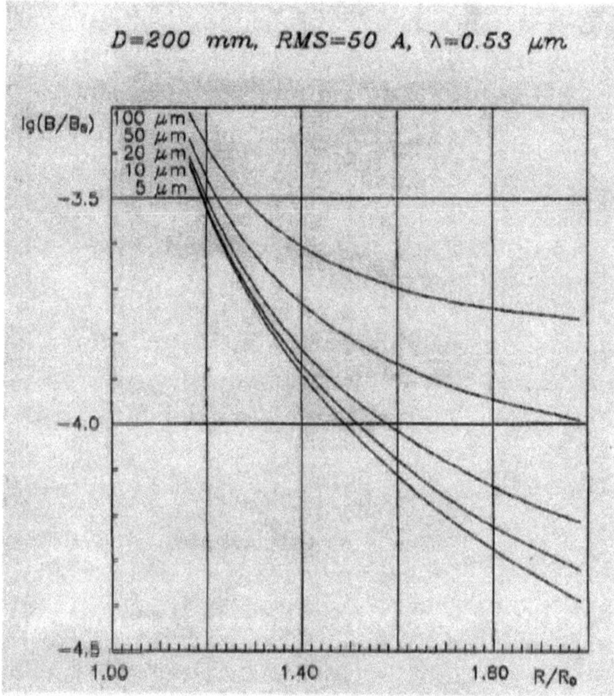

Figure 5. **Distribution of instrumental scattering versus distance from the optical axis in the 20 cm aperture telescope.**

- The phase relief depth is accepted to equal to the value of RMS microroughness of 50 Å.

It is seen that the significant level of scattered light near the limb prevents observations of the faint objects.

Recent advances in super-smooth polishing ($RMS \leq 10$ Å) revive the idea of a large-aperture reflecting coronagraph (Koutchmy and Smartt 1989, Smartt and Koutchmy 1999). Expected scattered light in the primary focal plane of a reflecting coronagraph versus distance is shown in Figure 6 (Kim et al. 1998). Diffraction at the entrance aperture and scattering at micro-relief of primary reflecting optics was taken into account. The following parameters were used:

- $D : F = 1 : 10$;
- $RMS = 10$ Å,
- correlative length of inhomogeneity $l = 20$ mkm,
- diameter of the solar disk of $959,63''$.

Limb darkening was not taking account. Calculations were made for the green coronal line Fe XIV, 5303 Å and diameter of entrance aperture of 200, 600, 1000 and 2000 . The input of scattered light caused by diffrac-

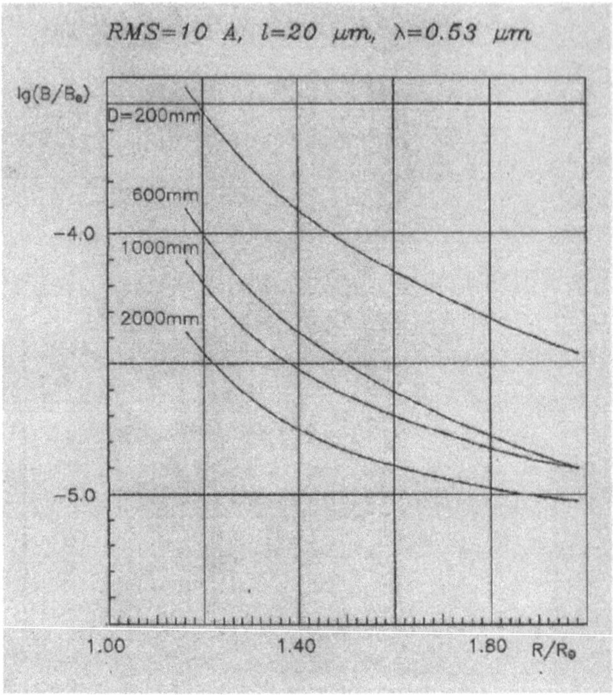

Figure 6. Scattered light versus distance in reflecting coronagraphs

tion at the edges of entrance aperture is dominant near the solar limb at distances $< 1.2R_\odot$. Employment of Lyot stop method can reduce this scattered light in 2 orders of magnitude. At distances $> 1.4R_\odot$ the scattered light is mainly caused by scattering at micro-relief. It is seen that employment of large aperture super-smooth reflecting primary optics combined with Lyot method allows to get the scattered light of $10^{-5} - 10^{-6}$.

Let us estimate the expected integration time (t_{exp}) to record V_0 for $B_{\prime\prime} = 20$ G, when a 2 m aperture reflecting coronagraph is used as a feeding optics. Lande factor g of Fe XIV, 5303 Å equals 1.5. Then $\delta\lambda_B \approx 3,9 \cdot 10^{-4}$. The average line width (FWHM - full width at half maximum) $\Delta\lambda$ equals 0.8 Å (Kim 1999).

$$\sqrt{I_{sc}} \leq 1,1 \cdot 10^{-4} I_0. \qquad (1)$$

Let

- $k_1 = I_0/I_\odot \approx 5 \cdot 10^{-5} I_\odot$,
- $k_2 = I_{sc}/I_\odot \approx 3 \cdot 10^{-5} I_\odot$,
- $f_\lambda = 1.88 \cdot 10^2$ergcm^{-2}s^{-1} Å$^{-1} \approx 5.02 \cdot 10^{13}$photonscm$^{-2}sc^{-1}$ - solar flux at the terrestrial atmosphere boundary,
- $S_\odot \approx 2.89 \cdot 10^6$ arc sec^2 - square of the solar disk,

- $s \approx 78.5$ arc sec^2 - square of the magnetograph aperture (diameter of $10''$),
- $t \approx 5 \cdot 10^{-1}$ - transparence of terrestrial atmosphere,
- $S = 3,14 \cdot 10^{42}$ - square of the entrance aperture of a feeding optics,
- $\tau \approx 10^{-2}$ - transmittance of "coronagraph+Fabry-Perot magnetograph" system,
- $\delta\lambda \approx 5 \cdot 10^{-1}$ Å - optimal width of instrumental profile($FWHM$) (Nikolsky et al. 1985),
- $q = 5 \cdot 10^{-1}$ - quantum efficiency of a detector.

Then $I_\odot \approx f_\lambda \cdot (s/S_\odot) \cdot t \cdot S \cdot \tau \cdot \delta\lambda \cdot q \cdot t_{exp} = A \cdot t_{exp}$ and

$$t_{exp} \geq \frac{k_2}{k_1^2} \cdot A \cdot 1,21 \cdot 10^{-8}.$$

The expected integration time $t_{exp} \geq 18$ c. The whole V-Stokes profile would be recorded during 3 min.

5. Conclusions

Several ways can be used for diagnostics of "weak" magnetic fields by direct methods in the upper solar atmosphere.

- During total solar eclipses the sky brightness is significantly reduced and reaches $(10^{-7} - 10^{-9})B_\odot$. Space conditions are imitated. The role of eclipse methods becomes significant in sense of testing of advanced instrumentation for spectral and magnetic (polarimetric) researches. Hanle analysis based on eclipse low-resolution coronagraphic polarimetric observations (Lebedev et al. 1997, Sykora 1998) looks very promising for prominences and the emission line corona "weak" magnetic field diagnostics.
- Recent advances in super-smooth polishing renew the idea of a large-aperture ($>$ 2 m) reflecting coronagraphs (coronagraphic telescopes) and allow to expect successes in high-resolution coronagraphic magnetic researches in prominences and the E-corona.
- Currently limb magnetic diagnostics in the chromosphere seems to be unsolving. The main preventing factor is significant level of scattered background near to the limb ($R < 0.007\,R_\odot$).

Certainly, we are encouraged to search for non-traditional approaches for the diagnostics of "weak" magnetic fields.

Acknowledgements

Researches were carried out in the frame of grants N 99-02-17753 of RF-BR, N 93-8-229 Russian State program for Astronomy, and N 1088 IR of INTAS-RFBR. My special thanks to I.A. Belenko and N.L. Kroussanova for assistance in preparing the figures.

References

1. Anzer, U. and Heinzel, P. (1998), *Solar Physics* **179**, 75
2. Anzer, U. and Heinzel, P. (1999), *Astron. Astrophys.* **349**, 974
3. Athay, G., Querfeld, C.W., Smartt, R.N., Landi Degl'Innocenti, E. and Bommier, V. (1983), *Solar Phys.* **89**, 3
4. Babcock, H.W. (1953), *Astrophys. J.* **118**, 387
5. Ballester, J.L. (1984) *Solar Phys.* **94**, 151
6. Baur, T.G., Elmore, D.E., Lee, R.H., Querfeld, C.W. and Rogers, S.R. (1981), *Solar Phys.* **70**, 935
7. Belenko, I.A., Bougaenko, O.I., Kim, I.S., Koutchmy, S., Matsuura, O.T. and Picazzio, E. (1997), *NATO ASI Series. Theoretical and Observational Problems Related to Solar Eclipses* (eds Z. Mouradian and M. Stavinsschi) **494**, 15
8. Bommier, V., Landi-Degl'Innocenti, E., Leroy, J.-L. and Sahal–Bréchot, S. (1994) *Solar Physics*, **154**, 231
9. Demkina, L.B. and Nikolsky, G.M. (1981), *Solnechnye Dannye* **N 12**, 105
10. Den, O.E. (1977), *Thesis*, IZMIRAN, Russia
11. Firstova, N.M., Boulatov, A.V. and Kashapova, L.K. (1999), *Solar Polarization*, Bangalore, India, 12-16 October 1998 (K.N. Nagendra and J.O. Stenflo eds.), 451
12. Gelfreikh, G.B. (1996), *Radio emission from the Stars and the Sun* (eds. A.R. Taylor and J.M. Paredes) ASP Conference Series **93**, 415
13. Gnevyshev, M.N., Nikolsky, G.M. and Sazanov, A.A. (1967), *Sol. Phys.* **2**, 223
14. Harvey, J.W. (1969), NCAR Co. Thesis, Univ. of Colorado
15. Heinzel, P. and Anzer, U. (1999), *Solar Physics* **184**, 103

House, L.L. (1970) *Solar Magnetic Field* (ed. R.D. Howard, Reidel Publ. Co., Holland)

16. Idlis, C.N., Karimov, M.T., Delone, A.B., Obashev C.O. (1956), *Izv. Astrof. Inst. Kaz. SSR, AN SSSR* **2**, 71
17. Ioshpa, B.A. (1962), *Geomagnetism i aeronomia (in Russian)* **2**, 149
18. Kim, I.S. (1989), *Lecture Notes in Physics* **363**, 49
19. Kim, I.S., Bugaenko, O.I., Bruevich, V.V. and Evseev O.A. (1995), *Bulletin of the Russian Academy of Sciences* **59**, N 7, 1239
20. Kim, I.S., Bougaenko, O.I., Belenko, I.A., Koutchmy S., Matsuura, O.T. and Picazzio, E. (1996), *Radiophysics and Quantum Electronics* **39**, N 10, p 869
21. Kim, I.S., Belinski, A.A., Lebedev, A.A., Folomkin, I. and Stroulea, I.L. (1998), *New Solar Cycle: Theoretical and Observational Aspects* (ed. V.I. Makarov and V.N. Obridko) S-Petersburg, PIYaF, 277
22. Kim, I.S., Folomkin, I.P and Stroulea I.L. (1999), 7th Symposium on Solar-Terrestrial Physics (ed. V.I. Makarov and V.N. Obridko) (in Russian), PIIYaF RAN, Gatchina, Russia, 77
23. Klepikov, V.Yu. (1999), Magnetic Fields of Quiscent Prominences, Thesis, IZMIRAN, Russia
24. Koutchmy, S. and Smartt, R.N. (1989), *High Spatial Resolution Solar observations* (ed. O. von der Lue) NSO/SPO Workshop N 10, 560
25. Lebedev, A.N., Belenko, I.A., Bougaenko, O.I., Kim, I.S., Matsuura, O.T., Picazzio, E., Silvi, R.P. and Livi, S.H. (1997), *New Perspectives on Solar Prominences* (ed. D. Webb, D. Rust and B. Schmieder) *ASP Conf. Ser.* **150**, 66

26. Leroy, J.-L. (1985), in M.J. Hagyard (ed.), NASA Conf. Publ. N 2374, 121
27. Lyot, B. (1930), *C.R. Acad. Sci.* **99**, 580
28. Molodensky, M.M. and Filippov, B.P.(1988) *Astron. Zh.* **65**, 1047
29. Nikolsky, G.M., Kim, I.S., Koutchmy, S. and Stellmacher, G. (1985), *Astron. Zh. (Russian)* **62**, N 6, 1147
30. Rust, D. (1966), NCAR Co. Thesis, Univ. of Colorado
31. Ryabov, B.I., Pilyeva, N.A., Akissandrakis, C.E., Shibasaki, K., Bogod, V.M., Garaimov, V.I., and Gelfreikh, G.B. (1999), *Solar Physics* **185**, 157
32. Sahal–Bréchot, S., Bommier, V. and Leroy, J.-L (1977), *Astron. Astrophys.* **59**, 223
33. Sazanov, A.A. (1968), Thesis, IZMIRAN, Russia
34. Smartt, R.N., Koutchmy, S. (1998), *SPIE* **33**, 614
35. Smolkov, G. Ya. and Bashkirtsev, V.S. (1973), *7th Regional Consultation on Solar Physics* (ed. J. Sykora) Slovak Academy Sci., 175
36. Stenflo, J.O. (1976), *Astron. Astrophys.* **46**, 61
37. Stenflo, J.O. (1999), Solar magnetism and the second solar spectrum, *Solar Polarization*, Bangalore, India, 12-16 October 1998 (K.N. Nagendra and J.O. Stenflo ed.), 1
38. Stenflo, J.O. (1994), Solar Magnetic Fields, Polarized Radiation Diagnostics, Kluwer Academic Publishers, Dordrecht/Boston/London, 385
39. Stepanov, A.I. (1989), *Thesis*, SAO RAN, Zelenchuk, Russia
40. Sykora, J., Badalyan, O.G., and Livshits, M.A. (1999), *Solar Polarization*, Bangalore, India, 12-16 October 1998 (K.N. Nagendra and J.O. Stenflo eds.), 363
41. Tandberg-Hanssen, E. (1970), *Solar Physics* **15**, 359
42. Zirin, H. and Severny, A.B. (1961), *Observatory* **81**, 155

SOLAR ACTIVITY

Z. MOURADIAN
Observatoire de Paris
92195 Meudon, France

Abstract. In this lecture, we start with a short introduction and give a few references for future study. Then we introduce two criteria to distinguish the quiet from the active sun: these are respectively the permanence of the structures and the presence of an inversion line of magnetic field background. In the third part we describe a few structures of the quiet sun that vary as a function of the solar activity cycle. We then discuss the two typical active sun sites: prominences and active regions. These sites usually evolve quietly until some transient event triggers an acceleration. Such transient events are the sudden disappearance of a prominence and, for active regions, flares and associated phenomena. At the end of the lecture we explain the activity indices and activity mapping, the 11-, 22- and 100-year solar activity cycles, and their effect on the quiet sun, with solar variability. We finish with a few comments on the solar-stellar relation.

1. INTRODUCTION

Solar activity is a special case of the general problem of stellar activity and, as such, the solar studies are of capital importance. The search for structures in stellar atmospheres is referenced to the Sun's, to the point where stellar activity sometimes appears to be an extrapolation of solar activity. In return, the stars provide a large sample of atmospheres with different physical conditions. Of course, solar activity is also important in understanding solar-terrestrial connection.

The first question, raised in solar physics is: What is solar activity? Empirically activity can be defined in terms of plasma movements or some energy absorption and realize process. But these concepts are too vague and they involve all solar features, for instance when we realize that nothing is mo-

J.-P. Zahn and M. Stavinschi (eds.),
Advances in Solar Research at Eclipses from Ground and from Space, 85–104.

tionless in the solar atmosphere, because of the continuous energy deposit coming from underneath. Some author, knowing that solar activity is related to the magnetic field, have pointed out the possibility of delimiting regions of activity by their magnetic field strength; but this is not satisfactory because it overlooks the activity of features with weak magnetic field, even if other structures with stronger field are quiet.

Solar activity is a vast and diversified domain. For this reason, the present lecture will give an overview of principal fields, and indicate some relations and interactions among their components, and simply give directions for further study. The lack of a theory on solar/stellar activity made the explanation of solar activity toilsome.

We will first begin by explaining some of the structures of the quiet sun, then those playing a role in solar activity and the solar cycle. We review here the principal form of activity. At the end we speak of the extension of solar activity to stars.

A number of good reference books can help to deepen our subject: Stix [36], Wilson [41], Zirin [42], and for further information see: Cox et al. [4], Sonett et al. [35].

There are some www sites concerning the sun:

http://wwwssl.msfc.nasa.gov/ssl/pad/solar/	(General)
http://solar-center.stanford.edu/	(General)
http://www.nso.noao.edu/	(General)
http://mesola.obspm.fr/	(Ground-based obs.)
http://sohowww.nascom.nasa.gov/	(SOHO satellite)
http://www.lmsal.com/YPOP/	(YOHKOH satellite)
http://vestige.lmsal.com/TRACE/	(TRACE satellite)

2. SOLAR ACTIVITY

We propose two criteria to distinguish the quiet from the active sun.

- The first is *the permanence of the structures, which is indicative of the quiet sun* . Permanence means that, while individual structures change, the global system does not. These structures exist independently of the state of solar activity or of the solar cycle. Permanence can be considered a characteristic definition of the quiet sun, and consequently non-permanent structures as that of the active sun.
- Secondly, *activity has a spatial characteristic: it is always located at the boundary of unipolar magnetic regions* , which are zones effective for magnetic field emergence. So, if activity exists, then it is located precisely at that boundary. Moreover, the transient elements of active sites are also exactly on this boundary.

With these two rules in mind, we can characterize solar activity as a set of structures formed between two opposing magnetic background regions containing transitory events accompanied by an energy release due to the action of the magnetic field. Fig.1 illustrates the sun in two periods, one quiet and the other active. In the upper row we can see the active structures (active regions and filaments) at work in a quiet sun environment. The bottom row shows the quiet sun at minimum of activity. The obvious question arises as to how this activity is modulated by the solar cycle.
We know that the solar *activity structures obey the solar cycle; but the cycle cannot be considered as characteristic only of activity, because it is also present in quiet sun structures.* So, the cyclicity is more general phenomenon than the activity.
The solar atmosphere can be sub-divided into active and quiet regions, according to the permanence of structures; but it can also be stratified vertically into three sub-atmospheres, in respect to altitude variation of temperature: the photosphere, the chromosphere, and the corona (Fig.1). All these three layers are equally present in the quiet and active regions. All solar activity is based on the presence of an active magnetic field, whereas in the quiet sun the magnetic field remains passive. The quiet/active classification is not a strict one, often the limits are loose.
All we know of solar activity concerns the Sun's atmosphere. We know very little about the roots of activity in the convective zone, where some forerunner conditions certainly exist. However, study is beginning in this direction using locally the helioseismological technique [14], [5]. The solar activity may quite plausibly be mainly an atmospheric phenomenon. For example, sunspots seem to be formed by a superficial agglutination of a magnetic field and plage is formed of fine photospheric tubes which expand in the photosphere-chromosphere transition.
Aside from solar rotation, there are mainly two phenomena structuring the solar atmosphere: convection, which is important in the photosphere; and the magnetic field, which predominates in chromosphere and corona. Both of these transport mass and energy from inside out. In order to see which - plasma pressure or magnetic pressure - plays the prime role at a given point of the atmosphere, we refer to the plasma β parameter:

$$\beta = \frac{(\text{gas pressure})}{(\text{magnetic pressure})}$$

Generally, in the photosphere, the plasma β is greater than unity and the mass motion carries and shakes the magnetic field tubes, whereas in the chromosphere and corona β is less than unity and the magnetic field then drives and guides the mass motions and energy.

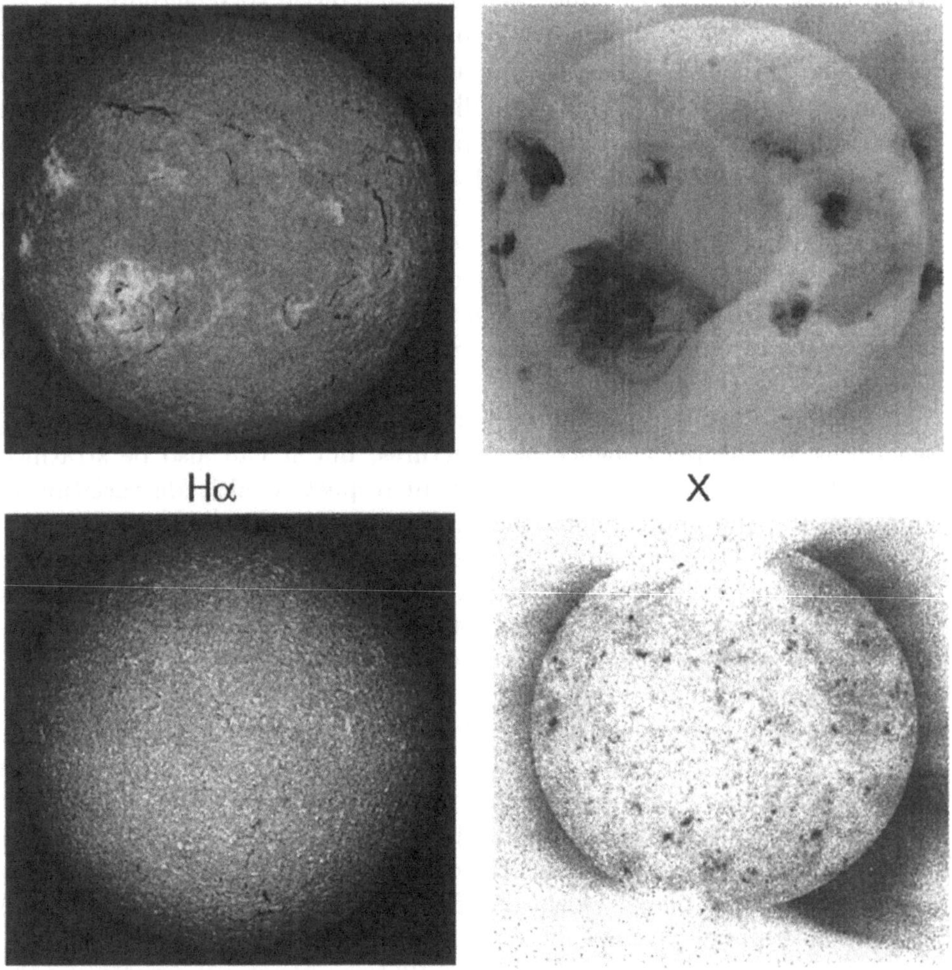

Figure 1. Two sets of daily solar images, at top row the active sun (October 26, 1991) and at bottom the quiet sun (5 October, 1996). From left to right: the photospheric magnetic field (Kitt Peak National Solar Observatory); the photosphere - chromosphere transition in CaII K1v line; the chromosphere in Hα line (Paris-Meudon Observatory); the corona in soft X-ray - negative image - (YOHKOH/SXT).

3. OUTSTANDING STRUCTURES OF THE SOLAR ATMOSPHERE

We shall point out only those specific features which are best related to our topic.

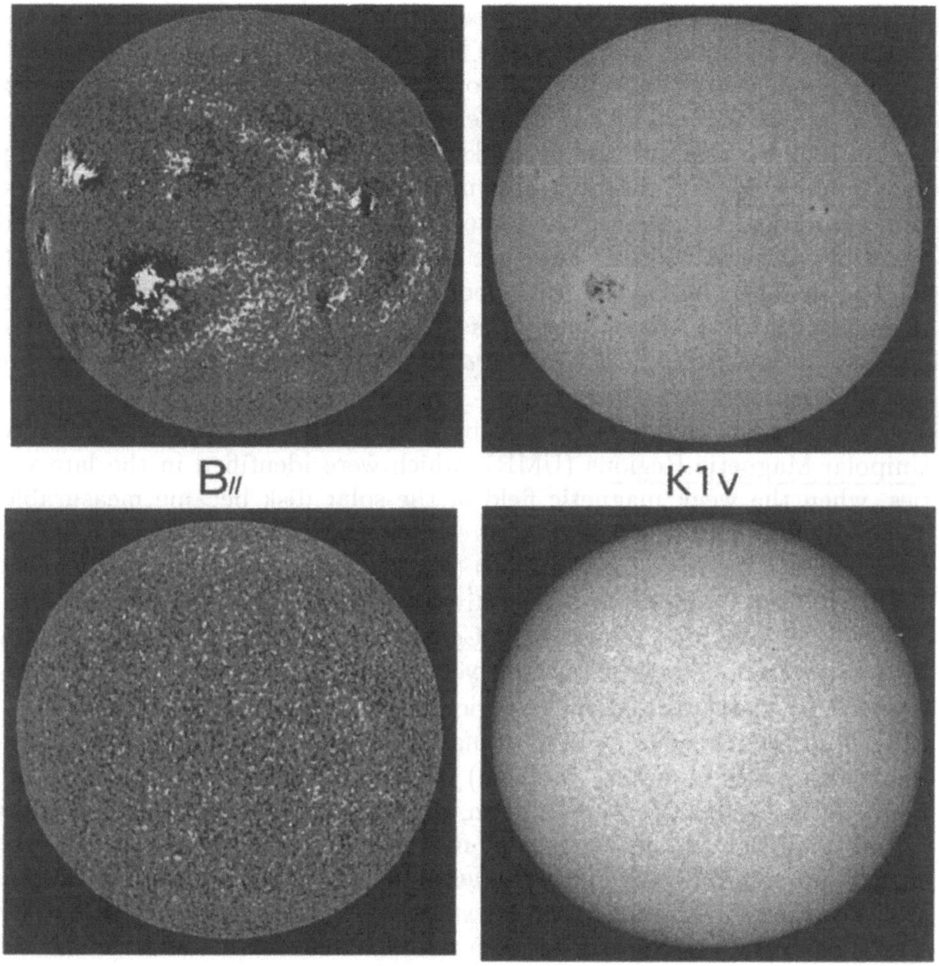

3.1. QUIET SUN STRUCTURES

The quiet sun structures are always present in the solar atmosphere. They are permanent, but undergo some variations in phase with the solar cycle. Granules are visible on the solar disk (photosphere), under good seeing conditions. Their dimension is about 1000 km and their life-time is 5 to 10 minutes. Granules are the smallest convective element of the solar atmosphere. In the intergranular space, we find fine foot-points of magnetic tubes which expand into the chromosphere and corona. The average number of granules in an area of 10 $\times$ 10 arc second changes during the solar cycle, from 37 at maximum to 32 at minimum [28], i.e. granule dimensions decrease as activity increases. The same type of variation is found at the approach of sunspots, near sunspots the dimension of granules is lessened. Super-granules are convective cells of 30 - 40,000 km diameter with a lifetime of about one day. They are permanent structures and were discovered in 1962 by Leighton and co-workers. Super-granules are surrounded by the Chromospheric Network, whose magnetic field is about 50 G. Spicules, long cylindrical structures, are rooted in the chromospheric network and penetrate the low corona. They are the mass supply for coronal material. The distance between spicule foot-points, the Chromospheric Grains, changes with the solar activity [6], as does the distance between granules, because the magnetic field tubes of spicules and chromospheric grains are rooted in intergranular space.

The whole solar surface is covered by a magnetic field background called Unipolar Magnetic Regions (UMR), which were identified in the late sixties, when the weak magnetic field of the solar disk became measurable. The magnetic field of these structures is not strictly unipolar, but one of the magnetic polarities is prevalent (Fig.2). Their dimensions may reach 10^4 to 10^5 km or more, and they are lengthened by solar differential rotation. UMR are also called Giant Cells, to imply that they are probably convective cells, though this is not yet clearly proved. The UMR are surrounded by filaments and active regions.

Fig. 2 shows the synoptic map of magnetic field observations at the maximum (top) and minimum (bottom) of activity cycle 22. The UMRs can be easily identified. At the minimum, the UMRs are almost substituted by mixed polarities region, whereas at maximum, they cover the entire solar surface and exhibit a turbulent character. The fractal dimension of UMR is D = 1 at the minimum and then grows to D = 1.30 at maximum of activity. The time evolution of UMR between minima and maxima was sketched by Mouradian and Soru-Escaut [23].

The North and South solar pole zones are UMRs of opposite magnetic polarity, very often accompanied by Coronal Holes. The fundamental characteristic of coronal holes is their open or divergent magnetic field. The temperature and density are less than in the "normal" corona ($T = 1.5$ instead of $2 \times 10^6 K$ and n_e is 1/3 of the quiet corona). Coronal holes may sometimes exist also at lower latitudes, but never overstep limits of a UMR.

3.2. ACTIVE SUN STRUCTURES

The active sun consists of two main structures: prominences and active regions. These are non-permanent structures located at the edges of UMRs. There exists a deep relation between UMRs and the activity features. *Active*

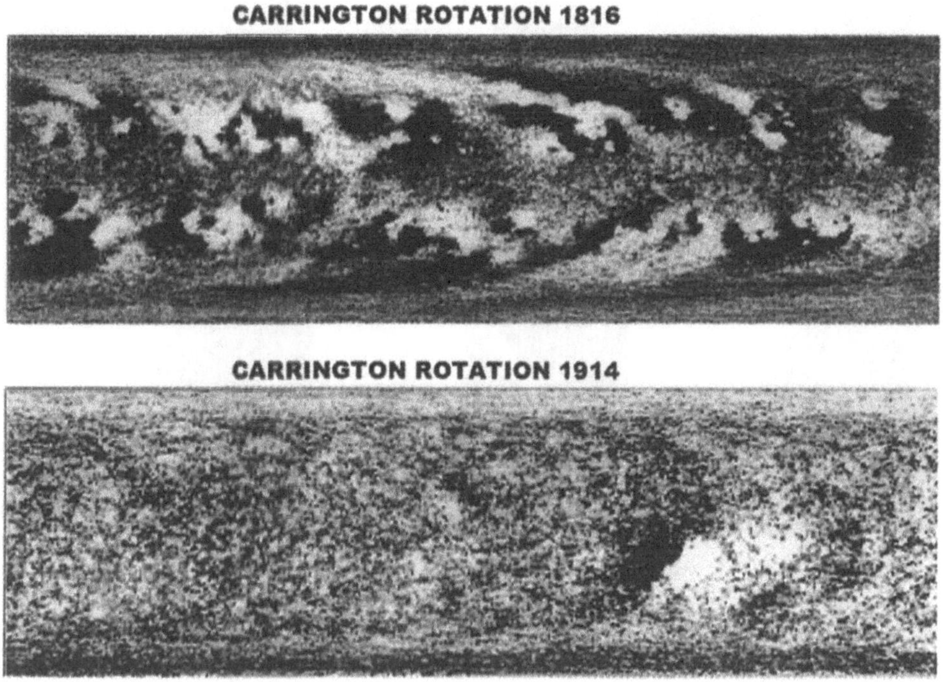

Figure 2. Two synoptic maps (Kitt Peak National Solar Observatory), of observed magnetic field at the maximum of activity (Carrington Rotation 1816; May 25.64 to Jun 21.84, 1989, at the top) and at minimum (Carrington Rotation 1914; Sep 18.39 to Oct 15.67, 1996, at the bottom).

sun structures all have a characteristic in common: their gradual evolution suddenly interrupted by a transitory event with energy realized generally followed by dramatic restructuring. The transient is triggered by an instability in the magnetic field and consequently, after the transient, the active site morphology changes. The number of active regions and prominences are subject to the 22 year solar cycle variation.

3.2.1. *Prominences*

Prominences, or Filaments when perceived on the solar disk, are clusters of many magnetic tubes containing a cool plasma ($T = 10^4 K$ and $n_e = 10^{10} - 10^{11} cm^{-3}$ [11]), forming a curtain-like structure in the hot coronal environment ($T = 10^6$, $n_e = 10^8 - 10^9$) [38]. A thin transition zone separates prominence from the corona. The prominence body is supported against gravity by Lorentz force. The geometrical dimensions of filaments

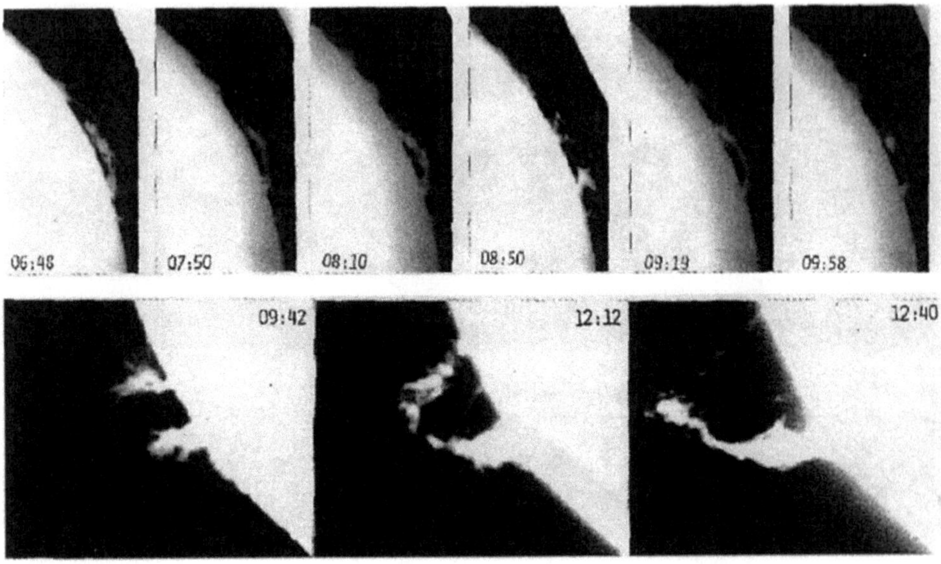

Figure 3. Sudden disappearance of prominences. Top: thermal disappearance (April 15, 1989); bottom: dynamic disappearance (August 16, 1988).

are extremely variable, but we mention some average figures: 100,000 km in length, 50,000 km in height and 5000 km thick, with possible variations of up to 100%. Prominence lifetimes may be a few days or months, but the plasma and magnetic tubes are always renewed. Even if a prominence is qualified as quiescent, it is in continual transformation and movement [29], but its general morphology does not really change.

A prominence may occur within an active region, in which case it is called an Active Region Filament, or it may occur in the quiet sun and is then called Quiescent Prominence or Quiescent Filament. Prominences are currently observed in the hydrogen Hα line, and are seen as emissive when viewed at the limb and absorbent when viewed on the disk.

During its lifetime, a prominence may suddenly disappear in a phenomenon called "DB" (for Disparition Brusque in French). This is a process in which the prominence disappears from the visible spectrum. The process may be of two types: thermal DB or dynamic DB [27]. DB is the transient process of prominences (filaments), occurring in both quiescent and active region filament (Fig.3).

Thermal DB (DBt) is due to the ionization of hydrogen when the energy input into the prominence grows significantly. The prominence disappears in the Hα line, the usual observation wavelength, but becomes emissive in the hotter lines of the EUV spectrum [21]. The DBt can be as rapid as 10 minutes. During this process, the altitude and the magnetic field structure of the prominence do not really change (Fig.3 top), so that, after the heating stops, the filament becomes visible again in Hα images, with the same shape and position as before; no Coronal Mass Ejection is triggered during DBt.

Dynamic DB (DBd) is an eruption of a quiescent or active region prominence, with plasma and magnetic field ejection into the corona. During this process, the magnetic field is entirely distorted. When a magnetic flux emerges close to one prominence foot-points, it disconnects the prominence, by magnetic reconnection, and the body of the prominence starts an ascending movement (Fig 3 bottom). The duration of a DBd is up to a few hours. This upward motion is accompanied by a Coronal Mass Ejection (CME) [8] and followed by a soft X-ray arcade forming a double-ribbon flare in the chromosphere [26], [12]. Some DBd precursors are visible before eruption.

A Filament Channel, clearly visible on both sides of the filament in Hα solar disk images, separates the filament from the quiet sun. Filament channel becomes visible by orientation of spicules parallel to the filament [19].

3.2.2. *Active Regions*

The active regions are multi-component magnetic structures that are formed at particular points on the Sun's surface, called Pivot Points [22]. During a few preceding Carrington rotations, before the formation of the active region, these pivot points exhibit rigid rotation with Carrington rate, regardless of the latitude. Pivot points are located at the boundary between two UMRs, and are deeply rooted in the convective zone.

The Active Region is the most important activity site. It comprises stable structures - a plage, a sunspot group - along with an occasional filament (Fig.4). Inside active region, transient structures do occur: flares, surges, spray, accelerated particle beams, among others [33].

Active regions are located in latitude bands from 5 to 40°, in either hemisphere, thus forming two parallel belts to the equator. At the beginning of the cycle, active regions appear at high latitudes and they progress toward the equator during the cycle, describing the so called Butterfly Diagram. As in the case of the prominence, the active region is also surrounded by spicules oriented in a Vortex system strengthening the whole active region. Toward the end of the life of an active region, a filament very often appears, between polarities of the plage.

The first component of an active region that appears in the solar chromosphere is the Plage. This begins as a compact, brilliant area sitting astride an inversion line in the magnetic field background. We find at least one

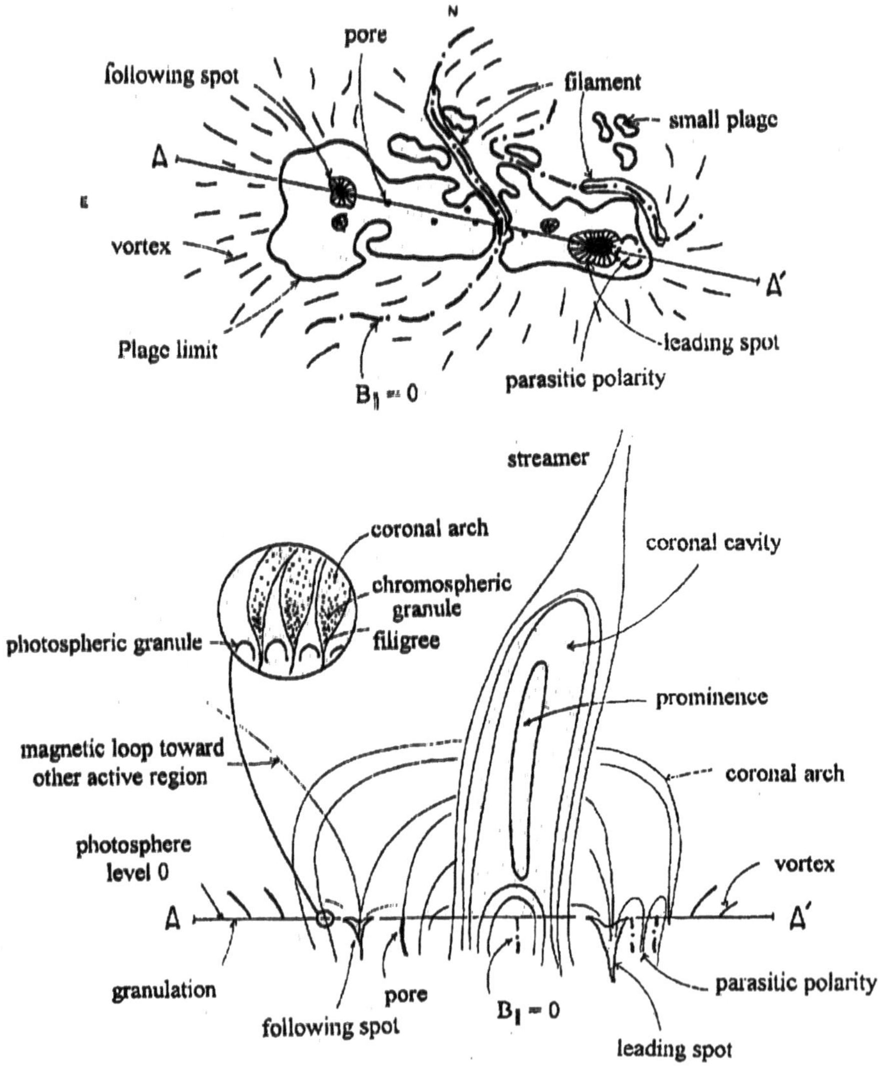

Figure 4. A schema of a quiet active region on the disk center and on the limb [33].

magnetic inversion line crossing an active region, dividing the plage into two opposite magnetic polarities. A few arches (Arch Filament System, AFS) join the two polarities at the birth of the active region [18]. With time, the plage grows and spreads out generally after two or three weeks, or more. The plage is hotter then the quiet sun (Fig.5), due to the moderate magnetic field (2 to 300 G) which transports and deposits the energy in the

chromosphere and corona.
The Sunspot Group is a magnetic bipolar group of spots inside a plage, and the magnetic polarities of the spots correspond to that of the plage polarities. The sunspot group appears after the plage and disappears before it. Brants and Steenbeek [1] found that the formation and disappearance of sunspots are due to local convergent and divergent motions respectively, of the magnetic tubes carried by photospheric movements. Generally, a group comprises a few Spots and Pores (small sunspots) oriented approximately East-West and may extend as much as 20 heliographic degrees and more. The Western spot is called Preceding (p) or Leading spot, whereas the eastern spot is called Following (f) or Trailing spot. Generally, the preceding spot is bigger and lasts longer. This is the classic ordering of sunspot groups, but often very complex groups are formed where the classic bipolar structure is no longer evident. These complex groups are composed of a mix of opposite-polarity spots and occur during the high activity period.
At the present (1999), in the growing phase of cycle 23 (see below), the p spots in the northern hemisphere are of north magnetic polarity, whereas those in southern hemisphere are of South polarity. This means that, in northern hemisphere, the western UMR of an active region, is of North polarity and the opposite in the southern hemisphere. During the following cycle, the magnetic situation will be contrary to that of the present cycle. This is the Hale's law. Consequently, we find the same magnetic configuration two cycles later, which doubles the 11- year cycle to 22 years. Note that in each hemisphere, in the growing phase of a cycle, i.e. before the maximum of activity, the magnetic polarity of p spot is always the same as the magnetic polarity of its solar pole [24]. Short after the maximum of activity, the magnetic polarity of solar poles is changing and consequently the polarity of p spot, in decreasing phase becomes opposite to the magnetic polarity of solar poles.
Sunspots are dark photospheric structures visible in white-light images. Their principal characteristic is a strong magnetic field between 1000 and 3000 G. This magnetic field prevents the energy transport to the surface, so that the area of the spot becomes cooler (1500- 2000 K less than the photosphere, Fig.5), and consequently less bright, so they appear dark by contrast. We note that this cooling of the atmosphere favors the formation of molecules in the spot atmosphere. The central part of a sunspot, the Umbra, is dark and surrounded by Penumbra of intermediate brightness, with a radial filamentary structure. Mean spot diameter may be larger than the Earth's, but pores are sometime less than 1000 km.
The active region magnetic field arches form a dome-shape structure above the chromospheric plage, which extends into a coronal streamer. The chromospheric temperature grows to $\sim 2 \cdot 10^6$ K in the corona (Fig.5).

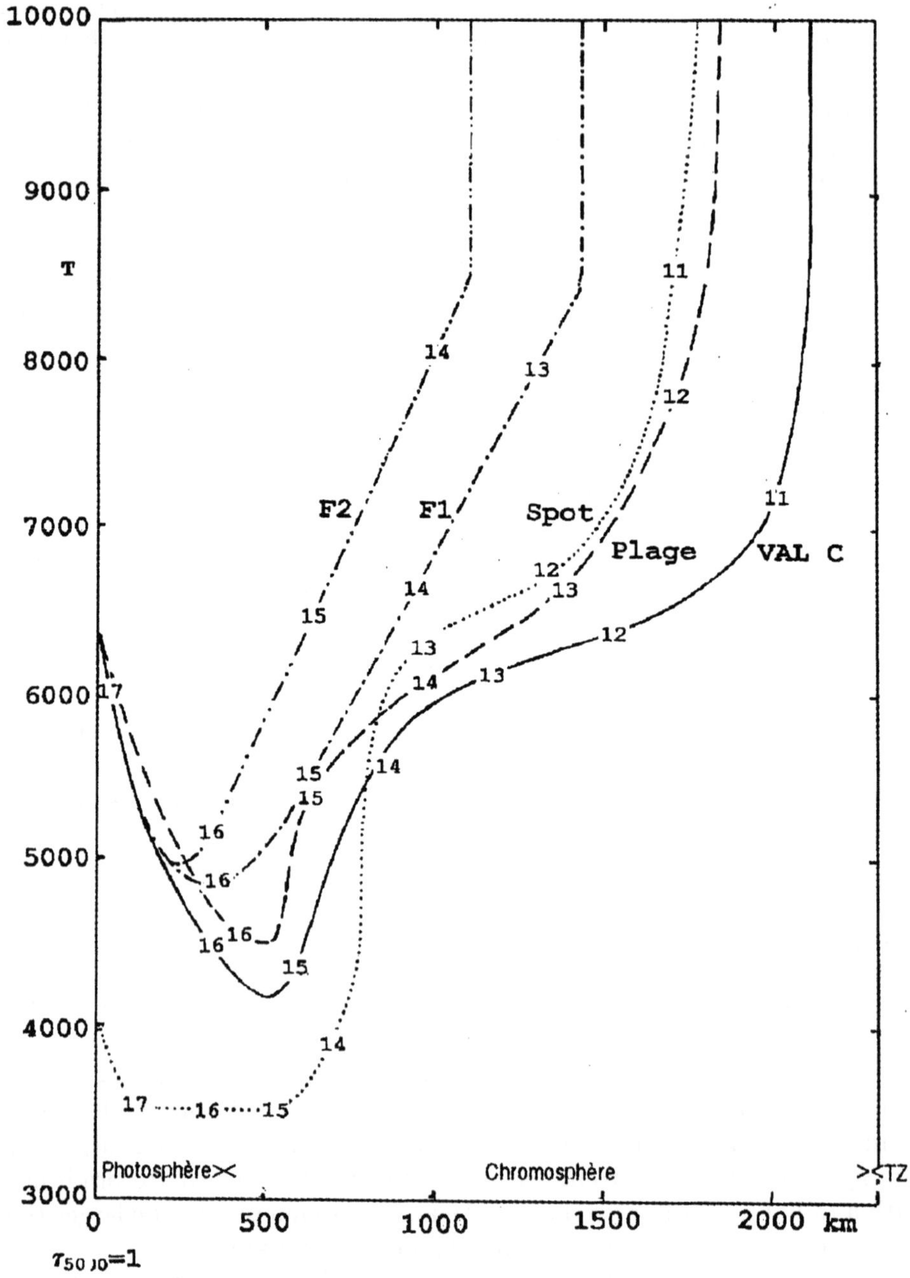

Figure 5. Temperature and total number of hydrogen atoms from structure models of active region. In abscissa the altitude in solar atmosphere [20].

Recent observations by TRACE satellite [31] show that cool and hot structures coexist tightly packed in the corona, and that the flow in arches is mostly a successive train of plasma package instead of continuous streaming.

A Complex of Activity, or Active Nests [7], is a group of active regions in space and/or in time, showing the existence of a common cause in deeper layers. The tendency of active regions to cluster in Active Longitudes has been known for a long time. One direct consequence of this clustering is the complexity of certain sunspot groups or complex spots, due to interpenetration and interaction of old and new flux.

Small active regions can evolve quietly without any transient events, but there are very few like this, and generally occur at the beginning or end of an activity cycle.

The Flare is the most energetic and spectacular phenomenon in the Sun's atmosphere (Fig.6) [39]. The energy released in the few minutes of the impulsive phase can reach 10^{32} erg. The entire spectral domain from elementary particles, γ and X-rays and to radio wavelengths participate in the event. A flare generally occurs inside an active region on the magnetic inversion line. A few flares can be found outside active regions, consequent to a quiescent filament DBt.

Flares, observed on the disk are classified by area in five classes: sub-flare, and importance (Imp.) 1 to 4. A brightness suffix can be added to this, as well as special soft X-ray classes. According to statistics for the 27 years, from 1967 to 1993 [25], the number of sub-flares account for 91.4% of the total number of flares, Imp. 1 for 8.0%, and strongest flares of Imp. ≥ 2, for only 0.7%.

A flare consists of a number of Flaring Arches [37] joining two magnetic polarities of an active region. Flaring arches may reach altitudes of several 10^5 km in a few minutes (Fig.6). Often their foot-points show up as a Two-ribbon Flare in Hα images. If the active region has a filament, often the flare may start at the filament DBd, and a two-ribbon flare occurs spanning the filament position. At the same time, a CME is triggered. The magnetic complexity and velocity field of an active region are likely to generate flares [15]. Solar flares can be broken down into two classes, F1 and F2 (Fig.5) [17]. F2 flares are hotter than F1, showing intense soft X-ray emission, accelerated particle beams producing polarized Hα light [9] and sometimes white light emission. Note that the first flare observed, in 1859 by Carrington, was a white light flare. We can distinguish two phases in a flare: the impulsive phase at the beginning, followed by a gradual phase [37]. Generally, during the short impulsive phase of a few minutes, all flare energy is released and high energy radiation (γ and X-ray) is emitted. The temperature in F2 flares can exceed 10^7 K, and so we can find Fe XXV

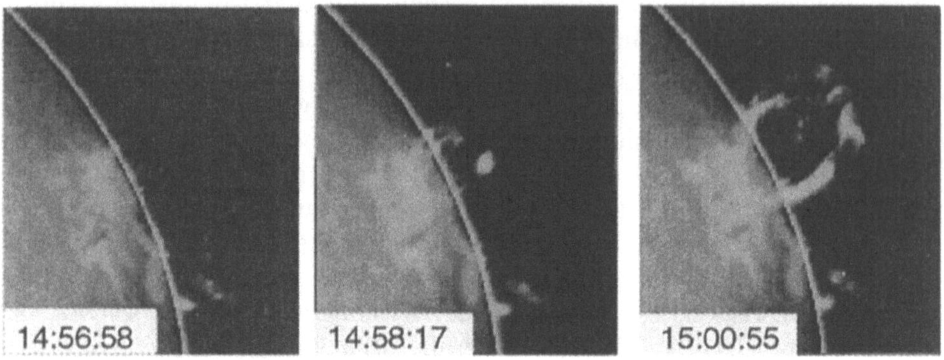

Figure 6. A limb flare observation of June 20, 1989 (Paris-Meudon Observatory).

lines in the flare spectrum, for instance. The gradual, or thermal phase, is a long cooling time when Post-flare Loops can be seen [30]. A flare model is given in the lecture of Y. Uchida [40].
High spatial and temporal observation by TRACE satellite [31] shows that a great number of spikes (small flares < 1 arc second) occur continually during a flare.
Flares are often preceded by a Surge, a magnetic arch emergence with flowing plasma inside [3].

4. ACTIVITY INDICES AND MAPPING

The use of indices comes from the need to qualify a states of solar activity by a number. The difficulty consists in choosing a representative quantity which is unique dependent on the phenomenon to be monitored. A good index is one whose response amplitude is sufficient to allow an easy reading. There are a certain number of indices measuring the global level of solar activity. Those most used are the sunspot number and the radio flux at 10.7 cm.

- An empirical index for the measure of global solar activity, based on the number of sunspots, was initiated by R. Wolf: $R_Z = k \times (10g + f)$, where g is the number of sunspot groups and f is the number of individual spots, for the whole visible hemisphere. k is a personal coefficient of the observer for normalization purposes. R_Z is established daily by a number of observers all round the world. It has been possible to reconstruct the Wolf Numbers back to 1749 (www.oma.be). Averaging R_Z numbers, Solar Geophysical Data [34] publishes the Monthly and

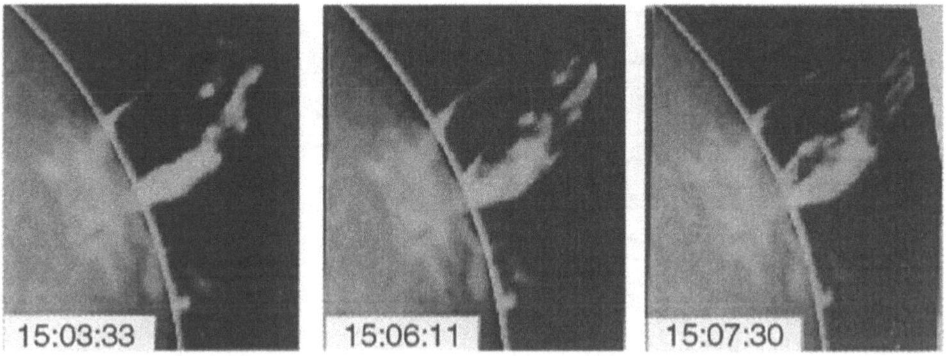

Annual Mean of Sunspots Number. A systematic real difference does exist between sunspots number in two hemispheres called the north-south asymmetry.

- The Radio Flux at 10.7 cm , i.e. 2800 MHz, (www.drao.nrc.ca) is a closer indicator of solar activity, because it is the direct measure of the radio flux of the entire active region corona on the disk. This index is less sensible to meteorological conditions than optical observations.

A few statistical linear relations can be established for these two parameters, as a function of the observed magnetic field norm integrated over the disk ($B_{||}$):

$$R_Z = 10.2 \times B_{||} - 43 \quad \text{and} \quad F_{10.7} = 9.67 \times B_{||} + 3$$

Other indices of activity are also in use. For example, there are the number of filaments; the total magnetic field flux ($B_{||}$); X-ray flux; Ca II K line plage area; coronal lines intensities; and others. Indices are usually represented by their time variation. The advantage of integrated flux indices of the sun is that they can be compared directly with analogous stellar indices for solar-stellar studies [16].

The indices mentioned above are one-dimensional representations of a given parameter as a function of time. In solar physics, two-dimensional representation is also used in the synoptic maps established for one Carrington rotation. A synoptic map is a longitude-latitude image of the distribution of a given structure or index, in cylindrical coordinates, through a solar rotation. Consequently for a given map, the East and the West ends represented structures are separated by a time interval of nearly one month (Fig.7), because one can never observe the sun entirely at the same time. Successive synoptic maps provide a complete space-time view of a given parameter, within the limits of space and time resolution.

Another kind of synoptic map is published by Paris Observatory, dedicated

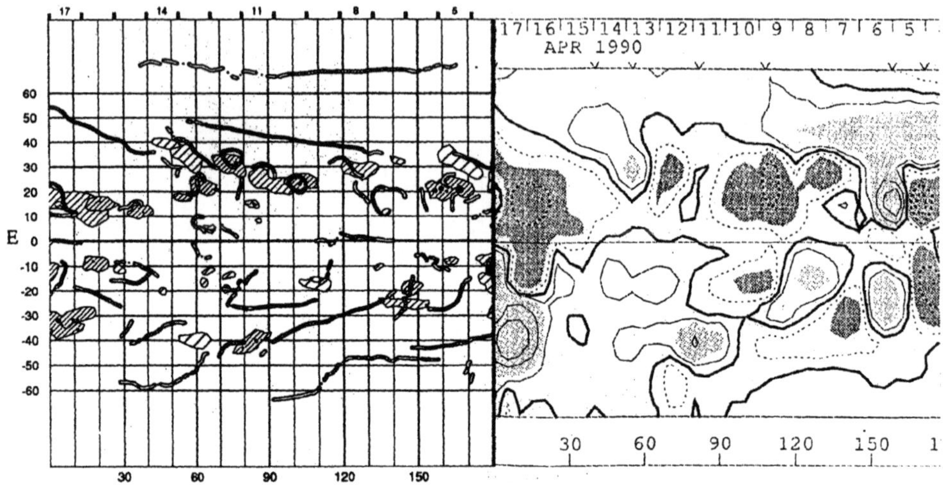

Figure 7. Partial synoptic maps of solar activity (left) and of weak photospheric magnetic field (right) respectively from Paris-Meudon Observatory and Wilcox Solar Observatory.

to solar activity (http://mesola.obspm.fr). These maps give a synthesis of each active region and each filament during their passage over whole disk (Fig.7).

5. THE SOLAR ACTIVITY CYCLE

The solar activity cycle is the time history either of an activity index or of the number of a given structure. The basic periodicity of the solar cycle is around 11 years for sunspot numbers, but it doubles to 22 years, if we refer to the magnetic field polarity changes. The maxima, as well as the minima of activity cycle show a long-lasting cycle of around 100 years. During the cycle, the number of active elements first grows fast up to the maximum and then decreases slowly toward the minimum. An activity cycle lasts longer than the time laps between two successive minima. Indeed, the new cycle starts before the end of the old one. For instance, the cycle 21 began in 1975.9 and ended in 1987.5, which gives 11.6 years of real duration, whereas the minima interval is only 9.9 years (1976.6 to 1986.5).

Different kinds of representations have been adopted to show the advance of the activity cycle. The most popular is the time dependence of the number of a given structure, for instance the R_Z number. Another is the latitude variation, again in time dependence, in which case the structures sketch the so-called Butterfly Diagram. This shows that activity starts at ± 30 to 40 degrees of latitude and migrates toward the equator during the cycle.

6. SOLAR VARIABILITY

The solar cycle duration and phase can also be seen in the global solar irradiance, where the quiet and active sun contribute to integrate disk emission variation. Solar irradiance is proportional to the Sun's radiative energy flux. Satellite measurements show that, at the activity minimum the irradiance is 0.15% lower than at maximum, but daily irradiance variations are due to spots, which weaken the radiation. Hudson [10] gives the relations of solar irradiance to the activity indices $F_{10.7}$ and R_Z. Other measurements of irradiation for solar chromospheric lines during the cycle are given by Livingston [16].
Not only the irradiance, but also some quiet sun structures show cyclic variation. The solar granulation dimensions change during the cycle. There are 14% more granules during the maximum than in the minimum [28], i.e. granule dimensions lessen with higher activity. At the chromospheric level, we find the same tendency in granules of the chromospheric network [6]. Even if granules or network are considered to be components of the quiet sun, they are influenced by the solar cycle.
The solar diameter also varies [13] with time. During a solar cycle, it grows from: 2×959.32 at the maximum, to 2×959.57 arc sec at the minimum, for a solar distance of one Astronomical Unit (AU).

7. SOLAR - STELLAR RELATION

The stellar rotation and variability, and the magnetic field detection capacities are opening paths of investigation of stellar activity [2]. Already long ago, eclipsing binary star observations have revealed the existence of spots on the central star, but the new Zeeman-Doppler Imaging technique [32] can map star spots with their magnetic polarity and strength.
The stellar flux spectra can also provide some important information concerning the atmospheric structures of the star. Different structures give particular signatures in the spectral lines [20]. So it is possible to detect spots, plages prominences, flares, CME, and wind.
It should be noted that stars allow the study of activity and cyclicity in shorter and longer time periods, or in different physical condition, than the sun.

8. CONCLUSION

Solar activity is conditioned by the existence of UMR boundaries, where the activity sites occur and they are non-permanent structures of solar atmosphere. Activity sites on the Sun are of two types: prominences and active regions. These evolve gradually but magnetic instabilities trigger certain transient phenomena such sudden disappearance of prominences and active region flares. The indexes of activity show that activity sites and transients obey 11 year cycle.

9. ACKNOWLEDGMENTS

The author is grateful to Mrs. I. Bualé for her assistance in preparation of figures.
This lecture is illustrated with observations obtained by Kitt Peak National Solar Observatory, Paris-Meudon Observatory, Wilcox Solar Observatory and Yohkoh/SXT Satellite.

References

1. Brants, J.J. and Steenbeek, J.C.M. (1985) Morphological evolution of an emerging flux region, *Solar Phys.* **96**, 229-252.
2. Cameron, A.Collier, and 17 co-authors (1999) Multisite observations of surface structures on AB Doradus in 1994 November, *MNRAS* **308**, pp. 493-509.
3. Canfield, R.C., Reardon, K.P., Leka, K.D., Shibata, K., Yokpyama, T., Shimoto, M., (1996) Hα surges and X-ray Jets in AR 7260, *Ap.J.* **464**, 1016-1029.
4. Cox, A.N., Livingston, W.C., Matthews, M.S. (eds.), (1991) *Solar Interior and Atmosphere*, The Univ. of Arizona Press, Tucson.
5. Donea, A., Lindsey, C., Braun, D. (1999) Helioseismic holography - a technique for studying seismic signatures of solar flares, Poster publication of this lectures in *Romanian Astro. J.* **9** Suppl., (in press).
6. Fang, Ch., Mouradian, Z., Banos, G., Dumont, S., Pecker, J.C. (1984) Structure and physics of solar faculae: IV-Chromospheric granular structure, *Solar Phys.* **91**, 61-70.
7. Gaizauskas, V., Harvey, K.L., Harvey, J.W., Zwaan, C. (1983) Large-scale patterns formed by solar active regions during the ascending phase of cycle 21, *Ap.J.* **265**, 1056-1065.
8. Gosling, J.T., (1993) The solar flare myth, *J. of Geophys. Res.* **98, No. A11**, 18937-18949.
9. Hénoux, J.-C. and Vogt, E. (1998) Application to astrophysics of collisional polarization spectroscopy, *Physica Scripta* **T78**, 60-67.
10. Hudson, H.S. (1988) Modeling of total Solar irradiance variability: an overview, in G. A. Chapman (ed.), *Solar and Middle Atmosphere Variability*, Adv. Space Res. **8 Nb. 7**, pp. 15-20.
11. Jensen, E., and Wiik, J.,E. (1990) Plasma parameters in quiescent prominences, in V. Ruzdjak and E. Tandberg-Hanssen (eds.), *Dynamics of Quiescent Prominences, IAU Coll. No 117, Lecture Notes in Physics.* **363**, Springer-Verlag, Berlin, pp.298-301.

12. Khan, J.I., Uchida, Y., McAlister, A.H., Mouradian, Z., Soru-Escaut, I., Hiei, E. (1998) A flare-associated filament eruption observed in soft X-rays by Yohkoh on 1992 May 7, *A&A*, **336**, 753-768.
13. Laclare, F., Dalmas, C., Coin, J.P., Irbah, A. (1996) Measurements and variations of the solar diameter, *Solar Phys.* **166**, 211-229.
14. Lindsey, C., and Braun, D.C. (1999) Helioseismic holography, *Ap.J.* **485**, 985-903.
15. Liu, Y., Akioka, M., Yan, Y. and Sato, J. (1998) A flare due to the interaction of a small loop and a large loop, *Solar Phys.* **180**, 377-392.
16. Livingston , W. (1993) Surrogates for total solar irradiance, in E. Nesme-Ribebs (ed.) *The Solar Engine and Its Influence on Terrestrial Atmosphere and Climate,* NATO ASI Series I **25**, pp.145-162.
17. Machado, M.E., Avret, E.H., Vernazza, J.E.,Noyes, R.W. (1980) Semi-empirical models of chromospheric flare region, *Ap.J.* **242**, 336-351.
18. Malherbe, J.M., Schmieder, B., Mein, P., Mein, N., van Driel-Gesztelyi, L. and von Uexküll, M. (1998) Arch filament systems associated with X-ray loops, *Solar Phys.* **180**, 313-329.
19. Martin, S.F. (1998) Filament chirality: A link between fine-scale and global patterns, in D. Webb, D. Rust and B. Schmieder, (eds.), *New Perspectives on Solar Prominences*, IAU Coll. 167, ASP Conf. Ser. **150**, pp. 419-429.
20. Mouradian, Z., (1995) Spectral signatures of solar structures, *Astrophysics* , **38**, 508-518.
21. Mouradian, Z., Martres, M.J., Soru-Escaut, I.(1987) The heating of filaments as a disappearance process in A.I. Poland (ed) *Coronal and Prominence Plasmas*, Proc. of Workshops in 1985 and 1986, NASA Conference Publication no. **2442**, pp. 221-223.
22. Mouradian, Z., Martres, M.J., Soru-Escaut, I.,Gesztelyi, L. (1987) Local rigid rotation and the emergence of active centers, *A&A* , **183**, 129-134.
23. Mouradian, Z., Soru-Escaut, I. (1991) On the dynamics of the large-scale magnetic fields of the sun and the sunspot cycle, *A&A* **251**, 649-654.
24. Mouradian, Z., Soru-Escaut,I.(1993) On solar activity and the solar cycle. A new analysis of the Butterfly Diagram of sunspots, *A&A* **280**, 661-665.
25. Mouradian, Z., Soru-Escaut, I. (1995) The flare productivity of the sun related to the 11 - and 80 - year activity cycles, *Solar Phyis.* **159**, 173-179.
26. Mouradian, Z., Soru-Escaut, I., Hiei, E., McAlister, A.H., Shibasaki, K., Ohyama, M., Khan, J.I., Uchida, Y.(1998) The filament disappearance of 1992, May 07 (the EBI), *Solar Phys.* **180**, 313-329.
27. Mouradian, Z., Soru-Escaut, I., Pojoga, S. (1994) On the two classes of filament-prominence disappearance and their relation to coronal mass ejections, *Solar Phys.* **158**, 269-281.
28. Muller, R.(1989) The solar granulation, in G.A. Chapman (ed.) *Solar and Middle Atmosphere Variability*, Adv. Space Res. **8, No.7**, pp 159-167.
29. Pojoga, S., Molowny-Horas, R. (1999) The transverse velocity field of EUV solar prominence, *Solar Phys.* **185**, 113-125.
30. Schmieder, B. Heinzel, P., Wiik, J.E., Leman, J., Anwar, B., Kotrec, P., Hiei, E. (1995) Relation between cool and hot post flare loops derived from optical observations and Yohkoh/STX, *Solar Phys.* **156**, 337-361.
31. Schrijver, C.J. and 16 co-authors (1999) A new view of the solar outer atmosphere by the transition region and coronal explorer, *Solar Phys.* **187**, 261-302.
32. Semel, M. (1989) Zeeman-Doppler imaging of active sun. I Basic principes,*A&A* **225**, 456-466.
33. Semel, M., Mouradian, Z., Soru-Escaut, I., Maltby, P., Rees, D., Makita, M. (1991) Active regions, sunspots and their magnetic fields, in A.N., Cox, W.C., Livingston, M.S., Matthews, (eds.), *Solar Interior and Atmosphere*, The Univ. of Arizona Press, Tucson, pp. 844-859.

34. Solar Geophysical Data, World Data Center A for Solar-Terrestrial Physics, NOAA, Boulder.
35. Sonett, C.P., Giampapa, M.S., Matthews, M.S. (eds.) (1991) *The Sun in Time*, The Univ. of Arizona Press, Tucson.
36. Stix, M. (1989) *The Sun*, Springer Verlag, Berlin.
37. Švestka, Z. (1989) Solar flares: the gradual phase, in B. M. Haisch and M. Rodono (eds.), *Solar and Stellar Flares*, Kluwer Acad. Pub. Dordrecht, pp. 399-417.
38. Tandberg-Hanssen, E. (1995) *The Nature of Solar Prominences*, D. Reidel Publ. Co., Dordrecht.
39. Tandberg-Hanssen, E. and Emslie, A. G. (1988) *The Physics of Solar Flares*, Cambridge Univ. Press, Cambridge.
40. Uchida, Y. (2000) New findings about flares and active regions by Yohkoh. (this issues).
41. Wilson, P.R. (1994) *Solar and Stellar Activity Cycles*, Cambridge Univ. Press, Cambridge.
42. Zirin, H. (1988) *Astrophysics of the Sun*, Cambridge Univ. Press, Cambridge.

NEW VIEWS OF ACTIVE REGIONS, FLARES AND CME'S FROM SPACE

Y. UCHIDA
Physics Department, Science University of Tokyo
Tokyo, Japan

Abstract. In the present article, we treat (1) the problem of how the active region coronae are heated in extra and supplied mass, and (2) the problems of how the small but high energy loop flares, and the large and dynamic arcade flares, sometimes accompanying massive coronal mass ejections, occur, based on the new observational results derived from the recent satellites, Yohkoh, SOHO, and TRACE, that revolutionized our knowledge about solar, and therefore, stellar activity. We also discuss possible models we propose to interpret those.

CONTENTS:

1. Introduction

Some of the recent observational results from space about the superhot, high energy, and mass-loss events in the solar atmosphere as observed by Yohkoh, SOHO, and TRACE, are described in the following, together with some modeling efforts. Since the results from those satellites range over very wide area, and the analyses are being done by a large number of researchers, I, of course, do not dare to cover all of them. Rather, I will confine myself only to those topics that my research group at the Science University of Tokyo has actually been working on. Those are (1) the problem of what are

J.-P. Zahn and M. Stavinschi (eds.),
Advances in Solar Research at Eclipses from Ground and from Space, 105–147.

active region coronae, why they are heated in extra and the density is high, or, how their energy and mass are supplied, (2) the problem of flares, how those are prepared, and how those develop in time after they start, and (3) how do those involve in the mass-loss processes from the Sun.

Our works are based on two approaches: One is the observational approach based on the analyses of the data from the Satellite Yohkoh, mainly the soft X-ray telescope, SXT, supplemented by the data from other more recent satellites, SOHO and TRACE, whose data are provided for our use from respective groups according to the NASA and ESA principles. We deeply acknowledge for those here. The other is the modeling based on the MHD simulations we have developed for years using supercomputers and 3D image processing.

Yohkoh/SXT is equipped with high sensitivity CCD detector, probably one of the first among the astrophysical satellites, and can take an unlimited number of pictures with high cadence, with wide dynamic range by the combination of computer-controlled rotating filter-wheels. These help us to go deep in the observations into, say, pre-event features which, by the causality principle, are very essential in telling us about the origin of certain physical event, but could not be obtained by the previous satellites. Therefore, we aimed at looking into such faint structures and their changes to pursue the mechanisms of the above-mentioned targets, and tried to make clear what physical processes are in operation in them by the MHD simulation approach.

In this Summer School lectures, the results in some of the already published papers, reports to some recent conferences, and the material in preparation, from our group, are briefly summarized. Audience interested in more details should take a look into those, either according to the references given, or by waiting for their publication in the near future in the case of unpublished material.

2. What Are Active Region Coronae?

We first ask a fundamental question of why the active region corona is heated in extra and has higher density. The answer to this question is vital also from the point of view of why the enduring component of X-ray emissions from other active stars are produced. The problem of the heating of the weaker background corona has been discussed for many years, but not yet solved. Our approach is from the brighter version of it, rather than from the basic weak end, the normal corona. Our hope is that if the problem of active region corona is made clear, then it may give clues to that of the normal corona, too.

The Sun is a modest small star, but still has most kinds of typical activities, and what is important, it is *the only star* on which the details of those events can be observed in the spatially-resolved form, though the magnitudes of them are one or two orders of magnitudes smaller than those very active stars emitting strong X-rays, both enduring and flaring components.

The special factor about active regions is known to be the presence of strong magnetic field, but the mere presence of a strong field can not explain the heating and the mass supply. There must be energy supply from somewhere, most likely from the photosphere below, and furthermore, there must be a process thermalizing it efficiently.

2.1. PREVIOUS VIEWS ABOUT ACTIVE REGION CORONAE

There were several views proposed as the mechanism of the coronal heating, in general, eg., due to thermalization of shocks of acoustic waves produced in the turbulent convection below the photosphere (eg. Ulmschneider 1996), or due to dissipation of Alfvén waves, which are produced under the photosphere, through phase discrepancy in propagating along the field in an inhomogeneous medium (eg. Sakurai & Granik 1984), or due to dissipation of electric current, which is assumed to be steadily flowing along magnetic loops (eg. Ionson 1985) by an excitation of anomalous resistivity (Galeev et al. 1981). In contrast to the steady current dissipation which turned out to be difficult, a mechanism of explaining the heating as well as mass supply by considering a magnetodynamic process of current injection from below was proposed by Uchida and Shibata (1988).

Observational testing of those models, however, could not be made before Yohkoh. The high sensitivity, wide-dynamic range, and high cadence observations by the Soft X-ray Telescope (SXT) aboard Yohkoh has allowed us for the first time to look into what is actually happening inside active region corona.

2.2. YOHKOH FINDINGS ABOUT ACTIVE REGION BEHAVIOR

Most remarkable of what were found about active regions by Yohkoh may be (1) that there exist a large number of strong loop brightenings inside active regions (Shimizu et al. 1992), with stronger and more frequent brightenings in more active active regions, and those are found to be giving a large part of the source of brightness of the active region corona in X-rays, and (2) that the outermost layers of active regions are intermittently (almost continually in the activity maximum period) expanding away (Uchida et al. 1992). Previously, probably due to the available low cadence observations, it was considered that the active regions are heated steadily, and are held tightly by the strong magnetic field in a more or less magnetohydrostatic

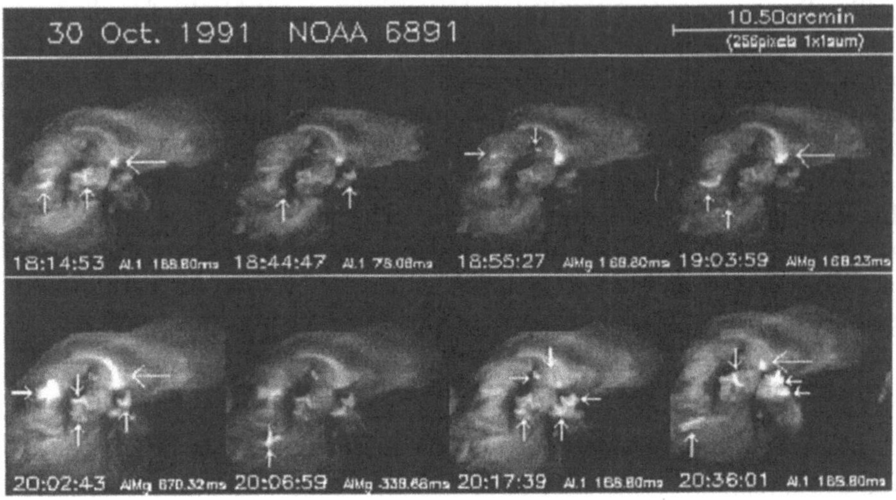

Figure 1. Active Region Transient Loop Brightenings (Shimizu et al. 1992): Time development of transient loop brightenings in a typical active region (the faint area on the back extending in the frame) obtained with a thick filter. One of the footpoints brightens, and the already heated plasma is injected into a pre-existed magnetic loop in the active region. Re-injection of the heated gas is seen in one and the same loop, suggesting that this is not a "reclosing of opened up magnetic loop" type mechanism. No obvious changes of loops nor a mass leakage from the loop is seen.

equilibrium. Both of these presumptions turned out to be not the case by the Yohkoh observations.

2.2.1. *Active Region Transient Loop Brightenings*

Soon after the start of observations by the Yohkoh/SXT in 1991, we noted that a dominant part of the brightness of active regions are due to the appearance of bright loops. By using thick filters to examine the bright features, differentiating from the less bright background active region, Shimizu et al. (1992) found that those bright loops come up transiently with already heated mass injected into the loop from one of the footpoints of that loop with a velocity of a few hundred km/s or larger.

It is remarkable that the gas injected is *already heated* from the beginning. The size of the brightening loop varies, but typically is of the length of magnetic loops connecting two opposite polarity magnetic sources in active regions, and the width is generally small, not wide in the middle as expected from the calculated potential field connections between them. The frequency of appearance differs from one active region to another, but more frequent in more active active regions.

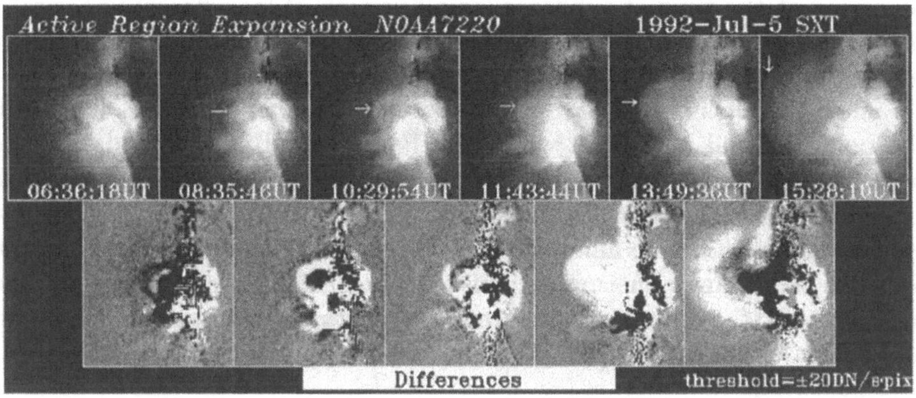

Figure 2. Observed Active region Expansion (Morita et al. 2000a): An example of active region expansion observed at the limb. The lower panels are the time-lapse difference of the frames above. The white part is the part brightened while the dark part is the part that lost the brightness during the time lapse between the successive frames. It is seen that the expansion is a smooth expansion by keeping the smooth round shape of the loop, not driven into a tipped shape by the push of a local agent like a rising blob or a jet. The velocity of expansion is about 10 km/s.

2.2.2. *Expansion of the Outermost Structures of Active regions - Active Region Expansions*

Active region expansion (Uchida et al. 1992) are, in contrast, faint dynamic features that were recognized in the movie format of the data with the thinnest filter. It can be analyzed by using a faint-feature-enhancing techniques like the "unsharpened image masking" method, or the "time-lapse differencing" method.

One point to be noted is that the mass motion occurs as an expansion keeping the loop shape, and likely to be Lorentz force-driven, very different from the case of the Parker wind that goes parallel, even in the case in which the magnetic effect was introduced for extra acceleration by the variation of the cross-section of the flow pipe.

The active region expansion (ARE's) has an average velocity of 10-15 km/s, and combining the Yohkoh data with the SOHO/EIT and SOHO/LASCO images, they seem to be accelerated with height to several tens km/s at several solar radii, and may explain the so-called "slow solar wind" concentrated to the active equatorial zone if they are further accelerated up to two hundred or so km/s.

It was made clear by Morita et al. (2000a) that the start of ARE's is correlated with the active region loop brightening (ARLB's) described in the above. Another noticeable point may be that a similar loop expansion occurring at the time of subflares (small version of loop flares) is *basically the same* as that associated with these fainter transient loop brightenings, but with only a bit larger velocity, say, 20 km/s or so.

Since the density of active region corona is higher than the density in the polar coronal hole from which the Parker wind is supposed to come by a factor of order two, the mass flux from the active region expansions, though slower in the velocity, and smaller in the area from which the outflow is provided, is shown to amount to 10-20 percent of the mass loss from the Sun due to Parker's thermal expansion wind. This may predict that the mass loss of this type may dominate in active stars having very much larger and active active regions like RSCVn stars (Uchida & Sakurai 1985).

2.3. OUR PROPOSED VIEWS ABOUT ACTIVE REGION BEHAVIOR

2.3.1. *Our Proposed Model for Active Region Loop Brightenings by Dynamic Current Injections*

Active region loop brightenings (ARLB) are newly discovered entity, and there was no model for them. The acoustic heating model obviously can not explain the correlation with the magnetic field, and the models with simple shear Alfvén waves can not explain the hypersonic injection of heated mass. Models with interruption of a steady electric current can not explain the repetitive occurrence of hypersonic injections of heated mass. A suggestion of applying the "reclosing of the once opened magnetic arcade" process proposed for arcade flares to this is not likely to apply for those small scale simple loop events embedded among the similar loops.

The observed behavior reminds us of the dynamic current injection model we proposed (Uchida & Shibata 1988). Therefore, we have extended it into a full 3D MHD treatment (Uchida et al. 2000a). This seems to provide us with a most natural answer to the question posed at the beginning of section 1, about the excess heating and mass supply to the active region coronae, if we consider the observation of active region loop brightenings found by Yohkoh. This idea is quite different from the previous coronal heating models, just reflecting the fact that what Yohkoh discovered is *very different* from what have been imagined prior to the Yohkoh observations.

We assume that a fraction of the kinetic energy in the (rotational) motion in the convection under the photosphere is given to the passive magnetic flux tubes in the form of the magnetic energy in the twisted field. The magnetic twist propagates as a torsional Alfvén wave train (TAWT), equivalent to a coaxially closed MHD current packet, moving along the flux tube with low Alfvén velocity. It may come up along the magnetic flux tube, and may emerge toward the photospheric surface, and comes out into the upper atmosphere where the flux tube protrudes above the surface, ie.,

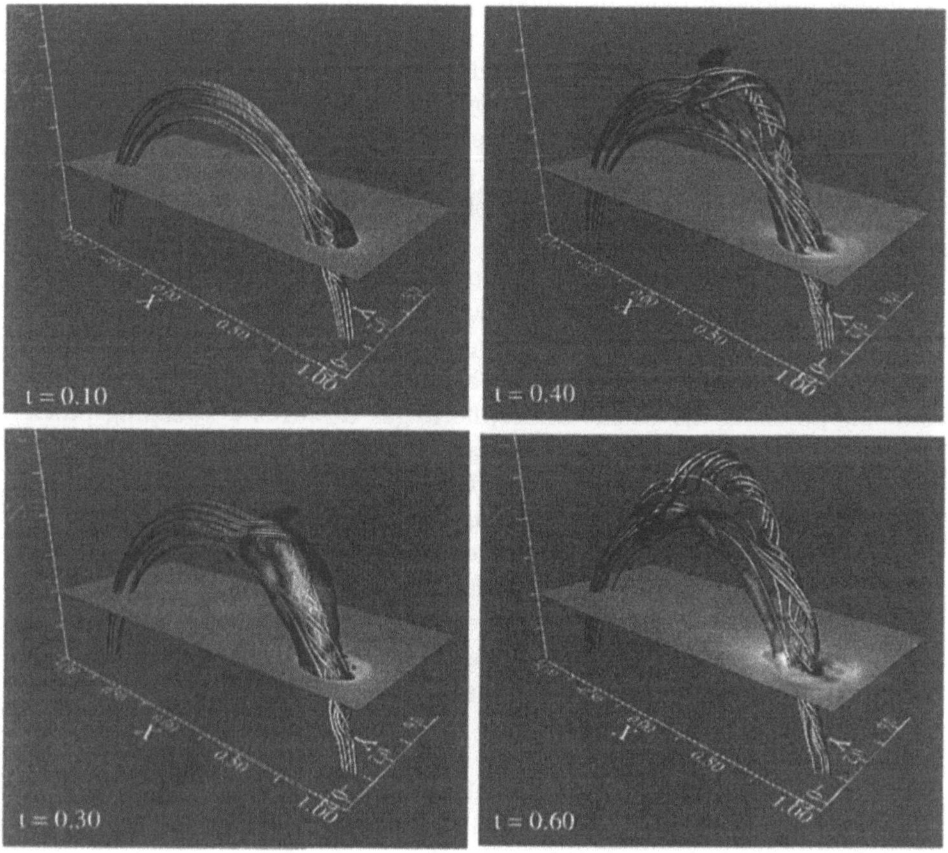

Figure 3. Our MHD Model of the ARLB (Uchida et al. 2000a): The leak of the magnetic twist (TAWP which is equivalent to a coaxially self-closed MHD current) drives the gas in the high chromosphere into the corona as a spinning jet along the loop. It is seen that a bow-shock type MHD fast shock goes ahead of it, followed by the mass flow with a MHD slow shock heating the gas, explaining the ARLB's (see text).

active regions. According to our picture, the active region corona is a *disposing site* of the MHD twist produced, either randomly or systematically, in the rotational gas motion in the convection under the photosphere.

As we said earlier, the mere presence of magnetic field, however strong it may be, can not explain the heating of the plasma by itself. In our proposed model, the excess energy is carried in as the magnetic energy of the twist packets (equivalent to a coaxially self-closed MHD current) from below.

These magnetic twist propagate with small Alfvén velocity under the photosphere along the flux tube, but coming up closer to the photospheric surface, the magnetic buoyancy will accelerate the twist packet to emerge quicker in a bunch. Thus, if it is propagating along the near-vertical part of the flux tube, the twist is transferred to the coronal part of the loop in a much more effective way than propagating with the normal Alfvén velocity. The following mechanism of enhancing the Alfvén velocity in a non-linear way will help it effectively: Namely, as the twisted part emerges to the photosphere from below, the quasi-magnetohydrostatic equilibrium with the dominant gas pressure below the photosphere changes over into another equilibrium without appreciable gas pressure in going into the corona. What happens then will be a pinching near the axis of the twist, and this pinching drives the atmospheric gas upwards ("sweeping pinch", Uchida & Shibata 1988). It was shown that this process squeezes out the gas near the photosphere, and the Alfvén velocity there gets greater due to the decrease in density in this non-linear process to allow a quicker supply of the twist to the coronal part of the loop.

This squeezing progresses up along the loop, pushing the gas to form a jet (heated to several times of the coronal temperature by pinching, and spinning due to the unwinding of the twist), driving the gas to the hypersonic Alfvén velocity along the coronal part of the flux tube. The incidence of the magnetic twist may also cause some distortion of the loop by helical instability. An MHD fast shock goes ahead of the jet with a form of a bow shock, together with a slow shock which is confined in the loop. We have proposed this "sweeping pinch" model to explain the active region loop brightenings (Uchida et al. 2000a, fig. 3).

2.3.2. *Mechanism We Propose for Active Region Expansion (ARE)*

As for the active region expansion, this is again something not known before, and no mechanism has been discussed in the past. We, however, think that this is potentially very important in dominantly magnetically active stars, if the mass-loss rate amounts to 10-15 percent of the Parker wind value in the case of the relatively weak active Sun.

If our remark about the observed indication that an active region loop brightening (ARLB) is found to occur at around the start time of active region expansion (ARE) is correct, it is thinkable that the incidence of ARLB exerts mechanical effect. It is seen that the heated mass in the ARLB itself seems to be confined in the flux tube itself and is not supplied to the expanding structure (with a shape of a paraglider), the effect must be exerted through the pressure (gas plus magnetic) increase effect across the field lines.

The incidence of the magnetic twist packet certainly increases the pressure in the active region corona due to the introduction with it of the additional magnetic pressure due to the toroidal magnetic field (the core of the loop is pinched, but the outer part expands). The fast shock front seen to propagate in front of the driven mass indicates this. So, it is possible that the outermost layer of the active region corona is pushed outwards by this. Then, how can we understand the continued expansion following this?

Our proposed explanation for this is that there is an outer boundary for the stable confinement of the magnetized plasma of the active region corona. This is why there is certain finite size for active region corona. The flux tube placed outside of this region may get unstable, and lost. Our tentative explanation is that the structure with weakened tension will be pushed outwards due to "melon seed effect" (a drive by the magnetic pressure gradient in a magnetic structure acting on an isolated magnetic feature) overcomes the line-tying effect. We will discuss this elsewhere.

3. How Does A Loop Flare Occur?

Loop flares are tiny (5000-15000 km) loop-shaped flare but emitting a considerable amount of energy, like a good fraction of big arcade type flares to be discussed in section 4, and emit higher energy photons than the arcade flares do.

3.1. PREVIOUS VIEWS ABOUT LOOP FLARES

Since the spatial resolution and the time cadence of images were not sufficient in previous satellites, the previous description of loop flares was not a very exciting one. Namely, "a small bright flare with an elongated shape which is suggestive of a loop, decays without any motion" (Pallavicini & Vaiana 1976). It was recognized that the top part of the seemingly loop structure was most strongly emitting. The observation in the initial years of the Yohkoh observation was just like that, too. Strangeness in these observed results is obvious: A very substantial energy release occurs, *without any motion.* We come back to this point in the next section.

Even with the disappointing observational counterpart like this, modeling works for loop flares have been done. Based on the fact that the energy-liberating location seemed to be the loop top, and the mass involved in a loop flare was larger than the mass pre-existed in the surrounding coronal region, those models concentrated on how the mass could be provided to the loop from the chromosphere with the energy source given at the top of the loop. There were two ideas in this: One assumed that a superhot region was created at the loop top, and the heat conducted down "evaporates" the chromospheric material into the loop (Nagai 1980, and many others). The other assumed an accelerated particles somehow created at the top part of the loop, and the down streaming of them bombarded the chromospheric gas to "evaporate" (eg. Brown 1986, many others). All these models assumed an energy source was somehow created at the loop top, and dealt with the supply of heated gas by heat conduction or by particle bombardment. The most important point of why and how the energy source was created there was *not* dealt with in them except for eg. Galeev et al. (1981) in the context of anomalously dissipating the steadily flowing electric current.

3.2. YOHKOH FINDINGS ABOUT LOOP FLARES

One of the exciting findings by Yohkoh is that of the dynamical processes hidden in the initial still faint phase of a loop flare, thanks to the advantage of the large dynamic range, high cadence observation of SXT, together with HXT (Uchida et al. 2000b).

The motionless point-like feature turned out to correspond to the later phase loop top source in which the dynamic source has settled down after the motion, and the *real dynamic phase was hidden in the still faint pre-event to the initial phase of the flare.* It turned out that, within the first one minute or so when the loop flare is still faint, there occurred quite an interesting dynamical development. Fig. 4 shows an example of those in the initial dynamical phase of the 1992 Feb. 17 event. These could not be seen with the previous instruments by the limitation of the dynamic range of the instruments. Also the high time cadence observation by Yohkoh was very important not to miss these crucial frames of the first one minute or so.

The dynamical process in which the bright blob (likely to be a shock) moves up, bounces back downward, and again reflected upwards, and eventually settles down at the loop top, is seen for the first time.

It is also of extreme interest that the shape of the loop (note contours of high values) is distorted into a "question mark" shape, but restores a smooth loop shape with somewhat reduced height, as the bright region settles down at the loop top, and increases its brightness further by one order of magnitude after that.

Another point to note is that the hard X-ray source moves down with the bright blob of soft X-ray, and a very strong hard X-ray burst took place at around 15:42UT when the blob touched down the foot. This suggests that, unlike the previous views, the high energy particles are not injected into the chromosphere as a free stream of individual particles, but dumped into the chromosphere together (probably trapped in the magnetic shock front and dumped simultaneously)(Uchida et al. 1996b, 2000b).

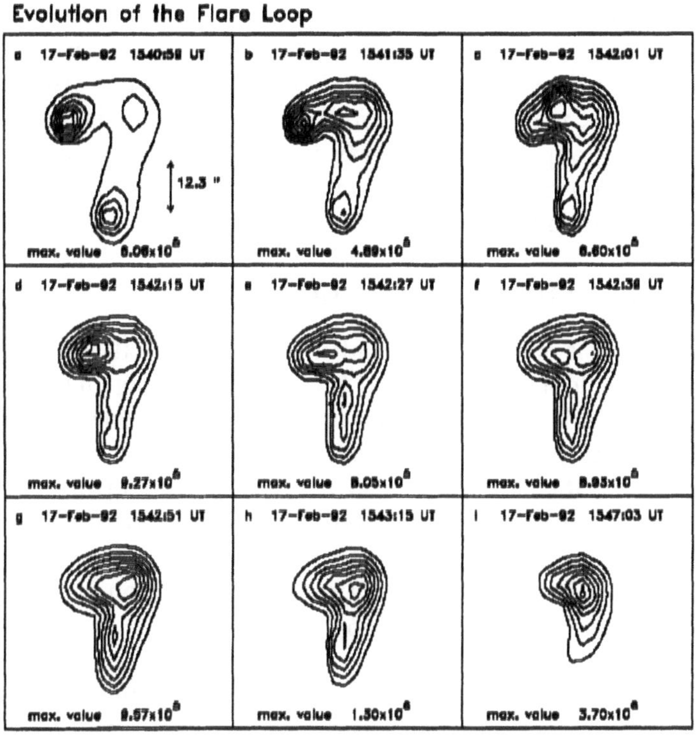

Figure 4. Initial Dynamical Phase of a Loop Flare (Uchida et al. 2000b): The initial dynamical development of a loop flare of 1992 Feb. 17. Two footpoints brighten first, and the bright part moves to the loop top (the south one is less obvious). The bright region seems to be bounced back towards the footpoints (north one is more conspicuous), and then retreats again, and finally settles down at the loop top. The loop top brightens one order of magnitude *after* that. The shape of the loop is distorted into a "question mark" shape, but restores a smooth loop shape *before* the loop top brightens one order of magnitude to the maximum phase.

Another interesting finding by Masuda et al. (1994) with other event, is that a hard X-ray source is found somewhat detached above the soft X-ray loop, and some very faint trace of the soft X-ray emitting source is found to rise above a loop flare (Shibata et al. 1995) with a considerable velocity. Those authors give explanations by bravely applying the traditional "re-closing of the once opened up simple magnetic arcade" model also to the loop flare case, and try to identify the "over the loop top source" with the injection point of the reconnection jet down onto the rigid loop below, and the rising source to correspond to the rising X-ray blob (probably the rising dark filament whose surface is heated) found in arcade flarings.

We, however, prefer another view partly because the situations of small loop flares and large arcade flares are quite different. Further, the precise relationship of the soft X-ray loop shape and the hard X-ray sources at the loop top and both footpoints by Masuda et al. (1994) has crucial questions to answer. Namely, while the loop top hard X-ray source is outside (upside) of the soft X-ray loop, the hard X-ray footpoints are located inside of the soft X-ray loop in a most clearly observed case. In the "open up-reclosing" picture, this is difficult to explain.

3.3. OUR PROPOSED MODEL FOR LOOP FLARES

We thought that it will be most natural for us to extend our model for the active region loop brightening (ARLB) to the loop flare case, as one might guess. A merit to consider the magnetic twist packet (or a coaxially-self closed current packet) by extending the model of active region loop brightenings described in section 2.3 to the case of loop flares is that the magnetic twist is created, transported, and destroyed purely magnetohydrodynamically (unlike the unchangeable current with huge inductance of the circuit under the photosphere (an erroneous notion !) in the previous "current circuit model" (Ionson 1985)).

If we take a horizontal flux tube in the convection zone, lying between the rising part and the falling part of the convective motion, a pair of opposite sense twist packets will be produced at each end of the rolled up part of the tube (pair creation). Such pair of packets can be annihilated if they are allowed to encounter with each other later (pair annihilation), with *direct* conversion of magnetic energy into kinetic energy of the spinning in the unwinding motion, canceling each other of the pair of magnetic twists (Uchida & Shibata 1988). This magnetic energy transport and conversion, which have not been considered in the flare problem (or in any other astrophysical problems), *does not depend upon the electrical resistivity at all, and can take place in the dynamic time scale* !

We consider such injections of self-closed MHD current packets (magnetic twist packets) from both footpoints of one and the same loop. One possible way for this is that both are physically related as proposed in the first proposal of this mechanism (Uchida & Shibata 1988), in which the subphotospheric flux tube, rolled up between the upgoing and downgoing parts of convective motion, may float up as emerging flux tube. The twists produced in this process may be stored near both ends of the floated up part of the loop, and may come up sometime later from both footpoints. Another possibility is that the second independent one may come up into one and the same loop while the first one is still propagating along it. Roughly speaking, the probability for this to occur will be the square of

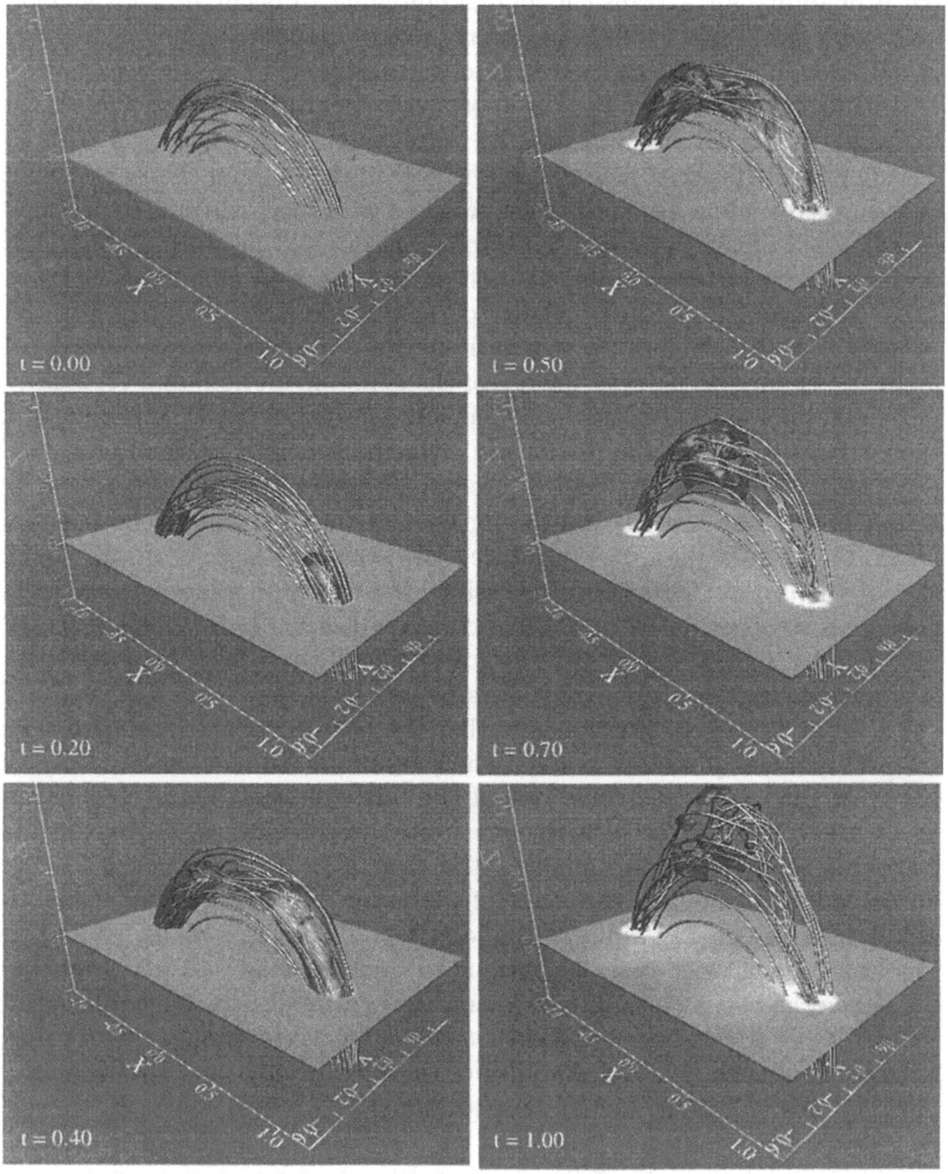

Figure 5. A 3D MHD Simulation of Current Injection Model of Loop Flares (Miyagoshi et al. 2000): The time development in the case of the coincidental injection of two TAWT into a loop. (A symmetrical case was dealt with only for simplicity, though in reality, likely to be asymmetric.) MHD fast bow-shocks goes together with the slow shocks in the tube with helical deformations as in the single case *until* the collision of the fast shocks takes place. The locus of the colliding part of the fast bow-shocks shows a ring structure expanding in time (see other figures in Miyagoshi et al. 2000). These may be our model counterpart of the over-the-loop-top hard X-ray thermal source by Masuda et al. (1994), and the rising faint emission by Shibata et al. (1995).

the probability for a transient loop brightening to occur in a given loop (Wheatland & Uchida 1999), and pretty much smaller than the probability of ARLB to occur, as expected from observations.

Incident packets drive the chromospheric gas in the form of hypersonic (because $V_A \gg a$) spinning jets by "sweeping pinch" effect, as seen in section 2.3, and produce fast and slow shocks ahead of them. Those pinching and shocks heat the gas, and those progress with near Alfvénic hypersonic speed. The collision of those hypersonic shocks will produce a very high temperature gas (up to 10^8K) at high part of the loop (Uchida & Shibata 1988; Miyagoshi et al. 2000). Other part of energy will be dumped in the process of cancellation of the magnetic twists, through direct conversion of magnetic energy into kinetic energy of the spinning motion, and eventually into heat, in case the packets have opposite sense of twists. The latter gives the SXT source, while the the former gives the superhot thermal plus non-thermal source above the SXT source. The former source appears higher in our model because the untwisting part of the flux gains height in the unwinding, while the latter produces the strongest source on the lower side of the initial twisted flux tube since the despinning motion heats the part with stronger field in the interaction (Miyagoshi et al. 2000) (Fig. 5). Aschwanden's (1995) delay time for lower energy particles is explained also in a natural way in Miyagoshi et al.'s paper.

A point to be remarked may be that the bending of the loop by helical instability takes place for both of the packets in 3D simulation, and the shape of the loop is distorted out of the initial vertical plane (Miyagoshi et al. 2000). When the contribution from the collision of the hypersonic jets fades after one or two minutes, the remaining source is the soft X-ray emitting lower loop, and the shape of the source restores a low and smooth shape with the energy input at the loop top. This part remains as long as the wake of the twist packets continue to come in, may be 10-20 minutes.

When the two packets have the same sense of twists, the twists strengthen each other's in collision, and if the toroidal field exceeds the critical value, a large amplitude helical instability of the loop occurs, and the loop may erupt, possibly leaving a small cusped loop.

4. How To Solve the Dilemma of Classical Model(s) of Arcade Flares?

Energetic solar flares were first known as an increase of white light in a small area near sunspot groups. Lyot's interference filter allowed the researchers to observe the sun in monochromatic images, and the images in Hα line revealed that the disappearance of a dark filament seen on the magnetic field polarity-reversal line leads to the appearance of a brightening in Hα

in the form of double-ribbon flare in the chromosphere. The bright patches in the white light (white light flare) were specially bright small parts in this double-ribbon flare, called "flare kernels".

Introduction of radio and X-ray observations later revealed that the Hα double ribbons are the loci of footpoints of the X-ray-emitting high temperature arcade structure existed in the corona above them, and the problem of explaining a double-ribbon flare became the problem of explaining an arcade flare.

4.1. PREVIOUS VIEWS ABOUT ARCADE FLARES AND ARCADE FORMATIONS

A model of mass suspension in a dipped magnetic-field configuration has been discussed by Kippenhahn and Schlüter (1957), and many authors have *assumed* that such a mass suspension could occur due to sagging of the field lines, even at the top of a convex magnetic field arching over the field-polarity reversal lines in the photosphere (see Tandberg-Hanssen 1974; Priest 1989). A model for arcade-type flares (Carmichael 1964; Sturrock 1966; Hirayama 1974; Kopp & Pneuman 1976) was developed based on this type dark-filament suspension model: When a dark filament is somehow destabilized and flies away, it drags the arcade field with it into interplanetary space. Magnetic reconnection occurs between the antiparallel fields at the leg parts of this stretched field. A reclosed arcade in the low corona and a detached closed loop field surrounding a flying blob will be formed as the result of magnetic reconnection. The energy released in the flare is interpreted as being the difference in the energies in the stretched and reclosed magnetic fields, a substantial part of which is liberated in the reclosed arcade during the process. The magnetic energy is converted into kinetic and thermal energies of the gas ejected downward from the reconnection site and the energy of high-energy particles accelerated in the reconnection process, and transported down to the chromosphere along the magnetic fields. The soft X-ray emitting mass in the reclosed arcade is considered to be chromospheric mass "evaporated" by being given excess energy via heat conduction and/or electron bombardment from above. Some researchers (eg. Shibata et al. 1995) refer to this model as the CSHKP model, taking the initials of the above authors.

A difficulty with this model of dark filament came to light in the observations by Leroy et al. (1983). They showed that the direction of the field perpendicular to the dark filament is *opposite* to the direction expected from the simple connection of the bipolar field below. A solution of this dilemma has been proposed, for example, by introducing the Kuperus and Raadu (1974) type model with a field configuration having an inverted circular sub-connection at the top of the convex arcade (cf. Priest 1989).

However, no sufficiently persuasive physical explanation has been given for the creation and stability of such an artificial configuration.

A more basic question about this classical model of arcade flares, however, was the one raised by Uchida (1980) concerning the energy of the rising dark filament. In order to stretch the strong field (whose magnetic energy ultimately should explain the large energy of flares in the reconnection process) and eventually cut it open, the energy in the rise of the dark filament should be *larger than* the energy of the flare itself. This point, which was the motivation for the work of Uchida and Jockers (1979), and Uchida (1980), was later formulated in a clearer form by Aly (1991). It should be pointed out here that there would be *no magnetic neutral sheet* in the "classical model" *if the process of cutting the field open can not actually take place due to the energy difficulty.* This is a severer dilemma. If the classical picture is correct, what requires a more serious investigation is the dark filament rise, not the flare itself, because the flare is then merely a repairing process of a more energetic break-up caused by the rise of the dark filament.

Some researchers, accordingly, have tried to explain the dark filament as a highly-stressed structure with a highly sheared field (cf., Moore & Roumeliotis 1992). The dark filament in this type model is an extremely energetic entity, but the observers know that the actual dark filament is a passive entity, and such a model is not plausible.

In order to avoid such a difficulty in energy, Uchida (1980), based on Uchida and Jockers (1979), proposed that the magnetic structure involving a dark filament may be due to a quadruple array distribution of the magnetic field (elongated regions of +,−,+,− polarities side by side, referred to as A, B, C. and D) instead of a simple bipolar region in the photosphere below. In that case, there appears a "neutral line" (the locus of the "neutral points", which will become a vertical "neutral sheet" if the photospheric footpoints are squeezed towards the central line) in the planar component of the magnetic field, $\boldsymbol{B}_\perp$ (the field components in the xz-plane perpendicular to the central polarity-reversal line in the photosphere which is taken along the y-axis), in the corona above the central polarity reversal line. We refer to this symbolically as the "neutral sheet" in the following, although it is actually a current sheet with the y-component of the field lying in it. The longitudinal component along the y-axis, B_y, is dominant in the "neutral sheet", because $\boldsymbol{B}_\perp$ is close to zero there. In our model, B_y, which may be relatively week initially, but will become of the same order of magnitude as $\boldsymbol{B}_\perp$ at certain distance on both sides when squeezed, is what is suspending the dark filament gas lying in the thin, high and long vertical "neutral sheet". B_y is supported, in turn, by the gradient in $\boldsymbol{B}_\perp$ as discussed later in section 4.3.

The quadruple-array source model can thus sustain the dark filament mass in the thin neutral sheet with "barbs", explaining the observed dark filament very well.

The quadruple source model, however, was not accepted too well when proposed in Uchida (1980), but Martin (1990) supported it from the Big Bear observation of dark filaments.

Many years after our proposal, the quadruple source characteristics have attracted increasingly greater attention, and have been discussed Demoulin and Forbes (1992), Antiochos et al. (1994), Sciffer (1997), Antiochos (1998), and Kuijpers (1997).

4.2. YOHKOH FINDINGS ABOUT ARCADE FLARES

Yohkoh, for the first time, observed an arcade flare on 1992 Feb. 21 as seen from the direction of the axis, as an impressive cusped feature standing up with a dark tunnel below (Tsuneta et al. 1992). This is a noticeable observation because this revealed several points that support the magnetic reconnection models (Note, plural !!). Some people thought that "the classical model for arcade flares (Carmichael 1964; Sturrock 1966 and others) was proved by this", probably because it looked just like the "cartoons" drawn for that, but it was too hasty to say this.

4.2.1. *Findings of Faint Structure in the Pre-event and Initial Phases*

Examining the faint stages in the pre-flare phase along our strategy of examining physical clues in the still faint phases of that event, we noted (Uchida et al. 1996a) that there were two large loop structures connecting the top part of the cusp back to the photosphere on both sides, and that there was a low-lying structure brightening up in the pre-flare to the initial phases at the axis of the dark tunnel below the cusp (Fig. 6(a)). Those features have *no model counterparts* in the widely conceived classical model proposed by Carmichael (1964) and independently by Sturrock (1966), and discussed by some others (Hirayama 1974; Kopp & Pneuman 1976; Priest & Forbes 1990; and others). We confirmed the same features in another flare of 1991 Dec 2, that there also occurred a cusp with dark tunnel below, and there existed a low-lying structure brightening at the axis of the dark tunnel in the same way. There were high back connections to both sides in this flare, too (Fig. 6(b)). Therefore, we think that the features found in them were quite basic in arcade flares, when seen along the axis, essentially in support of the quadruple magnetic source model advocated by Uchida (1980; see Fig. 11, dated as 1977 Jul 11).

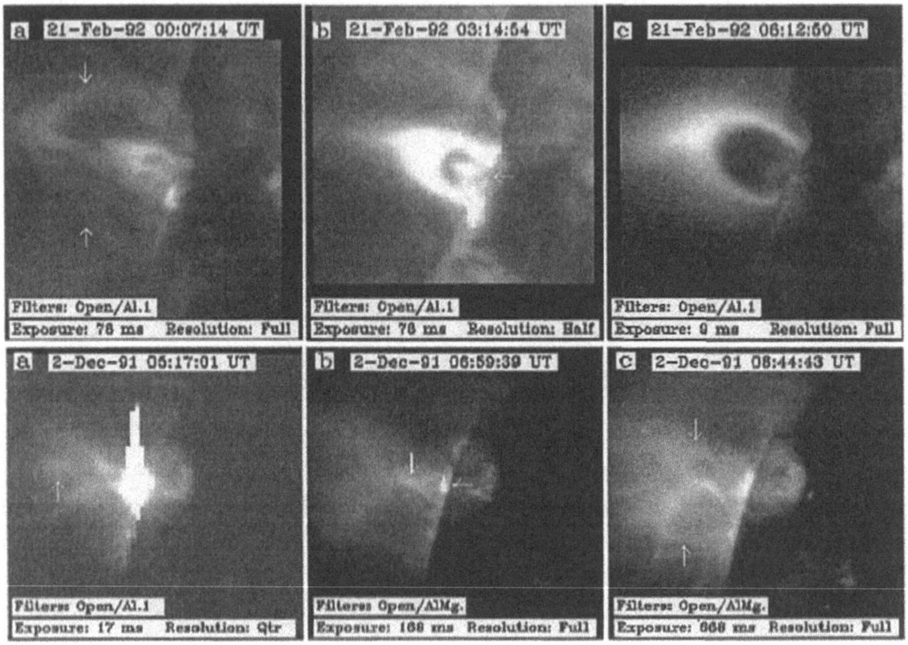

Figure 6. Features Indicating the Quadrupolarity of Arcade Flares (Examples of (a) 1992 Feb. 21, above, and (b) 1991 Dec 2, below, Uchida et al. 1996b): High connections back to the photosphere on both sides and a brightening low at the axis of the tunnel below are seen just in the same way in both of these flares, indicating that these are fundamental features in arcade type flares.

4.2.2. *3D Structure Derived from Homologous Flare Series*

Morita et al. (2000b), based on the idea that the flares occurring consecutively on Feb. 24, and Feb. 27 after the cusped flare of 1992 Feb. 21, constitute a homologous flare series (the shapes of them are basically the same, though, of course, not in a mathematical rigor), tried to derive the 3D structure of these flares by utilizing the information obtained from three different viewing angles. They succeeded to derive a 3D structure, and found very important facts: (1) The magnetic sources involved are 4, not 2 as the classical model assumed, (2) the cusp is *not* the X-ray arcade itself, but a bright arch diagonal to the body of the X-ray arcade upcoming later, leading its formation, (3) Hα double-ribbons widen their separation as noted before, but not in a simple way, but rotating its orientation, keeping the end-points of the above-mentioned diagonal bright arch as pivots. These are quite different from what the classical model supposed.

It was also reported by Tsuneta et al. (1992) that there was a faint X-ray emitting small blob rising in the pre-event phase in both of these flares,

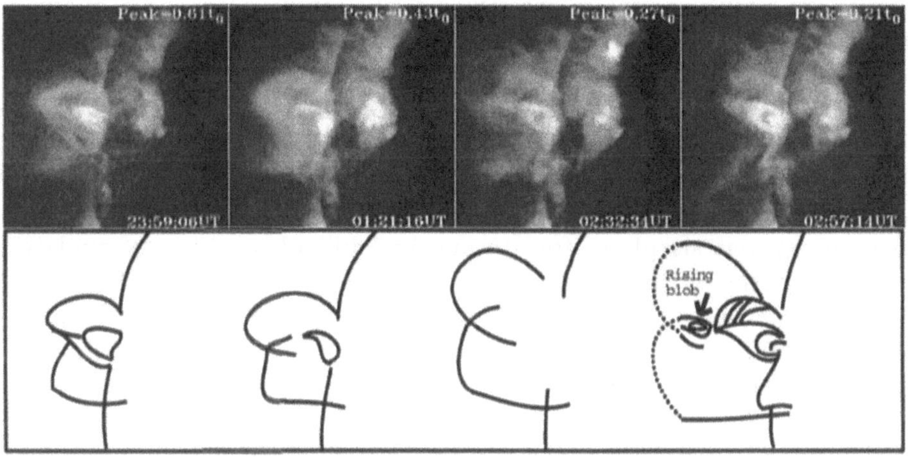

Figure 7. Evidence for Passive Rise of a Blob (Dark Filament with Its Heated Periphery), Rather than Cutting the Simple Magnetic Arcade Open (Morita et al. 2000b): It is seen that the expansion of the high loops on both sides pulls out a dark blob surrounded by bright periphery (likely to be dark filament whose surface has been heated), instead of the blob actively pulling up the loops and cutting them open. Cartoons above are the sketches of relevant structures.

and the blob was considered to be related to the rising dark filament seen on axis. Morita et al. (2000b) examined the 1992 Feb. 21 event in detail, and found that the blob was *not pulling* the overlying loops as supposed in the classical model in which the rising dark filament was supposed to stretch the magnetic arcade and eventually cut it open, causing the magnetic reconnection behind. What Morita et al. found, instead, was that the blob *came out of the valley* between the two back-connecting large loops mentioned above (the valley coincided with the tip of the flare cusp), by *being pulled up* by the upward expansion of those two high loops (Fig. 7). In other words, what were found are different from what one expects from the classical "reclosing of the once opened up simple magnetic arcade" model by Sturrock (1966) and others, in the sense that the dark filament rise is *not actively cutting* the magnetic arcade open, but it is *passively pulled out* from the valley between the two high loops connecting the top of the cusp back to the photosphere on both sides. These findings are very severe evidence against the classical Sturrock model and other similar models based on the magnetic bipolarity.

What the observation revealed made us recur to the quadruple source model suggested by the early cartoons by Sweet (1958), and elaborated by one of us (Uchida & Jockers 1979; Uchida 1980) years ago. Noting the existence of a theoretical difficulty in the classical "reclosing of the once opened up simple magnetic arcade" model, Uchida proposed a quadruple source model in which the magnetic field has a "neutral sheet" structure due to the effect of another pair of sources from the beginning.

In opening up the magnetic arcade in the classical model, the rise of the dark filament should do an amount of work, equal to or even greater than the magnetic energy providing the energy of the flare. If one dares to assume that the dark filament has such an energy, the difficult problem of flare energy was simply shifted to another difficult problem of explaining the large energy of the dark filament rise. It is, however, concluded that the energy of the rising blob is much smaller than that of the flare (Ohyama & Shibata 1997).

In Uchida's idea (1980), the strong part of the field is already open by the influence of another opposite pair of magnetic sources, say, on both sides of the pair of sources, and the passive dark filament is stably suspended in the "neutral sheet" with $B_{||}$, *prohibiting* the reconnection between the oppositely directed (to $+z$ and to $-z$ directions) field on both sides of the neutral sheet. The removal of this dark filament with $B_{||}$ allows the thinning of the neutral sheet region, and the contact of the opposite polarity $B_{\perp}$. Then, the magnetic reconnection starts, releasing a large amount of magnetic energy stored in the distorted magnetic field configuration prior to that release. We found that Uchida's idea seems to be favored by observations. We come back to the detail of the model later in section 4.3.

4.2.3. *Observation of an Arcade Flare by TRACE*

More recently, the satellite TRACE is giving wonderful data in higher resolution. Alan Title provided us with the TRACE data, noting that the data obtained by TRACE seem to be very favorable for our quadruple magnetic source model. Thus we start examining the case of an arcade flare of 1999 Jul 19 by using the TRACE data, and obtained a quite clear evidence supporting our quadruple source model (Fig. 8).

First, it is seen that the dark filament, sustained above a polarity-reversal line of the photospheric field, flies away. Then a bright arcade starts to form between the two magnetic regions on both sides of the polarity reversal line, B and C. We further note that there are two more regions outside of B-C involved in the flaring, and call these as A and D. Those form a quadruple magnetic sources, + - + - in the order of ABCD from the left.

It is seen that the structure covering a wider area between the sources A and D is gone with the rise of the dark filament. An arcade flare whose width gradually expands, occurred between B and C, but the width was limited just between the sources B and C, though the structure between A and D is gone. In this process, the structures connecting A to B, and

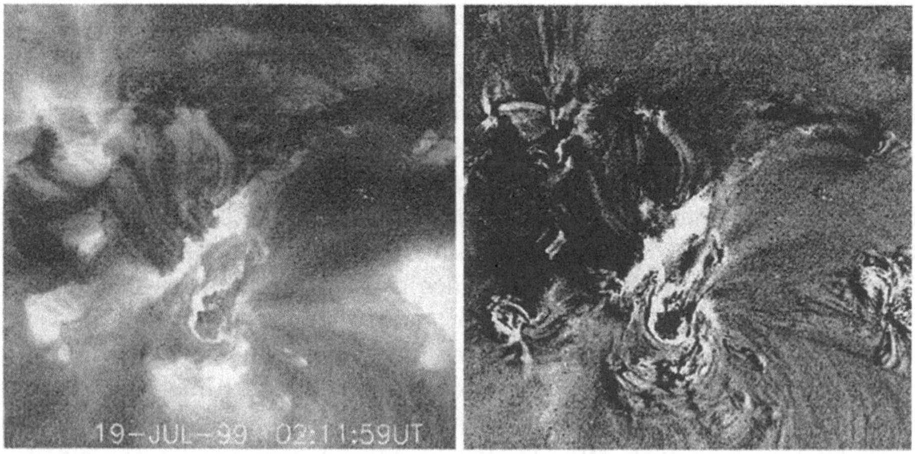

Figure 8. Observational Results of the Arcade Flare of 1999 Jul 19 by TRACE (Uchida et al. 2000c): A TRACE observation (Title et al. 1998) supporting the quadruple magnetic source model (Uchida et al. 1999c). It is seen that a wide structure covering the sources A to D has been unveiled by the rise of the dark filament which caused an arcade flare whose width gradually expands but limited within much narrower width between the source B and C. The structures connecting the source regions A to B, and C to D are clearly seen to be remaining, while the structure connecting A to D was peeled off by the rise of the dark filament. Connections A to B and C to D are coming closer as the flare progresses. These are exactly what are expected in the quadruple magnetic source model.

C to D *remained unaffected*, while the loop structure connecting A to D disappeared. The structures connecting A to B, and C to D, slowly came towards the center line as the flare progresses. This is exactly what is seen in the quadruple magnetic source model (movie version of the simulation of Fig. 13 shows this). The involvement of the sources A and D in the flare is very definitive in view of the flare material flowing to them in the process.

The situation with the 1992 Feb. homologous flares was very similar to this. Therefore, from these, it is seen that the actual observations are clearly in support of our quadruple magnetic source model.

4.2.4. *Yohkoh Observations of High Latitude Arcade Formations*

How about the fainter high latitude weak field versions of the same process, namely, the high latitude X-ray arcade formations?

It was known that there are larger scale, weaker version of double-ribbon Hα flares outside active regions. Actually, very large X-ray arcade formation events have been found by Yohkoh even at high latitudes (though the Hα brightening may be too faint to see in them). Those arcade formations were recognized from the qualitative characteristics as belonging to the same category as arcade flares, but occurring in weaker, larger scale magnetic fields.

This case of high latitude arcade formations was actually a challenge for our quadruple source model, because it has been always reported that "they have dark filament lying just above the field polarity reversal line in the photosphere, *between two unipolar* regions". This was actually the basis for the idea of simply-connected magnetic arcade model for the dark filament suspension between two unipolar regions, and therefore, the basis for the classical "reclosing of the once opened up simple arcade" model for arcade flarings. The model of the dark filament suspension above the convex magnetic arcade would never have been considered without this observation, because it is readily noted that the mass loading on the convex loop is simply difficult (at least before a large enough amount of mass has been loaded. And even after loaded, it will be unstable for the sideward shaking. It will be readily seen that the application of the treatment by Kippenhahn and Schlüter (1957) to the present problem of the dark filament suspension will not be valid.)

Anyway, it has been a challenge for the quadruple source model to be confronted, since it is reported to occur at the border of two extended unipolar regions. The problem was where can the rest of the two sources for the quadruple magnetic source model be found in this case.

Pre-event "Dual Arcades" Structure in the Location of High Latitude Arcade Formations.

The high latitude arcade formations themselves are not bright events, and the even fainter pre-event features of them require some special technique to examine with. We use a faint structure-enhancing technique known as "unsharpened-image masking" method, in addition to selecting the frames with the thinnest filter and the longest exposure as available.

What we found prior to the high latitude arcade formation in the same location was a pair of arcade-like structures consisted of the X-ray loops, lying side by side, with the inside legs crossing with each others' (Fig. 9a). By carefully overlaying images by considering the differential rotation, the height of the dark filament etc., it is found that the locus of the top of the dark filament coincides with the locus of the crossing region of the X-ray threads, and lies above the field polarity reversal line in the photosphere (Uchida et al. 1999a).

The situation here is again quite different from what would be expected from the classical models. In the classical picture, what is expected is a magnetic arcade simply connecting the unipolar field regions on both sides, with the dark filament material suspended at the top of it with a certain sagging of the convex arcade there. Some of the previous observations by

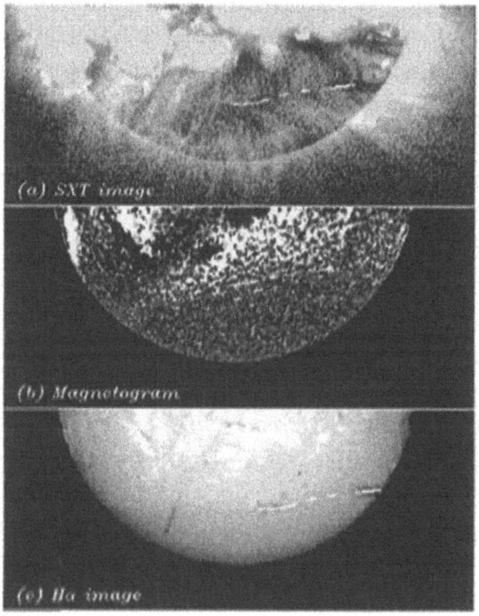

Figure 9. Pre-event Coronal Structure at the Site of an Arcade Formation of 1992 Jan 25 (Uchida et al. 1999a): "Dual arcades" structure is found in the location of high latitude arcade formation events in the pre-event phase. White lines in the panels are the locus of the top (south side) of the dark filament in the third panel, rotated to the time of the X-ray picture, and down-projected to the photosphere in the case of the second panel. Interpretation of this unexpected "dual arcades" structure is discussed in the text.

radio with lower resolution suggested the presence of a dark lane there, and it was interpreted as due to absorption by the cool dark filament material suspended in the sagged simple arcade. Now with higher resolution of Yohkoh-SXT, it turned out that this is not the case, and the loop-like features in the "dual arcades" structure have landing points in the opposite side of the "polarity reversal" line (the angle of the crossing seen from above is too large to keep the shape without landing footpoints). This structure seems to be compatible with the report on Hα dark filament configuration by Martin (1990; Martin & Nichols 1994).

What is the "dual arcades" structure then ? One characteristic feature we found there is that the magnetic field distribution in the photosphere below is *not at all* a clean bipolar regions with the dark filament sitting above the clear border line between them. There is a belt of highly mixed polarity having a width of tens of thousands of kilometers (Fig. 9(b)). There, many strong field patches of the wrong polarity exist in the normal polarity area on both sides of the (imaginary) field polarity reversal line.

Previous descriptions of two large unipolar regions of opposite polarities in contact seems to have missed this point, or ignored this fact. Further, a strange thing is that the loop-like features coming from one polarity seem to be connected to the patches of the *same* polarity on the opposite side, though the resolution does not permit a definite statement. To connect the same polarity sources is, however, *not possible for normal magnetic field lines*!

The answer we arrived at for this riddle is that the "dual arcades" structure may not be usual magnetic arcades consisted of ordinary flux loops, but those may be the critical lines of force composing the "separatrix surfaces" (Fig. 12(a)). It will be shown later in section 4.3 that a "separatrix surface" can look as if connecting the same polarity flux sources (see Fig. 12(b)). The intersection of the two "separatrix surfaces" is a "neutral line" (or sheet when squeezed) in the case of the elongated quadruple source situation. It is reasonable that the "separatrix surfaces" may well be preferentially heated if a slower energy release may be taking place in the neutral line (or sheet) already in the pre-event phase, and may well be visible in X-rays. This explains the fact that the only feature seen there is the "dual arcades" structure, while there must be many other normal magnetic field lines inside and outside. They may not be seen because they are cooler.

Evolution of High Latitude Arcade Formation Events.

We have also investigated the time evolution of high latitude arcade formations themselves (Fujisaki et al. 2000). A typical example is the 1993 Feb. 24-25 event. The pre-event structure is a typical "dual arcades" structure. The X-ray event started sometime after the dark filament eruption as a bright thin horizontal thread ("spine") rising from the location where the dark filament was in the pre-event configuration. As this "spine" rose, there appeared a number of thin loops connecting the "spine" to the photosphere on its north-western side. It is likely that there were similar threads on the south-eastern side, but may be shadowed by the "spine". Finally, the west-most part of the "spine" ballooned up to make a large cusp-like shape, and then entire structure faded away in ten hours time (Fig. 10(a)).

Another example of the same category was the one that led to the so-called "Giant Cusp" event on 1992 Jan 25 (Fig. 10(b)). The pre-event structure was also a very clear "dual arcade" structure (shown in Fig. 9). The event took place closer to the west limb, and the final product, the "Giant Cusp", occurred somewhat beyond the limb. The event developed from a smaller cusp, and grew into a large cusp. This "Giant Cusp" has been taken to be a large scale version of the 1992 Feb. 21 cusped arcade flare, occurring in weaker magnetic field, and Hiei et al. (1993) noted that

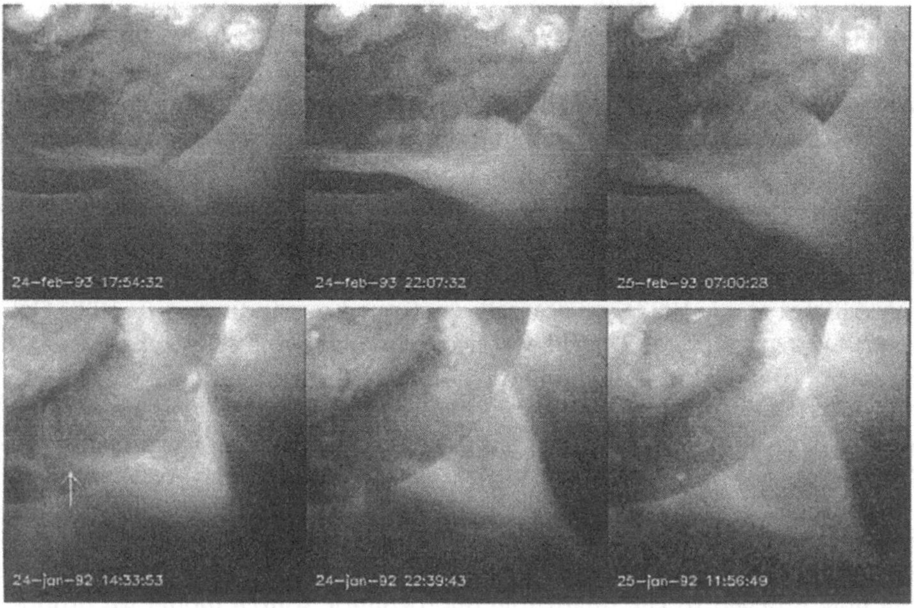

Figure 10. Evolution of High Latitude Arcade Formations (Fujisaki et al. 2000): (a) (upper panels) In this event of 1993 Feb. 24-25, the western part of the rising spine finally balloons up to form a "giant cusp"-like feature. (b) (lower panels) The "Giant Cusp" event, which was considered as an example of the classical "reclosing of the once opened up simple magnetic arcade" in a large scale, is, in our view, an event of the same category as the 1993 Feb. 24-25 event, and to be interpreted like that (see text).

some subsidiary features were pealed off downward, somewhat suggestive of the behavior of the reconnected flux in the classical model.

We, however, do not agree to the view that this large cusp-like structure can be interpreted as a large scale version of the "reclosing of the once opened up simple arcade" model. Our argument is based on the fact that this "Giant Cusp" event and the 1993 Feb. 24-25 event are very similar with each other, and apparently belong to the same category. We are convinced in this because, in the early development of the "Giant Cusp", we found that the cusp, while still small, had a very long "tail" of 10-15 degrees in longitude eastward from the southern foot (arrow in Fig. 10(b)-1), while the main body of the cusp was at or beyond the limb. This can not be explained in the classical model, but it is *readily identifiable* with the "spine" as seen in Fig. 10(a). Then, the "Giant Cusp" is very likely to correspond to the large hump of the "spine" at its west-most part of Fig. 10(a)-3, and not the "reclosing of the once opened up simple arcade".

The observed behavior may be given a tentative explanation that the rising "spine", coming up after the dark filament flew-away, may be identified with the remaining half of the dark filament once compressed down as S-shaped structure when the upper half of the dark filament is squeezed out. Such a splitting of the dark filament at its rise has been reported in Hα by Tang (1987). The remaining half may or may not come up later from below the X-ray arcade that was produced through the major magnetic reconnections taking place at the neutral sheet (see section 4.3 below), and pulling them up into a shape of a "tent"but stopped as in 1991 Sep 28 event (McAllister et al. 1992), or scarcely continued expanding as in 1996 Oct 5 event (Watari et al. 1997). Theoretical explanation is still to be sought for for such a delayed rise of the lower half of the dark filament.

4.3. OUR PROPOSED MODELS FOR ARCADE FLARES AND ARCADE FORMATIONS OUTSIDE ACTIVE REGIONS

4.3.1. *Quadruple Magnetic Source Model for High Latitude Arcade Formations – 2.5D Case*

We now proceed to the quadruple photospheric source model we proposed years ago (Uchida 1980), and we are now developing again (Uchida et al. 1999b, Hirose et al. 2000a), since many aspects of it have turned out to be supported by the new findings from Yohkoh as described above.

Equilibrium Configurations in Quadruple Magnetic Source Model.

Uchida and Jockers (1979) originally dealt with the quadruple source model, and advocated that in the Skylab Workshop on Flares (Uchida 1980). Our point was that the Sturrock's model (1966) and the similar ones seemed to have a problem that, in order to cut the strong magnetic arcade open, the dark filament rise should have an energy equal to or greater than that of the flare itself, and the flare is then merely a repairing process. The problem of large flare energy is simply shifted to that of the dark filament rise, and is not solved by their model. This point about the energy was raised again later by Aly (1991) in a different form.

Uchida (1980) proposed a different way of explaining the arcade flares. The dark filament is suspended in the magnetic neutral sheet (precisely speaking, the neutral sheet of $\mathbf{B}_\perp$, containing the gas together with $B_{||}$) in a quadruple source field, and the thin curtain like structure of the dark filament with "barbs" can be very naturally explained. In his model, an arcade flare occurs when the dark filament having $B_{||}$ is passively expelled out by some disturbance, and then the reconnection, occurring in the naked neutral sheet after $B_{||}$ is expelled out, allows the system to liberate the accumulated magnetic stress. The enhanced rate of reconnection was considered to take place after a dynamical collapse of this neutral sheet into a "book

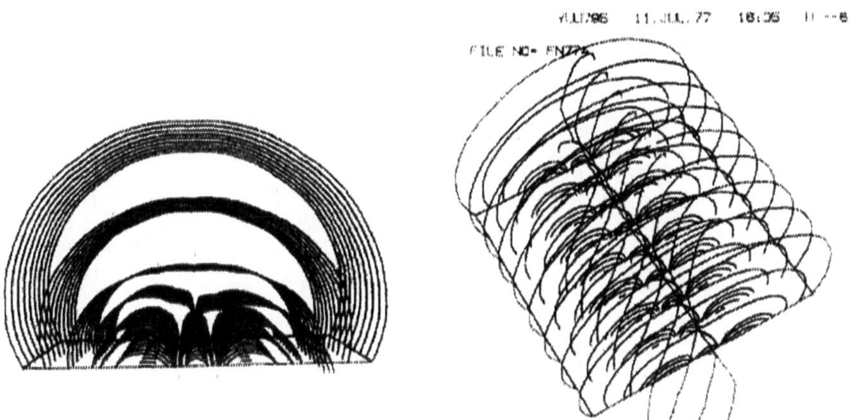

(a) (b)

Figure 11. Calculated Field Lines for a Quadruple Source Distribution (Uchida & Jockers 1979; Uchida 1980): (a) An example of quadruple magnetic source field after the photospheric footpoints are homologously squeezed towards the center line, and there is a current distribution corresponding to the magnetic stress stored in the deformation. (b) A birds-eye view of the same field. It is seen that the region of neutral sheet in $\mathbf{B}_\perp$ is dominated by $B_\parallel$. This seemingly strongly sheared configuration along the dark filament is, in this model, simply due to the presence of the null of $\mathbf{B}_\perp$, and due essentially to the quadrupolarity.

page"-like structure through interchange instability (Uchida 1980; Uchida et al. 1994).

The equilibrium configuration of the quadruple field was derived in a 2.5D (no dependence of physical quantities on the y-coordinate, but take into account of the y-components of vectors) approximation (Uchida & Jockers 1979; Uchida et al. 1999b), and the solution is like Fig. 11.

We now discuss briefly the comparison of the computed results of the pre-event magnetic field with the observed structure by Yohkoh, given above.

Comparison with the Observed Features Surrounding Dark Filaments.
In our quadruple source model, there exists a neutral sheet (of $\mathbf{B}_\perp$), which is the deformed (pressed) neutral line region located at the crossing line of the two "separatrix surfaces". Then the observed "dual arcades" structure in the pre-event area is very nicely explained as the "separatrix surfaces" for the elongated quadruple source case (Fig. 12(a)). The "separatrix surfaces" in a more realistic distribution of the field sources with the transported flux patches are given in Fig. 12(b), representing the observed

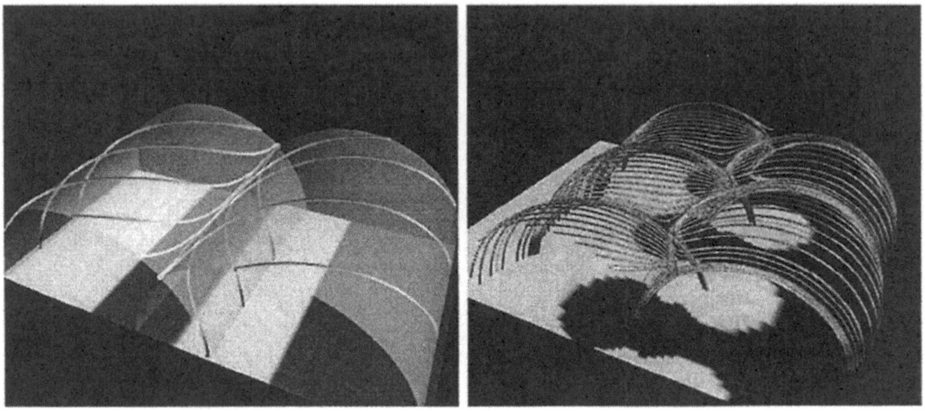

Figure 12. "Separatrix Surfaces" for a Quadruple Source Model in 2.5D, and More Realistic 3D Cases: (a) "Separatrix surfaces" for the 2.5D elongated quadruple sources as readily predicted from Fig. 11(a). (b) The version with more realistic 3D case with certain number of opposite polarity patches transported to the other sides, gives a good matching to our findings of the crossed legs of the "dual arcades" described above.

features very realistically (note, however, that these figures are for potential fields before pressing).

Other features of dark filaments such as their very thin curtain-like structure, and the leg structure called "barbs" observed in Hα (Martin 1994) (that have *never* been given explanation by any previous models, because their dark filament suspended in a dip of the otherwise convex arcade *did not* have any direct magnetic connection to the photosphere below it near the middle) have their intrinsic model counterpart in our model.

When the dark filament material is efficiently cooled by the radiation loss, further coronal gas is supplied along the field lines connected to the dark filament in the neutral sheet through $B_{||}$ connections due to the pressure difference (the siphon mechanism) in a very natural way in our model.

4.3.2. *2.5D MHD Simulations in the Quadruple Magnetic Source Model*

In the extension of the calculation of the equilibrium described above, we have performed dynamic simulations for the high latitude arcade formation in 2.5D approximation (Hirose et al. 2000a)

Initial Model and the Destabilization

We derived the initial conditions for arcade formation with accumulated energy by using the same simulation scheme to be used in the main simulation. Starting from a state with a current-free quadruple source magnetic field, we derive a state with accumulated magnetic energy eg., by moving slowly the photospheric sources homologously toward the central line (see Fig. 11). The following is assumed according to the observation: The configuration surrounding the high latitude dark filament is observed to be extremely stable, staying almost the same for a few solar rotations. The dark filament is destabilized when a small magnetic pair emerges nearby (Feynman & Martin 1995), and then erupts. This situation is taken into the modeling for the simulations. We modelize the destabilization of the dark filament as due to the new emergence of magnetic pair in the neighborhood with an injection of electric current (magnetic twist) into the dark filament via some newly created paths of the current by local reconnection with the new pair. The dark filament in which a current is flowing can be destabilized if a new current antiparallel to that is injected, and the upper part of the dark filament may be expelled out of the neutral sheet region upward, whereas the lower part of it may be expelled out downward and pressed, and may be seen in X-rays as an S-shaped structure below (see Fig. 6). Magnetic reconnection can now occur in the neutral sheet from which $B_{||}$ with plasma (dark filament) has been evacuated, since the opposite polarity main fields of $\mathbf{B}_{\perp}$ are now pressed together directly.

Results of 2.5D Magnetodynamic Simulation for High Latitude Arcade Formations.

The results of dynamic simulations thus made are shown in Figs.13 and 14 (Hirose et al. 2000a). It is seen that after an injection of an appropriate current (with opposite direction and a magnitude comparable to what is flowing in the dark filament due to the squeezing) as suggested from observations, a splitting of the dark filament occurs with the upper part which is squeezed out of the neutral sheet of $\mathbf{B}_{\perp}$ upwards, and the lower part which is pressed down and heated adiabatically.

The upgoing part, corresponding to the rising dark filament in Hα shows a remarkable behavior reminding the observed behavior of it (Fig. 14): The rising dark filament once hesitates to rise, and then restarts rising with remarkable acceleration. The restart of acceleration to greater velocity is found to correspond to the start of the reconnection in the magnetically neutral sheet created below after the dark filament gas with $B_{||}$ has been expelled out.

It is seen from the results of simulation that, as the dark filament gas together with $B_{||}$ is actually expelled out, the magnetic reconnection at the central part occurs (an anomalous resistivity has been included, but not effective until this phase, because the pressure of $B_{||}$ with dense gas *has prevented the thinning of the current sheet* until then), and the relaxation of the accumulated magnetic stress energy starts with restructuring that

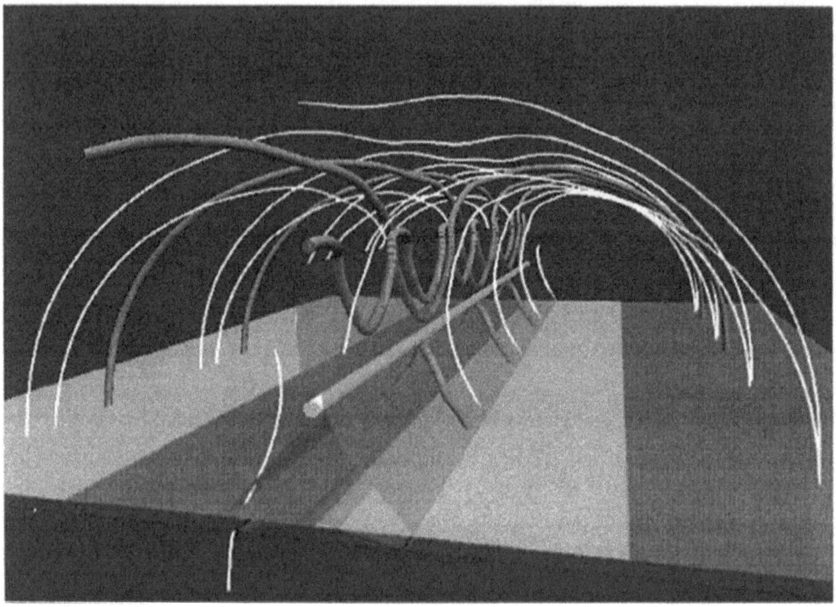

Figure 13. The Result of Simulation of 2.5D Quadruple Source Model (Hirose et al. 2000a): The transparent structure at the middle is an iso-density surface, corresponding to the dark filament with barbs. Relevant magnetic field lines are drawn. It is seen that there are helical longitudinal field in the dark filament, corresponding to the longitudinal field in a magnetic island formed in the dark filament material.

is allowed by the reconnection. The ejection of the dark filament occurs in two steps, explaining the observed "hesitation".

A most remarkable point here, made clear for the first time in the present *global* numerical simulation, is that the energy liberation is *not only* taking place in the neutral point region, *but rather, largely* takes place due to the overall expansion of the compressed field across the "separatrix surfaces" directly into the outer region of lower magnetic stresses (Uchida et al. 1999c). This point was not realized in any previous treatment in which the region near the neutral point was locally idealized into an X-type configuration. The magnetic energy converted to the kinetic and thermal energies in the neighborhood of the X-point is indeed a relatively small fraction of the total energy release. This seems to explain what is claimed about CME that the kinetic energy in CME's is *larger than* that of the "flare energy" itself, if the flare energy is thought to be the thermal and kinetic energies directly connected to the neutral point regions. We note that the total energy of a flare should include also the energy liberated through

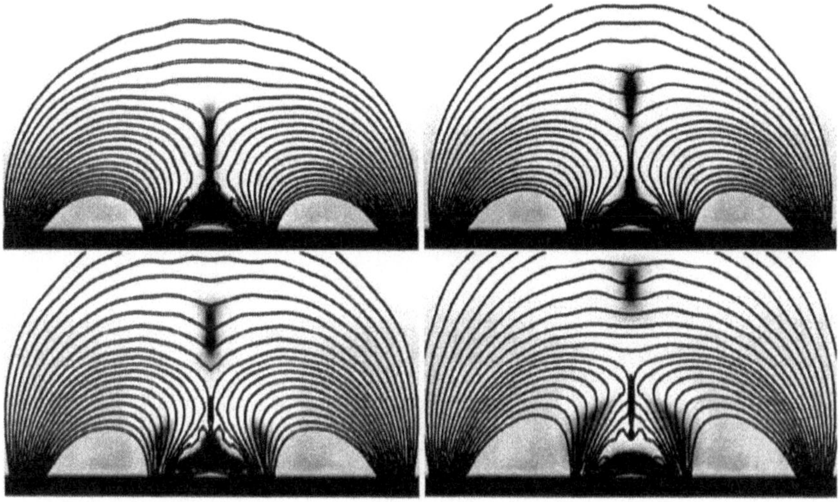

Figure 14. Time Development of the Result of Simulation of 2.5D Quadruple Source Model (Hirose et al. 2000a): It is seen that the upper part of the dark filament is squeezed out upwards, while the lower part pressed down. After the dark filament is squeezed out, magnetic reconnection starts, and the stressed field in the region below the separatrix surfaces can be relaxed (made possible by reconnection at the central neutral point) expanding across the previous separatrix surfaces. The dark filament, once hesitated to go up, is then pushed with strong acceleration.

the expansion of the thus-far squeezed magnetic field under the separatrix surfaces.

This liberation, however, is made possible by the magnetic reconnection at the neutral point after the dark filament is gone, and that in turn is induced by the destabilization of the otherwise very stable dark filament by the emergence of a small magnetic pair nearby in our model. It should be noted that many observed features have model counterparts in our model.

The presence of the dark filament plays an essential role here in allowing the accumulation of the magnetic energy by preventing continuous magnetic energy release. The cooled dark filament gas that came into the location through the siphon mechanism along $B_{||}$ plays the role of the weight to prevent the $B_{||}$ from being expelled out too easily.

In the above, the applicability of the quadruple magnetic source model in 2.5D was discussed, and shown that the model is very much more compatible with the new observational results by Yohkoh of high latitude arcade formation (2.5D-like) than any other previous models. The question whether there can be quadruple sources in the case of high latitude arcade formations *was* a crucial question, but it turned out that there is a mixed polarity belt in between the two unipolar regions, and the Yohkoh observa-

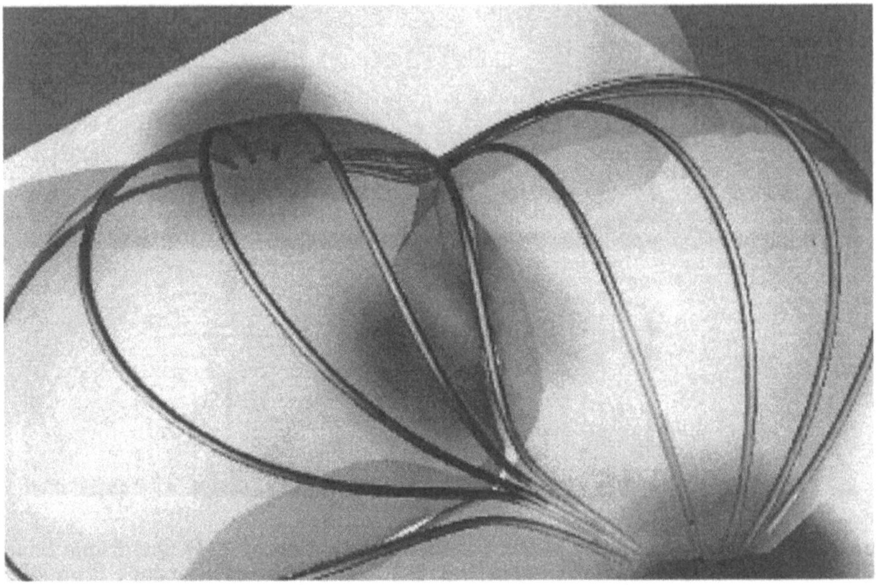

Figure 15. An Example of Separatrix Surfaces of Quadruple Source Situation in 3D (Uemura et al. 2000, a potential Field case): The contact line of separatrix surfaces is a neutral line. It is seen that this contact line becomes a surface if the two separatrix surfaces are pushed toward each other, and a current sheet along which the longitudinal field line will be created.

tions showed that the actual pre-event coronal structure makes use of those exchanged patches. The situation, on averaging along the y-direction, can be represented by a "quadruple source model" in the 2.5D approximation. The thin partition type shape of the dark filament with "barbs" can be explained for the first time by our model.

The physically most important point may be that the energy paradox in the classical model can be avoided, as well as the *new* energy paradox that the CME ejection caused by a flare has a larger energy than the flare energy itself. The definition of the flare energy as the thermal yield and the motions directly from the X-point was not viable. We should include in counting the flare energy the energy due to the work done by the relaxation of the stress from the overall region across the separatrix surfaces that is allowed by the reconnection.

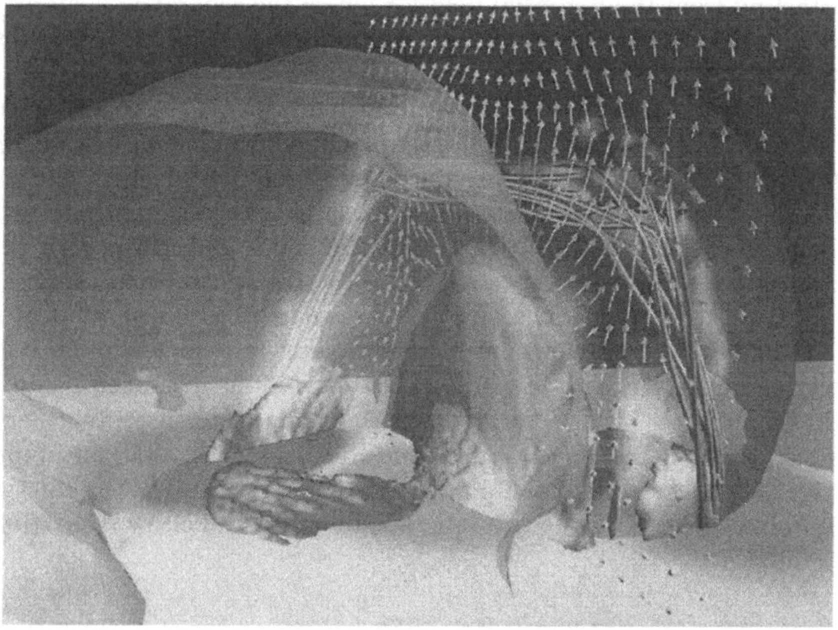

Figure 16. 3D Magnetodynamic Simulation of Quadruple Source Model (Hirose et al. 2000b): A realistic full 3D version of our quadruple magnetic source model corresponding to the separatrix surfaces as in Fig. 15. The behavior of the magnetic field passing in the current sheet together with the gas held by the field suggest that the result of this simulation can explain the observed behavior of the dark filament as observed by TRACE (Fig. 8). (Other field lines than the ones in the dark filament are suppressed in order to show the behavior of such special lines of force more clearly.)

4.3.3. *Results of 3D Magnetodynamic Simulations for Arcade Flares in Active Region*

Separatrix Surfaces in Active Regions in 3D.
If we do not simplify the sources in arrays, the separatrix surfaces naturally become fully 3D. Uemura et al. (2000) developed a method for deriving the separatrix surfaces, and showed nicely that the contact lines of the two separatrix surfaces give a neutral line, and the longitudinal field lines can lie along it (Fig. 15). The neutral line is easily deformed into a neutral sheet if the magnetic flux regions under the separatrix surfaces are pressed together.

Some Results of Full 3D MHD Simulations of Quadruple Source Model of Arcade Flares.

Starting from the potential field for which the separatrix surfaces have been thus derived, Hirose et al. (2000b) performed a full 3D MHD simulations.

They derived a configuration with magnetic stress stored by squeezing the photosphere in which the sources are embedded. The derivation of this state was made also by applying the same simulation scheme. The supply of the dark filament material is most likely occur along the lines of force by "siphon mechanism", but in this preliminary trial, we gave a certain appropriate mass in the magnetic field above the neutral sheet region after the system is squeezed. The start of the event is assumed to take place by continuing the squeezing here, before going into such a more realistic triggering action as the injection of MHD current as discussed before.

In the fully 3D MHD simulations with a "quadruple magnetic sources", it is seen that the results of simulations already suggest that this approach is extremely successful, explaining what are made clear by the observations with Yohkoh, TRACE, etc.. Details will be published elsewhere (Hirose et al. 2000b).

5. New Views of Coronal Mass Ejections

5.1. PREVIOUS VIEWS ABOUT CORONAL MASS EJECTIONS

CME was previously considered to have a form of huge loop structure expanding into interplanetary space (Hundhausen 1993). One of the arguments in support of this was that the locations of the two footpoints are fixed, and the CME loops are kept connected to those footpoints.

5.2. NEW ASPECTS OF CME'S FROM SOHO/LASCO, SOHO/EIT, AND YOHKOH/SXT

SOHO/LASCO is giving out a lot of interesting data about coronal mass ejections (CME's), and the observed results revealed various new information about CME's. For example, it was revealed that the frequency of occurrence of halo-type CME's turned out to be much higher than thought. Some people think that all CME's may be halo-type CME's, and some times look as if loop-type due to the effect of the length-of-the-line-of-sight across the structure.

We, however, do not agree to this due to the following reasons. In the following, we describe that there are special pre-existing connections from the flare and the footpoints, and those footpoints are special points in the case of loop-type CME's, telling that the loop-type CME's are *not* of a dome shape projected on the plane of the sky. Our arguments in the following are based on this.

5.2.1. *Observation of Loop-shaped CME's*

The relation of CME's with flares has been one of the focuses of discussions, and people naturally thought that they are caused by flares, because there are flares, either big ones or small ones, in association with them. Harrison (1995), however, noted in some cases that the CME had an earlier start time than the start time of the associated flare. Another fact, that the total energy involved in a CME turned out to be larger than that of the flare itself, made people think of another possibility that a CME causes a flare, not the reverse (Gosling 1993). We have given a plausible way out of such paradoxes in section 4.3.2.2.

Relation of Loop-Shaped CME's and Arcade Flarings.

It is actually confirmed by LASCO data that CME's take place with an arcade type flaring somewhere in between the footpoints of them that are often separated in the north and south high latitudes.

We wondered why the two footpoints separated by such a large distance can cooperate with the arcade flare occurring between them in lower latitudes, and examined this problem in a depth. We found by deep observations (long exposure time with thin filters, and use faint-structure-enhancing technique) of Yohkoh, that there are magnetic connections *pre-existing* between the footpoints of CME separated on both hemispheres, and the arcade flare between them! Therefore, there exists a physical reason for the cooperation of the footpoints to the arcade flare that seems to be isolated in between, and that those footpoints are special points in the CME event (Fig. 17).

In this context, we have found that there are interesting features in certain examples, in not rare cases. We looked into the cases of CME's having the associated arcade flare between the footpoints, and found the case in which the structure expanding into the interplanetary space behind the CME front (that is usually considered to be related with the erupted filament) has a deep dip right above the arcade flare in the early phase of expansion, like the MacDonald's logo (Fig. 18). This supports our claim above that is based on our finding of the physical connections between the two footpoints and the arcade flare between them as mentioned above.

Also, there are cases in which the CME has a rhombic shape (Fig. 17). Those were not seen in the SMM or in the earlier data. We see those in stronger event having greater acceleration, and are developing a magneto-dynamic model for those. If there are magnetic connections between the flare and the footpoints in the pre-event phase, it is possible that substantial magnetic twists are injected into the connections much in excess of the Krascal-Shafranov limit of helical instability. The twists are bounced from the two footpoints back along them. By the time they come back after

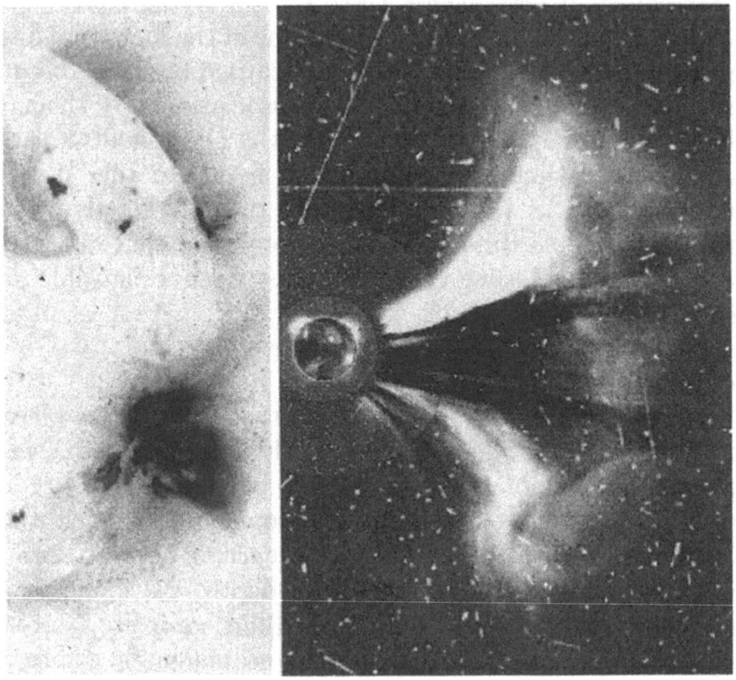

Figure 17. The Relation of the CME Footpoints and the Associated Arcade Flarings (Tanaka et al. 2000): (a)(left) The faint-structure-enhanced image from Yohkoh, and (b)(left) the LASCL C2-C3 image, of the CME of 1997 Nov 6, with the footpoints separated in the north and south hemispheres occurred together with an arcade flare in between. It has been a riddle why a CME whose footpoints are widely separated occurs in association with an arcade flare somewhere in between, but our finding made clear the physical explanation of this association.

bouncing, the two loops anchored at the flare region is reconnected to each other to make a large connection between the footpoints, and expanding. The helicity in excess in the two legs may deform both legs into the rhombic shape, and may drive the expansion of the long loop by the sweeping pinch effect somewhat like the model of loop flare in section 3.3 (Fig. 5).

5.2.2. *Halo-type CME's*

The halo-type CME's are those showing a roundish front, coming toward the earth, or, those with the front expanding beyond the occulting disk of the space coronagraph to both sides, or to all directions. This clearly has a dome-like shape, and definitely different from the loop-type CME's that have clearly defined footpoint regions.

The loop-type CME's have an arcade formation, or a dark filament eruption associated with them, but what observationally do the halo-type

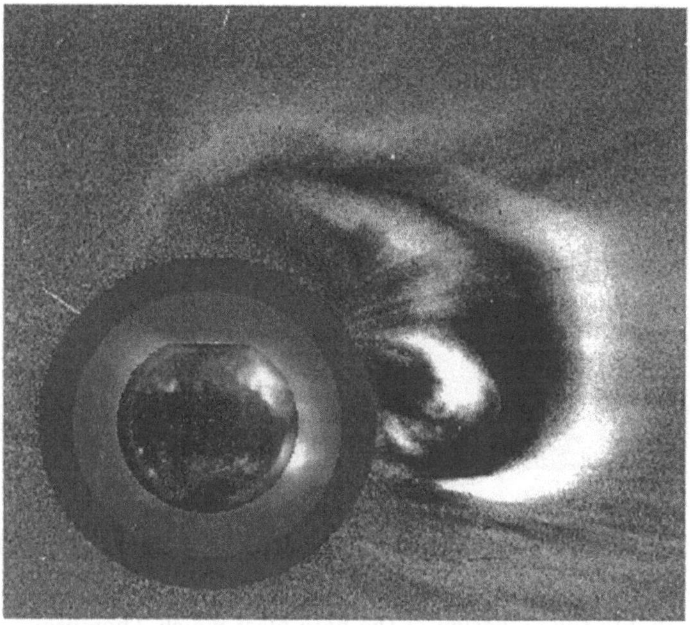

Figure 18. A CME with a Dip at the Associated Arcade Formation : The CME of 1997 Dec 6, clearly indicating an initial anchoring of the internal loop structure (which is related to the associated dark filament erupted from the arcade flare), is an example of such type found in LASCO data, supporting our claim in the text.

CME's have in association with them? Our proposed view is that the halo-type CME's are caused by the flare blast waves.

Relation of Halo-type CME's and Flare Blast Waves.

The flare blast wave were first noted by Moreton (1961) in the wings of Hα line profile. The disturbance he found propagates along the solar surface into a wide angle, with a velocity of 500-1500 km/s. When the disturbance, called Moreton wave, passes, the chromospheric material swings first downward, and then upwards.

Theoretical model should explain the propagation velocity as large as 1000 km/s, much higher than the acoustic and Alfvén velocities in the chromosphere, 10-30 km/s, the down-up swing, and the origin of the directivity of the disturbance, etc.. Uchida (1973) proposed, based on his treatment of the Type II radio source as due to MHD fast shock, that the Moreton wave disturbance is due to down refraction of an MHD fast shock front propagating in the corona. The Alfvén velocity in the corona is as high as 1000-2000km/s, and the calculation based on Eikonal method (WKBJ method in 3D) for MHD weak shocks showed that the Moreton waves can

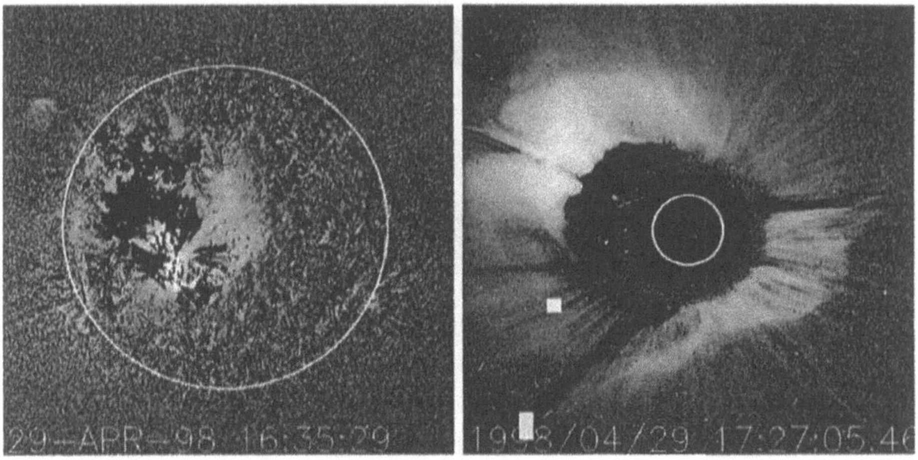

Figure 19. Halo Type CME's (Hata et al. 2000): Halo type CME's have been noted earlier, but it becomes clear that the frequency of the halo type CME is higher than thought before. We propose that these are related to the strong-shock version of the MHD blast produced at the explosive start of a flare, and this type CME's are most probably the extension of the flare blast which Thompson et al. (1997) discovered in SOHO/EIT analyses (see text). (a) Time-lapse difference image of EIT. The white part is the region increasing the brightness, and the dark part is where the brightness decreased in the successive frames, and (b) The same in the LASCO frames, of the 1998 Apr. 29 event.

be nicely explained by his model (Uchida et al. 1973), and also explained Type II radio bursts as well (Uchida 1974).

This flare blast wave was re-discovered by Thompson et al. (1977) in the SOHO/EIT observations, and was investigated in more detail. Pronounced difference of the disturbance observed in EIT and Hα as Moreton waves is its propagation velocity, 300-900 km/s vs 800-1500 km/s, but a most natural explanation at present is due to the difference in Alfvén velocity in different phase in the activity cycle. Figure 18 shows the EIT and LASCO pictures of a halo-type CME event.

5.3. OUR PROPOSED EXPLANATION OF CME'S

5.3.1. *Loop-type CME's*

Our proposed explanation for the loop-like CME's is related to our quadruple magnetic source model of arcade flares discussed in section 4.3.

The large scale connection between the magnetic sources in the north and south hemispheres will interact with the outer shell of the quadruple field of the pre-event magnetic structure of an arcade causing a flare. Then, a magnetic neutral point exists between them, if the outer shell part of the pre-event structure of arcade has a magnetic field component opposite to the overlying large connection of these two sources, and a similar magnetic structure including a neutral point (line, and sheet if squeezed) will be formed just like the one in the basic quadruple structure below, as discussed in section 4.3.

A large scale version of the process in Fig. 14 can take place, accordingly. Thus, the energy release in the form of the release of the magnetic stress in an extended magnetic structure, which is allowed to occur by the magnetic reconnection at the arcade flare can take place here by the same reason. It is noted that this occurs when the inner quadruple field starts releasing the upper part of flux tubes by reconnection at the low-lying neutral sheet. Therefore, the arcade flare occurring in between the footpoints of CME can cause a large scale arches connecting the two distant footpoints. The acceleration after that may be due to the acceleration mechanism due to magnetic buoyancy (melon seed effect) described in section 2.3.2, but the acceleration mechanism (driving by sweeping pinches) described in section 5.2.1.1 may work in a more powerful case.

5.3.2. *Our Proposed Explanation for Halo-type CME's*
In our tentative view, the halo-type CME's are driven by the flare blast. It was unfortunate that the wave hypothesis was discounted in the earlier phase of the CME research, because the halo-type CME were very rare in the observation period by earlier satellites. The argument for that was that the two footpoints were fixed on certain part of the Sun, whereas the wave front had none of such fixed footpoints. The wave hypothesis (eg. Wu 1988) for CME's should be paid proper attention *in the case* of the halo-type CME's Since a flare blast of the same explosion velocity will produce a stronger shock if it takes place in an atmosphere with a smaller Alfvén velocity, we may expect that the shocks due to the flare blast may tend to have higher Alfvén Mach number in the activity minimum. If this is the case, the treatment by Uchida with a weak shock approximation was better applicable in the activity maximum corona, but a stronger shock treatment is needed for the period of slower Alfvén velocity period, though the basic behavior of the fast-mode front may be represented reasonably well by the weak shock treatment. It is known that the basic behavior of the front will not be much different, but the mass carried by the shock front will be influenced appreciably by the strength of the shock. The faint dome-like front of the halo-type CME's suggests us of this possibility.

6. Conclusion

We discussed some of the new findings by Yohkoh obtained by using specifically its capability of detecting faint features that were not observable by the previous satellites, but can be crucial for the understanding of some basic phenomena of the solar activity. Those include, for example, some of the faint pre-event features that clearly showed contradictions with some of the previous ideas for theoretical explanations of flares and active regions.

Yohkoh has given various other outcome, and has contributed to the research in the field of solar and stellar active phenomena, and will continue to give remarkable findings in the rise phase of the new activity cycle in collaboration with other excellent satellites, SOHO, TRACE, etc., that are up in space, and hopefully with Solar-B to be launched in 2004.

The author acknowledges his indebtness to the colleagues of International Yohkoh Team for their wonderful cooperation in preparing and operating the Satellite Yohkoh, and for their collaborative data analyses and discussions. He also thanks Shigenobu Hirose and the graduate students in the Uchida Lab for their excellent works. The cooperations with Barbara Thompson, Alan Title, and Monique Pick are gratefully acknowledged. The use of excellent data from SOHO/EIT, SOHO/LASCO, and TRACE is gratefully acknowledged.

References

1. Aly, J.J. (1991) How much energy can be stored in a three-dimensional force-free magnetic field? *Astrophys. J.* **375**, L61.
2. Antiochos, S.K. (1998) The magnetic topology of solar eruptions, *Astrophys. J.* **502**, L181.
3. Antiochos, S.K., Dahlburg, R.B. and Klimchuk, J.A. (1994) The magnetic field of solar prominences, *Astrophys. J.* **420**, 41.
4. Aschwanden, M.J., Schwartz, R.A. and Alt, D.M. (1995) Electron time-of-flight differences in solar flares, *Astrophys. J.* **447**, 923.
5. Brown, J.C. (1986) Solar Maximum Mission Symposium (NSO), 431.
6. Carmichael, H. (1964) in Proc. of NASA Symp. *The Physics of Solar Flares*, (ed. W.N. Hess; NASA SP-50), 51.
7. Demoulin, P. and Forbes, T. (1992) Weighted current sheets supported in normal and inverse configurations – A model for prominence observations, *Astrophys. J.* **387**, 394-402.
8. Feynman, J. and Martin, S.F. (1995) The initiation of coronal mass ejections by newly emerging magnetic flux, *J. Geophys. Res.* **100**, 3355.
9. Fujisaki, K., et al. (2000), submitted to *Publ. Astron. Soc. Japan*
10. Galeev, A.A., Rosner, R., Serio, S. and Vaiana, G.S. (1981) Dynamics of coronal structures - Magnetic field-related heating and loop energy balance, *Astrophys. J.* **248**, 882.
11. Gosling, J.T. (1993) The solar flare myth, *J. Geophys. Res.* **98**, 18937.

12. Harrison, R.A. (1995) The nature of solar flares associated with coronal mass ejection, *Astron. Astrophys.* **304**, 585.
13. Hiei, E., Hundhausen, A. and Sime, D. (1993) Reformation of a coronal helmet streamer by magnetic reconnection after a coronal mass ejection, *Geophys. Res. Lett.* **20**, 2785.
14. Hirayama, T. (1974) *Solar Phys.* **34**, 323.
15. Hirose, S., Uchida, Y., Cable, S., Morita, S. and Torii, M. (2000a), in preparation.
16. Hirose, S., Uchida, Y., Uemura, S. and Cable, S. (2000b), in preparation.
17. Hundhausen, A. (1993) Sizes and locations of coronal mass ejections – SMM observations from 1980 and 1984-1989, *J. Geophys. Res.* **98**, 13177.
18. Ionson, J.A. (1985) The heating of coronae, *Solar Phys.* **100**, 289.
19. Kippenhahn, R. and Schlüter (1957)
20. Kopp, R. and Pneuman, G. (1976) Magnetic reconnection in the corona and the loop prominence phenomenon, *Solar Phys.* **50**, 85.
21. Kuijpers, J. (1997) A Solar Prominence Model, *Astrophys. J. Letters* **489**, 201
22. Kuperus, M. and Raadu, M.A. (1974) The support of prominences formed in neutral sheets, *Astron. Astrophys.* **31**, 189
23. Leroy, J.L., Bommier, V. and Sahal-Bréchot, S. (1983) The magnetic field in the prominences of the polar crown, *Solar Phys.* **83**, 135.
24. Martin, S. (1990) in *Dynamics of Quiescent Prominences*, Springer Lecture Notes on Physics **363**, 1-44.
25. Martin, S. and Nichols, C. (1994) in *Solar Surface Magnetism* (ed. R. Rutten and C. Schrijver; Kluwer), 339-346.
26. Masuda, S., Kosugi, T., Hara, H., Tsuneta, S. and Ogawara, Y. (1994) A loop-top hard X-ray source in a compact solar flare as evidence for magnetic reconnection, *Nature* **371**, 495-497.
27. McAllister, A., Uchida, Y., Tsuneta, S., Strong, K, Acton, L., Hiei, E., Bruner, E., Watanabe, Ta. and Shibata, K. (1992) Thickness variations along coronal loops observed by the Soft X-ray Telescope on YOHKOH, *Publ. Astron. Soc. Japan* **44**, L205.
28. Miyagoshi, T., Uchida, Y., Yabiku, T., Cable, S. and Hirose, S. (2000) *Publ. Astron. Soc. Japan*, submitted.
29. Moore, R. and Rumeliotis, G. (1992) in *Eruptive Solar Flares*, (ed. Z. Svestka, B. Jackson and M. Machado; Springer), 69.
30. Morita, S., Kamata, T. and Uchida, Y. (2000a) *Publ. Astron. Soc. Japan*, to be submitted.
31. Morita, S, Uchida, Y. and Torii, M. (2000b) *Publ. Astron. Soc. Japan*, to be submitted.
32. Moreton, G. (1961) *Sky and Telescope* **21**, 145.
33. Ohyama, M. and Shibata, K. (1997) Preflare heating and mass motion in a solar flare associated with hot plasma ejection: 1993 November 11, C9.7 flare, *Publ. Astron. Soc. Japan* **49**, 249.
34. Pallavicini, R. and Vaiana, G.S. (1976) The spatial structure of a solar flare in soft X-rays and centimetric wavelengths, *Solar Phys.* **49**, 297.
35. Priest, E. (1989) in *Dynamics and Structure of Quiet Solar Prominences* (Kluwer), 1.
36. Priest, E. and Forbes, T. (1990) Magnetic field evolution during prominence eruptions and two-ribbon flares, *Solar Phys.* **126**, 319.
37. Sakurai, T. and Granik, A. (1984) Generation of coronal electric currents due to convective motions on the photosphere. II - Resonance and phase mixing of Alfvén waves, *Astrophys. J.* **277**, 404.
38. Sciffer, M. (1997) Thesis, University of Newcastle, Australia.
39. Shibata, K., Masuda, S., Shimojo, M., Hara, H., Yokoyama, T., Nitta, N., Tsuneta, S. and Kosugi, T. (1995) Hot-plasma ejections associated with compact-loop solar flares, *Astrophys. J.* **451**, L83-85.

40. Shimizu, T., Tsuneta, S., Acton, L., Lemen, J. and Uchida, Y. (1992) Transient brightenings in active regions observed by the Soft X-ray Telescope on YOHKOH *Publ. Astron. Soc. Japan* **44**, L147.
41. Sturrock, P.A. (1966) *Nature* **221**, 695.
42. Sweet, P.A. (1958) in *Electromagnetic Phenomena in Cosmic Physics* (ed. B. Lehnert; Cambridge Univ. Press), 123.
43. Tanaka et al. (2000)
44. Tandberg-Hanssen, E. (1974) *Solar Prominences* (D. Reidel)
45. Tang, F. (1987) Quiescent prominences - Where are they formed? *Solar Phys.* **107**, 233.
46. Thompson, B., Newkirk, J., Gurman, J., St Cyr, O. and Stezelberger, S. (1977) *AAS/SPD* **29**, 01.30.
47. Title, A. et al. (1998) The TRACE mission: Initial scientific results, *AAS* **193**, 100.08.
48. Tsuneta, S., Hara, H., Shimizu, T., Acton, L.W., Strong, K.T., Hudson, H.S. and Ogawara, Y. (1992) Observation of a solar flare at the limb with the YOHKOH Soft X-ray Telescope, *Publ. Astron. Soc. Japan* **44**, L63-69.
49. Uchida, Y. (1973) in NASA Symp. *High Energy Phenomena on the Sun* (ed. R. Ramaty and R. Stone; NASA), 577.
50. Uchida, Y., Altschuler, M. and Newkirk, G. (1973) *Solar Phys.* **28**, 495.
51. Uchida, Y. (1974) *Solar Phys.* **39**, 431.
52. Uchida, Y. and Jockers, K. (1979) unpublished.
53. Uchida, Y. (1980) in *Skylab Workshop on Solar Flares* (ed. P.A. Sturrock; University of Colorado Press), 67 and 110.
54. Uchida, Y. and Sakurai, T. (1985) Magnetodynamical processes in interacting magnetospheres of RS CVn binaries, in *Unstable Current Systems and Plasma Instabilities in Astrophysics* (ed. M. Kundu and G. Holman; D. Reidel), 281.
55. Uchida, Y. and Shibata, K. (1988) Magnetodynamic mechanism for the heating of emerging magnetic flux tubes and loop flares, *Solar Phys.* **116**, 291.
56. Uchida, Y., McAllister, A., Strong, K., Ogawara, Y., Shimizu, T., Matsumoto, R. and Hudson, H. (1992) Continual expansion of the active-region corona observed by the YOHKOH Soft X-ray Telescope, *Publ. Astron. Soc. Japan* **44**, L155.
57. Uchida, Y., McAllister, A., Khan, J., Sakurai, T. and Jockers, K. (1994) in *X-Ray Solar Physics from Yohkoh* (ed. Y. Uchida, T. Watanabe, K. Shibata and H. Hudson; Universal Academy Press), 161.
58. Uchida, Y. (1996a) New aspects about solar flares revealed by the wide-dynamic range, high-cadence observations from YOHKOH, *Adv. Space Res.* **17**, (4/5)19-28.
59. Uchida, Y. 1996b, in *Magnetodynamic Phenomena in the Solar Atmosphere – Prototypes of Stellar Magnetic Activity* (ed. Y. Uchida, T. Kosugi and H.S. Hudson; Kluwer Academic Publishers), 295.
60. Uchida, Y., Fujisaki, K., Morita, S., Hirose, S. and Cable, S. (1999a) Soft X-ray coronal structure surrounding high-latitude dark filaments observed by Yohkoh, *Publ. Astron. Soc. Japan* **51**, 53.
61. Uchida, Y., Hirose, S., Cable, S., Morita, S., Torii, M., Uemura, S. and Yamaguchi, T. (1999b) A quadruple magnetic source model for arcade flares and X-ray arcade formations outside active regions – I. Dark filament suspension and the magnetic structure in the pre-event regions, *Publ. Astron. Soc. Japan* **51**, 553
62. Uchida, Y., Hirose, S., Morita, S., Fujisaki, K., Torii, M., Tanaka, T., Yabiku, T., Miyagoshi, T., Uemura , S. and Yamaguchi, T. (1999c), *Astrophys. Space Sci.* **264**, 145-199.
63. Uchida, Y., Miyagoshi, T., Yabiku, T., Cable, S. and Hirose, S. (2000a) *Publ. Astron. Soc. Japan*, submitted.
64. Uchida, Y., Khan, J., Doschek, G., Masuda, S., McAllister, A., Hirose, S., Feldman, U. and Cheng, C.C. (2000b) *Publ. Astron. Soc. Japan*, submitted.

65. Ulmschneider, P. (1996) in *Cool Stars, Stellar Systems, and the Sun* (ed. R. Pallavicini and A. Dupree), 71.
66. Watari, S., Watanabe, Ta., Acton, L. and Hudson, L. (1997) in *The Corona and Solar Wind near Minimum Activity* (ESA), 725-730.
67. Wheatland, M. and Uchida, Y. (1999) Frequency-energy distributions of flares and active region transient brightenings, *Solar Phys.* **189**, 163.
68. Wu, S.T. (1988) in *Developments in Theoretical and Applied Mechanics* **14**, 62.

MODELING THE EVOLUTION OF SOLAR MAGNETIC FIELDS

K. GALSGAARD
University of St Andrews, Department of Mathematics
North Haugh
St Andrews KY16 9SS, Scotland

Abstract. The solar atmosphere is a laboratory for studying complicated physical processes. The Magneto-Hydro-Dynamic (MHD) equations is the simplest theoretical approach that captures the complicated dynamical interaction between plasma and magnetic fields. These equations are therefore widely used for investigating the dynamical processes that takes place in the solar atmosphere. The MHD equations can also be used to provide information about the field line structure of the coronal magnetic field. Knowledge of the magnetic field topology is required before we can understand where the magnetic field tends to liberate it's free energy. To fully understand the energy release process, magnetic reconnection and wave dissipation processes have to be understood in detail. The MHD equations can also provide information about temperature profile along loops that are heated by, both static and time depend, heating profiles. This paper gives a review of magnetic reconnection, the basic idea behind magnetic field extrapolation, and the problems related to using temperature profiles for determining the underlying heating profile. A full understanding and utilisation of all of these disciplines within solar physics MHD is required to significantly advance our knowledge of the dynamical solar corona.

1. Introduction

Recent data obtained from satellites (Yohkoh, SoHO, and TRACE) and ground based observations provide us with information about density and temperature structures, dynamical properties and assumed magnetic topologies in the solar atmosphere. But, without proper theoretical models that can explain why and how the physics of the various phenomena works, we have little or no possibility of understanding why the solar atmosphere be-

J.-P. Zahn and M. Stavinschi (eds.),
Advances in Solar Research at Eclipses from Ground and from Space, 149–170.

haves and looks the way it does. It is generally accepted that the magnetic field plays a crucial role in determining the dynamics and structure of the solar corona, while in and below the photosphere the dynamics is dominated by the plasma. To build a consistent picture of the physical properties of the solar atmosphere, models have to be constructed that bridge the gap between the photosphere and the corona. The simplest theoretical description of a magnetic plasma is the *Magneto-Hydro-Dynamic* (MHD) equations [48]. The MHD equations consist of 6 coupled partial differential equations, that describing the macroscopic behaviour of an electrically conducting plasma:

$$
\begin{aligned}
\frac{\partial \rho}{\partial t} &= -\nabla \cdot (\rho \mathbf{u}), && (1)\\
\frac{\partial \rho \mathbf{u}}{\partial t} &= -\nabla \cdot (\rho \mathbf{u}\mathbf{u} + \underline{\underline{\tau}}) - \nabla P + \mathbf{J} \times \mathbf{B} - \rho g \mathbf{e}_z, && (2)\\
\frac{\partial e}{\partial t} &= -\nabla \cdot (e\mathbf{u}) - P\nabla \cdot \mathbf{u} - Q_{\text{Rad}} + Q_{\text{cond}} + Q_{\text{Joule}} + Q_{\text{visc}}, && (3)\\
\frac{\partial \mathbf{B}}{\partial t} &= -\nabla \times \mathbf{E}, && (4)\\
\mathbf{J} &= \sigma(\mathbf{E} + \mathbf{u} \times \mathbf{B}), && (5)\\
\mathbf{J} &= 1/\mu \nabla \times \mathbf{B}, && (6)
\end{aligned}
$$

where ρ, $\mathbf{u}$, e, g, $\mathbf{e}_z$, $\mathbf{B}$, $\mathbf{E}$, σ, $\mathbf{J}$, μ, $\underline{\underline{\tau}}$, $P = k_B/m\rho T$, k_B, m, T, Q_{Rad}, Q_{cond}, Q_{visc}, and Q_{Joule} are the density, velocity, thermal energy, gravity, unit vector in the z direction, magnetic field, electric field, electric conductivity, electric current density, magnetic permeability, viscous stress tensor, gas pressure, Boltzmann's constant, the mean molecule weight, temperature, radiative cooling, heat conduction, viscous, and Joule dissipation, respectively. The first equation describes how the plasma density changes as material is advected and compressed/de-compressed by the fluid flow field. The second equation is the momentum equation - Newton's second law - that describes how a plasma element is accelerated due to the different forces acting on it. The third equation is the energy equation, which in it's present form represents the plasma energy. The forth equation is the induction equation, that describes the changes in the magnetic field due to the advection and resistive processes described by Ohm's law, equation five. The sixth equation is Ampère's law that relates the magnetic field and current. The MHD equations are obtained by taking different moments of the Boltzmann equations and then adding the Maxwell equations [58]. The equations are therefore only valid when the plasma is well represented by statistical mean values, i.e. when the velocity distribution of the particles in the average volume is represented by a Maxwellian velocity distribution. For this to be fulfilled the length scale of the average volume represented by the MHD equations have to be much longer than the mean free path of the particles and the time scale much longer than the average time between particle collisions. These conditions for the solar atmosphere are generally fulfilled. The MHD approach is therefore a valid approach to investigate the structural and dynamical evolution of the various phenomena observed in the solar atmosphere.

This paper discusses some of the basic processes that are involved in releasing free magnetic energy stored in the magnetic field through reconnection, Section 2. To understand where in the solar magnetic field this can take place we have to know the topology of the solar coronal magnetic field. This information can at present only be retrieved by making magnetic field extrapolations into the corona based on photospheric magnetograms, Section 3. As different mechanisms are expected to contribute to maintaining the hot solar corona, Section 4, we have to investigate how they influence various parameters along the affected magnetic field lines - loops. Finally Section 5 investigates how observations of temperature profiles along loops may be used to probe the underlying mechanism that heats them.

2. Magnetic Reconnection

This chapter discusses magnetic reconnection, a mechanism by which a magnetic field can liberate some of the free energy it contains. Magnetic reconnection is expected to be one of the main mechanisms responsible for releasing magnetic energy in the solar corona and could as such be responsible for maintaining the hot solar corona (see Section 4). Let's assume that we are dealing with a magnetic field where both footpoints are rooted in the solar photosphere and the field lines between these extend into the corona. The magnetic energy is then given by the volume integral of $\frac{1}{2}B^2$ over the volume that the field occupies. Only a small part of this energy may be released, namely the amount that exceeds the energy of a potential magnetic field that lies within the same volume and has the same normal component of the magnetic field at the boundary. How can this free magnetic energy be released on a short enough time scale to maintain the hot corona? How much of the free energy is easily accessible on a short time scale?

To answer the second question, the amount of energy that may be released on a short time scale is strongly limited by the requirement that the *magnetic helicity* is conserved [6]. The magnetic helicity is a quantity that contains information on the twist and braiding of the large scale field line topology. Berger [5] showed that on a time scale much shorter then the decay time for helicity, a magnetic field can only release the amount of magnetic energy required to reach a constant α force free magnetic configuration defined by the same boundary conditions and helicity as the initial magnetic field. This puts a strong limitation on the actual energy release.

To answer the first question we have to take a closer look at the induction equation, Eq. (4). The change in the magnetic field is described by the curl of the two contributions defining the electric field, **E**, in Ohm's law, Eq. (5). The $\mathbf{u} \times \mathbf{B}$ term represents the effect of advection of the field with the plasma flow and the $1/\sigma\mathbf{j}$ term the effect of resistivity where field lines can both diffuse through the plasma and reconnect. The ratio of the advection term to the diffusion term defines the magnetic Reynolds number: $Rm = |\mathbf{v} \times \mathbf{B}|/|\eta\mathbf{J}| \approx vl/\eta$. This number can be determined for the solar atmosphere using typical values obtained from observations: it is of order $10^6 - 10^{12}$ [40] – very much larger then unity. This means the advection term in the induction equation determines the changes of the magnetic field. In this limit, where the resistive effects are unimportant for the time dependent development of the magnetic field, the magnetic field is said to be frozen into the plasma. This implies that when a given charged particle is identified to be associated with a particular magnetic field line in space at one time, then at a later time the same particle will still be associated with the same magnetic field line. Observations on the other hand provide evidence that the frozen in condition cannot always be true. Flares are large explosive events where free magnetic energy is found to be released on a short time scale. Therefore, situations exist where the resistive term in the induction equation becomes important for the development of the coronal magnetic field - this requires that the magnetic Reynolds number locally approaches unity. By ignoring the advection term in the induction equation an estimate of the resistive time scale for a loop, $t_r = l^2/\eta$, can be obtained. Then by using typical values for a coronal loop of 100 Mm and classical Spitzer resistivity [55], one finds the resistive time scale, t_r, to be on the order of 160 days. A similar time scale is found for the diffusion of helicity. To get the energy release to occur on a much shorter time scale, then the length scale l, on which the resistive process occurs, has to decrease. In flares, hard x-ray bursts, type III radio bursts and narrow band spikes indicate that the energy release takes place in many small events, on different length scales and with event times significantly below one second [4, 3]. To match these time scales with classical Spitzer resistivity a length scale on the order of meters or less is required! Even taking into account different types of anomalies resistivity, the typical length scale only increases by a factor of about 1000 and the required length scale remains significantly shorter than what present observations can resolve. Observations can therefore only provide indirect evidence that magnetic reconnection is taking place in the solar atmosphere ([59] and references therein). The following sections give an overview of the concepts of magnetic reconnection and the different ways that it may occur.

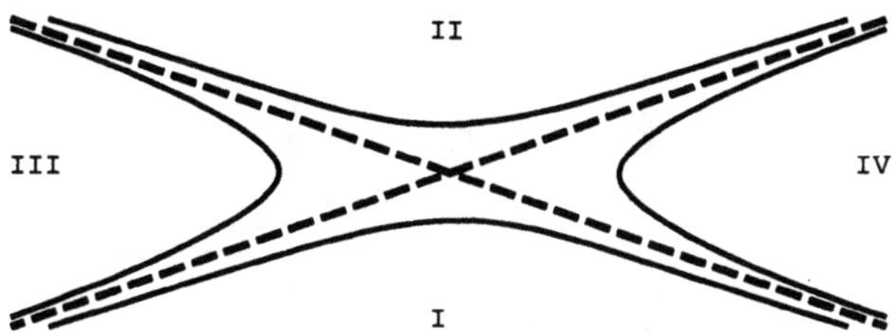

Figure 1. The x-point topology of the magnetic field hosts the locations in space where 2D magnetic reconnection may occur. The dashed lines represent the separator lines that divides space into four independent flux regions. The full lines indicate the field line topology.

2.1. 2D RECONNECTION

What is magnetic reconnection and where can it occur? This section is dedicated to 2D reconnection. A clear simplification of the real world, but it contains the basis to provide us with the concepts required to discuss 3D reconnection in the following section.

In 2D, reconnection can only occur at locations in space where the magnetic field vector vanishes and the magnetic field has an x-type topology, Fig. 1. At the x-point the magnetic field from regions of different connectivity gets infinitely close together and by reconnecting at the x-point the field line connectivity can change and the magnetic field evolves towards a lower energy state. For this to occur it is only required that the magnetic resistivity is non-zero in a local neighbourhood of the x-point. The magnetic flux from regions I and II can then reconnect in the diffusion region around the x-point and the new field lines move into region III and IV, with half of each of these field lines defined by the incoming field lines from region I and II (see Fig. 1). The question that remains to be answered is: How fast is this process? Can it process the incoming flux fast enough to account for the energy release processes that take place in the solar atmosphere? To investigate this we look in more detail at the classical example first described by Sweet [56] and Parker [42]. In this case the x-point has been collapsed into a tangential discontinuity of length L and width l, Fig. 2. By assuming steady state, applying conservation laws and using a force balance between the Lorentz force and the inertia term in the momentum equation along the sheet, it is possible to derive the ratio of the plasma velocity into the sheet and the external Alfvén velocity as a function of the external magnetic Reynolds number; $M = 1/\sqrt{Rm}$. The problem with this result is obvious: as the magnetic Reynolds number eaches a solar coronal value, then the plasma inflow velocity becomes very small and the amount of flux that can be processed approaches zero. This reconnection process is therefore totally inadequate to explain the fast energy release in solar flares.

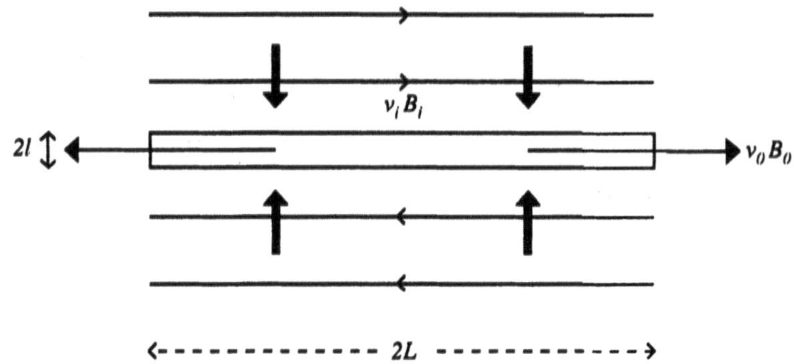

Figure 2. The layout of the Sweet-Parker current sheet. From [47]

Over the years reconnection processes with fast magnetic release have been found. The investigations culminated when Priest and Forbes [46] linearised the magnetic field around an x-point with a narrow opening angle and found that a family of solutions exit. The characteristic of the reconnection process in their solutions is controlled by the imposed boundary condition on the horizontal velocity at one corner of the domain. As this parameter is changed then the reconnection process evolves through a sequence of regimes with varying reconnection speeds, Fig. 3. It should be noticed how the velocity field advecting the magnetic field into the diffusion region changes character as the imposed boundary velocity changes from a compression type flow to a stagnation type flow. It is only with the stagnation type flow that the process reaches reconnection speeds that can become independent of the magnetic Reynolds number. The linearised solutions are only one particular class of solutions. If the opening angle of the outflow region increases a totally different regime is reached. These were first discovered by Biskamp [7], who made numerical experiments of magnetic reconnection and found that the results did not agree with the theory of the time. The reason for this was the difference in opening angle. Following this analytical solutions have been found for magnetic field configurations with large opening angles of the outflow region (For a recent review of reconnection see [47]).

The general situation regarding 2D reconnection can therefore be said to be well resolved today, although there are still problems that remain to be solved. These regard the understanding of the time dependent reconnection and to verify why some numerical experiments only give the expected result if special space dependent forms of the magnetic resistivity are used.

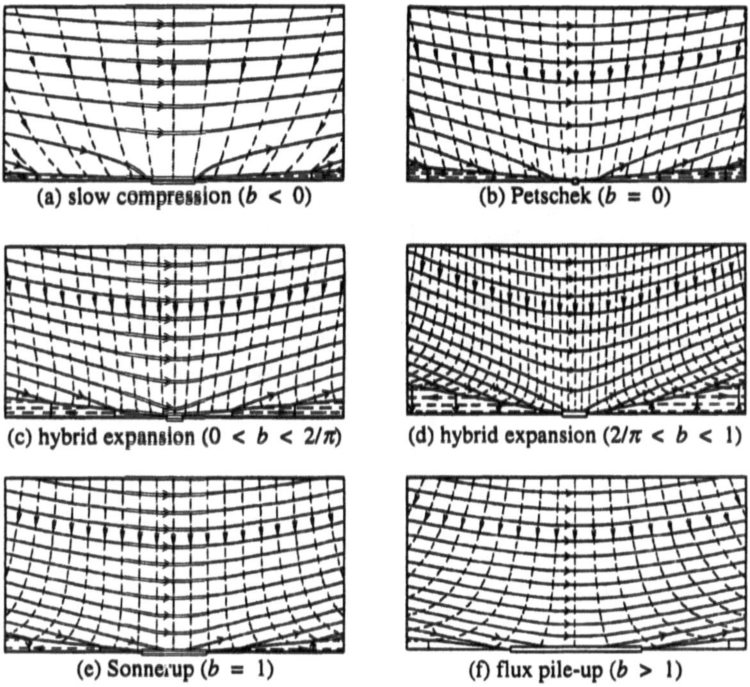

Figure 3. The panels show how the magnetic field and velocity structure of the reconnection regime changes as the boundary value of the horizontal velocity in the upper right corner of the domain is changed, controlled by the parameter b. Full lines are magnetic field lines while dashed lines are stream lines. The reconnection speed decreases going from top left to bottom right panel.(From [47])

2.2. 3D RECONNECTION

The situation becomes more complicated as we start investigating reconnection in three dimensions. Here reconnection is no longer constrained to field regions where the magnetic field strength goes to zero, as in 2D, but can also occur in field regions where the magnetic field is non-vanishing [22]. In the following we discuss both situations, starting with reconnection requiring magnetic singularities - null points.

2.2.1. *Null Point Reconnection*

The topology of magnetic null points in 3D is different from the 2D x-points discussed in the previous section. In three dimensions the null point consist of a fan plane [52] in which field lines ray out from the null point in a plane. The fan plane divides space into two independent flux regions. From each side of the fan plane a single field line, the spine axis, enters into the null point, Fig. 4. The local linear topology of the null point is determined by the eigenvalues and corresponding eigenvectors of the Jacobian matrix of the

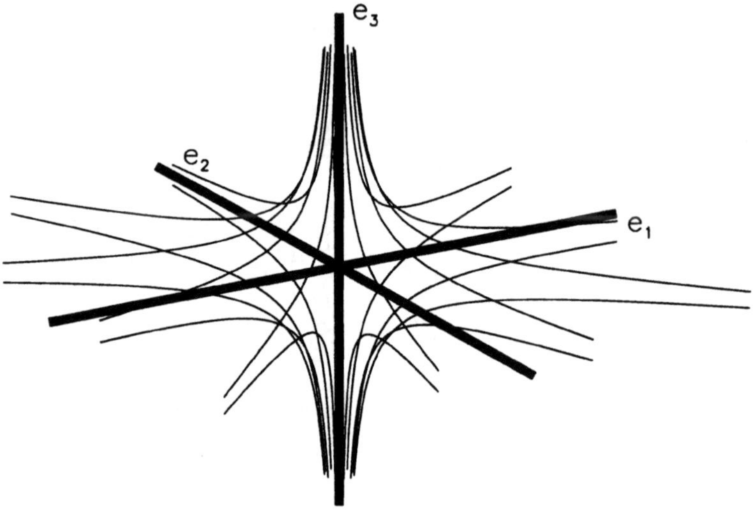

Figure 4. The structure of a single null. The eigenvectors $e_{(1,2)}$ define the fan plane, while e_3 represents the direction of the spine axis.

magnetic field at the null point (for more details see [45]). Null points do not exist alone but are formed in bifurcation processes, and will be formed in even numbers [20, 51]. When a double null point system is formed the nulls will be connected by at least one magnetic field line, the separator line. This field line is defined by the intersection of the two null fan planes, Fig. 5. Lau and Finn [32] and Priest and Titov [52] investigated the possible locations of magnetic reconnection in single and double null point systems using a kinematic approach. In the kinematic approach it is assumed that the magnetic field line structure is maintained constant in time. Different velocity functions are then imposed on the boundaries where the field lines enter the domain and by using the ideal Ohm's law, Eq. (5) with $\eta = 0$, the field line velocity in the domain may be determined. This shows that there are two possible modes of magnetic reconnection in three dimensional null points. Namely spine reconnection, which is associated with current accumulation along the spine axis, and fan reconnection, where current accumulates in the fan plane. In double null point systems there is further possibility for current accumulation along the separator.

The kinematic approach is the simplest possible and does not allow for any time dependent or plasma related influence on the investigation. One therefore has to investigate the results obtained by kinematic investigations by more advanced approaches, either analytical or numerical. Both of these have been used in this case. More recent investigations [9, 11, 10] have confirmed the kinematic findings using an analytical approach that

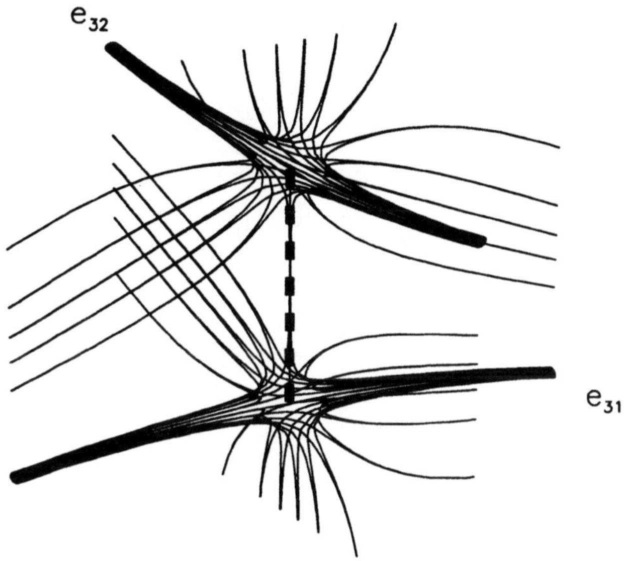

Figure 5. The structure of a double null system. The vectors $e_{3(1,2)}$ represents the spine axis, while the separator line is indicated by the dashed line.

takes into account the ideal induction equation and a simple form of the momentum equation written in vorticity form. They utilise the appearance of symmetry in the equations to find fully dynamical solutions. Doing this, they have been able to, not only describe the structure of the current and velocity flows, but also obtain scaling relations for the energy release. They show, not surprisingly, that the reconnection speed at the null point depends strongly on the topology of the null point and the imposed structure of the velocity field. Also numerical experiments of the full set of MHD equations have shown that current accumulations at the spine axis, the fan plane and along the separator are possible under given stressing conditions of the magnetic field [53, 15, 17, 16]. Both the analytical investigations [11] and the numerical simulations [15, 17, 16] indicate that the spine reconnection is insignificant. The simple reason is that the spine axis represents a single field line in 3D space and that current accumulations only occur if the field line topology is driven very precisely around the spine axis. In any other case the current will either accumulate in the fan plane or along the separator line in the double null point system. From the numerical experiment by Galsgaard and Nordlund [15], that contained 8 null points within the domain, is was found that only the separator seemed to attract current as the magnetic field was stressed by imposed boundary motions, Fig. 6. The field line topology along the separator has an x-type topology perpendicular to it. One could therefore expect that the collapse of the separator

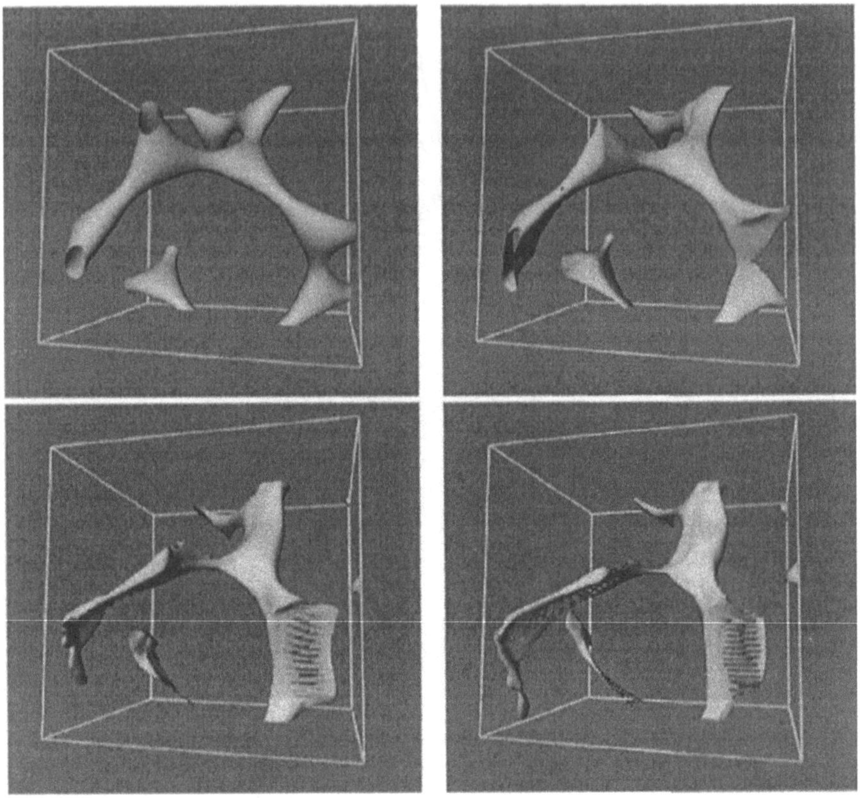

Figure 6. The panels show the collapse of the magnetic field along separator lines connecting nulls. Current sheets are formed at the locations of these collaps. (from [16]).

occurs due to the advection of flux towards the x-line as in 2D reconnection. Galsgaard et. al [16] investigated the process that produced the null points to collapse in the 8 null point experiment. Surprisingly it was found that the shear imposed by the boundary motion caused a shear stress across the fan plane of the nulls and this shear initiated the current sheet formation to occur along the separator line. Only when the reconnection starts to occur along the current sheet will the traditional stagnation flow become important for maintaining the current sheet during the proceeding reconnection process.

Despite advances over the last five to ten years, the true nature of 3D null point reconnection is still far from understood. We do not have a clear picture of the real changes in field line topology, nor are we sure about the speed at which the flux can reconnect at the null or how different structures of the null or velocity flow influences the process.

2.2.2. *Reconnection Without Null Points*

In reconnection without nulls there are a few characteristics that are seen in different numerical experiments and therefore seems to be a must for reconnection to occur. These are the need for the magnetic field to contain a fast changing tangential field component and a stagnation type flow advecting the plasma towards the centre region of the tangential field change. For the reconnection to be efficient, then the stagnation flow has to be set-up by the magnetic field itself. Otherwise the collapse of the current sheet may be stopped as the magnetic pressure builds up at the centre and the magnitude of the current may not become strong enough for fast reconnection to proceed. If a stagnation flow is initiated then it can be shown that the frozen in condition implies that the current magnitude will grow exponentially with time. In a plasma with a non-zero magnetic resistivity a balance between the advected plasma into the current region and the diffusion there reaches an equilibrium within a time that scales logarithmically with the value of the resistivity, η [14]. Therefore, even in a low resistive plasma, as the solar atmosphere, the current magnitude can build up to a sufficient high value for magnetic reconnection to proceed with in a finite time.

Parker [43] stressed that a three dimensional magnetic field can only exist without current singularities (read current sheets) if the field line structure is independent of the direction along the main direction of the magnetic field. This conclusion has been debated over the years and the problem has never been settled beyond any doubt for an ideal plasma [60, 61, 41]. On the other hand, numerical results [14, 34, 35, 21] strongly suggest that this is not the real issue to be debated. There is no such thing as an ideal plasma and the discussion of whether a true current sheet exists or not is not important. The importance is whether the length scales of the magnetic field can collapse enough for the current strength to reach values large enough for reconnection to become dynamically important. This in itself may be very difficult to prove analytically or numerically.

A different approach to 3D reconnection was taken by Hornig and Rastätter [25]. They investigated magnetic reconnection using four dimensional Minokovsky space. They found that the electro-magnetic field is conserved in 4 space and that for magnetic reconnection to occur in a non-zero magnetic field a stagnation flow has to exist that advects the magnetic field into the centre of the stagnation region in a finite time. For a traditional stagnation flow this is not the case, the time to reach the centre is infinite. But, in a non-ideal plasma the diffusion region will have a finite size and the advection time into this region will therefore be finite. The results from the flux braiding experiments mentioned above therefore fit very well with this result.

Priest and Démoulin [49] introduced another term into 3D reconnection, namely *Quasi-Separatrix-Layers* (QSL). QSLs are locations in space where the changes in the field line mapping is particular fast, but where no singularity is present. These locations should be particular susceptible to current build up as the field on either side of them is stressed. Further analytical investigations by Inverarity and Titov [27] and numerical experiments by Galsgaard [18] indicate that to initiate QSL reconnection is not that simple and that stagnation type flows are again required. The evolution of the QSLs therefore seems not be too different from the situation found in the flux braiding experiments, where there is also a large change in the field line mapping around the current concentrations. The initial discussion of the QSLs may therefore only be a different way of viewing the initial idea by Parker [43]. New analysis by Titov et al. [57] have shown that the previous definition of QSLs was not general enough and they suggested a slight redefinition of QSLs to make the measure independent of the chosen foot point.

In all of the above mechanisms it has been assumed that an infinite current density would be reached in an infinite time. If it, on the other hand, can be proven that the MHD equations are not regular, which means that for an initial smooth distribution of velocity and magnetic field vectors, then an infinite current density can develop in a finite time. This is difficult to prove or disprove analytically. Instead Kerr and Brandenburg [30] constructed a numerical experiment that indicate its existence. The characteristic of a finite time singularity in MHD is that the time integral of the sum of the max norm of the absolute values of the vorticity and the current will approach infinity as the time approaches the critical time. They use two interlocked magnetic rings that contract under their own tension force and create a current growth that seems to become infinite at a finite time. A similar finite time singularity may also have been found by Arber et al. [1], where the current build up of an internal kink instability is investigated using three different numerical techniques.

3. Magnetic Field Extrapolation

Before we can make any prediction of where and how magnetic energy release may take place in the solar atmosphere, we have to know the topological structure of the magnetic field. Presently it is not possible to make routine observations of the magnetic field at other locations than in the photosphere. If we therefore want qualitative information of the structure and field strength in the corona, we have to extrapolate the field into the corona using the photospheric magnetic field observations as boundary conditions. The general approach to determine the coronal magnetic field is to assume that the dynamics of coronal structures is dominated by the Lorentz force and that the magnetic field structure is in a nearly force free state. The problem then reduces to solving the steady state momentum equation with only the Lorentz force,

$$\mathbf{j} \times \mathbf{B} = \mathbf{0}. \tag{7}$$

For this equation to be fulfilled the current has to be proportional to the magnetic field, $\nabla \times \mathbf{B} = \alpha \mathbf{B}$. By taking the divergence of this equation, α can easily be shown to be a constant along field lines, but not necessarily to have the same value on different field lines. The solutions to Eq. (7) can be divided into three different types; $\alpha = 0$, a potential magnetic field; a constant α; and a space dependent α. There are various solution methods applicable to determining magnetic field extrapolations for these three types of force free fields. For a given set of boundary conditions a potential magnetic field solutions may always be found, however there are severe restrictions for the constant α magnetic field depending on the chosen boundary conditions and the imposed domain size. No simple solutions are found to determine non constant α force free solutions (for a more detailed discussion of the basic solution methods see [54]).

The extrapolation of solar magnetic fields has been applied to a number of different subjects, from the global solar magnetic field structure to the more detailed understanding of the field line structure in active regions and prominences.

The determination of the global solar field has been performed by Mikić et al. [39]. A problem in determining the three dimensional structure of the solar magnetic field is that we only have precise observational data of the part of the solar surface that is directly visible. The photospheric distribution of the magnetic field on the back side of the solar surface can only be estimated using models. Mikić et al. [39] have provided predictions of the structure of the solar corona in advance of several of the latest total solar eclipses (http://haven.saic.com/corona). It shows that the method they used works relatively well when the activity level on the sun is low. However as the complexity of the magnetic field increases, the predictions start to diverge from the observations.

Before the dynamics of active regions can be understood, the topology of the magnetic field has to be known in detail. Because good information on the magnetic field can only be obtained in the photosphere, magnetic field extrapolations of active regions are not easily verified. Different indirect techniques have been applied. Information from vector magnetograms in the photosphere; intensity contrasts in EUV/X-ray are used to reveal projected parts of magnetic field lines in the corona; and recently Lee et al. [33] have used radio observations for comparison. Radio observations are sensitive to both the strength and direction of the coronal magnetic field with different wavelengths being optical thick at different layers in the solar atmosphere. By combining the three techniques progress should be achieved in generating reliable field line structures of the solar coronal magnetic field.

Magnetic field extrapolations have also been employed to investigate the typical topology of field line structures in prominences. Mackay et al. [37] used field extrapolations in a closed domain to investigate the structural changes of the magnetic field line structure by comparing the extrapolated field with H_α observations before and after a prominence was formed.

Similarly Mackay et al. [38] investigated the influence of the structural form and strength distribution of flux patterns in the photosphere on the field line structure in the corona, in order to determine the requirements for achieving dipped field lines across the polarity inversion line that may support dense cold plasma in a prominence at coronal heights.

4. Coronal Heating

All satellite observations (Yohkoh, SoHO, TRACE) clearly show that the temperature of the outer solar corona is much higher than at the photosphere. The reason for this is not well understood. We are struggling to explain how the plasma temperature can rise from about 6000 K in the photosphere to a few million degrees in the corona. From Skylab an average radiative energy loss from different parts of the solar corona (open and closed regions) have been derived [64]. In active regions with closed magnetic field loops the radiative loss is estimated to be 1 $J/s/m^2$, if we assume a plasma column of $1m^2$ times $10^6 m$ containing a magnetic field with a field strength of 0.5 Gauss, a plasma density of $8.5\,10^{-14} kg/m^3$ and a temperature of 2 M Kelvin. If all of the plasma and magnetic energy could be released as input into the radiation, the plasma would be able to maintain the radiation for under half an hour, while the magnetic field would maintain the hot coronal plasma for just over a day. These estimates are much too large, as the coronal plasma cannot release all of the energy: a large fraction of the released energy is in fact conducted away from the release site and along the field lines. It is therefore obvious that a source of constant energy supply is required to maintain the hot corona. It has long been recognised that the magnetic field has a significant role in the heating process. The magnetic field is expected to work both as a transport medium of convecting energy from below the photosphere into the corona, and as a storage reservoir for building up free magnetic energy and finally as a provider for the release mechanism that liberates the free magnetic energy into the corona. There are two main classes of processes that are being discussed as heating mechanisms. The classification relates to the ratio of the time scale of the perturbation to the propagation time along the magnetic field structure. When the driving time is short compared to the crossing time the heating is classified as *Alternating Current* heating (AC), while when the driving time is longer then the crossing time it is known as *Direct Current* heating (DC). It is not our intention to go into details about the various mechanisms that have been suggested within these two groups, but just to outline the proposed methods. For the AC heating the main problems are: what is actually generating the high frequency wave energy to heat the corona, and how is the wave energy dissipated. The first question is not solved yet, but the energy is expected to arise from fast buffering of the magnetic footpoints of the loops in the photosphere. How to dissipate the energy is not obvious either, but there are ideas that seems to work. These are phase mixing [23] and resonant absorption [28, 19]. The phase mixing depends on neighbouring regions having different Alfvén velocity whereby a wave front is being stretched out across the boundary between the two regions. The fastest part of the wave will eventually catch up with the previous wave front and a large gradient in the magnetic field across the boundary is created. On the other hand then the resonant absorption relies on a density change along the line of propagation where the frequency of the wave then resonates with the eigen-frequency of the localised density perturbation and the wave amplitude may be significantly amplified.

A mechanism often mentioned as a possible DC heating mechanism is explosive events like flares. The problem with flares is that they do not supply enough energy to maintain a hot corona. Several investigations of the energy distribution of explosive events have found that they follow a power law [26, 13, 12, 31, 44]. Hudson [26] showed that the total energy output from an energy distribution that follows a power law is given by

$$P = \frac{f_0}{2-\gamma}\left(E_{max}^{2-\gamma} - E_{min}^{2-\gamma}\right), \tag{8}$$

where γ is the power index, f_0 a scaling constant and $E_{(max,min)}$ the energy range in which the power law exist. If γ is smaller than -2, then the small events make the main contribution to the energy release, while for γ larger than -2, the large events provide the main heating contribution. The power law distribution found for flare like events has a slope of about -1.6 and can therefore not provide the required heating of the solar corona.

Observations indicate that as the event energy becomes smaller then the slope tends to become steeper. But to derive the energy release of the events that approach the limit of present days observations implies that assumptions about their geometrical shape have to be made and these assumptions have profound influence on the result of the derived slope.

Krucker and Benz [31] and Parnell and Jupp [44] have both investigated events down towards 10^{24} ergs and obtained power indexes of -2.4 to -2.0. Recently Aschwanden [2] argued that there is a problem in going to too low energies in the individual events. They will eventually become so numerous that if they were to heat the corona then it would be physically impossible to host them in the solar atmosphere. It therefore seams that we are still a long way from understanding the mechanisms that heat the solar corona.

The flux braiding mechanism suggested by Parker [43] is also classified as a DC heating mechanism. Numerical experiments by Galsgaard and Nordlund [14] have shown that this mechanism may be able to heat the corona. Their experiments show the large scale driving cascades down to smaller length scales and that the energy release takes place on a range of length scales all the way down to the numerical resolution limit throughout the numerical domain.

5. Temperature Profiles

We have discussed how the energy may be released and how information about the magnetic field line structure in the corona can be obtained in the previous sections. But how does the location of the energy release influence the structure of the temperature profile along a single magnetic field line. The static temperature profile along a field line can be found by solving the energy equation and the hydrostatic equilibrium equation in one dimension,

$$\frac{dT}{ds} = VT^{-5/2}, \tag{9}$$

$$\frac{dV}{ds} = b\left(p^2\chi T^{\alpha-2} - H(s)\right), \tag{10}$$

$$\frac{dp}{ds} = -\rho g(s). \tag{11}$$

Here χ in Eq. (10) determines the radiative loss as a function of temperature. Different forms of this function have been derived [24, 8]. Despite differences in actual values, they agree on the basic shape of the function. The heat deposition along the loop is given by $H(s)$ and $g(s)$ is the value of gravity along the loop parallel to the field line. These equations can be solved as a boundary value problem when a shape of the loop has been chosen. A simple way to represent the possible solutions is to use a phase diagram, Fig. 7. In the phase diagram each line represents a solution with a given set of boundary parameters. The diagram in Fig. 7 represents a fixed uniform heat input along the loop. The choice of base temperature (x axis) and heat flux (y axis) through the base then determines which solution the curve represents. If the heating along the loop changes then the phase diagram should be three dimensional with the value of heat deposition given

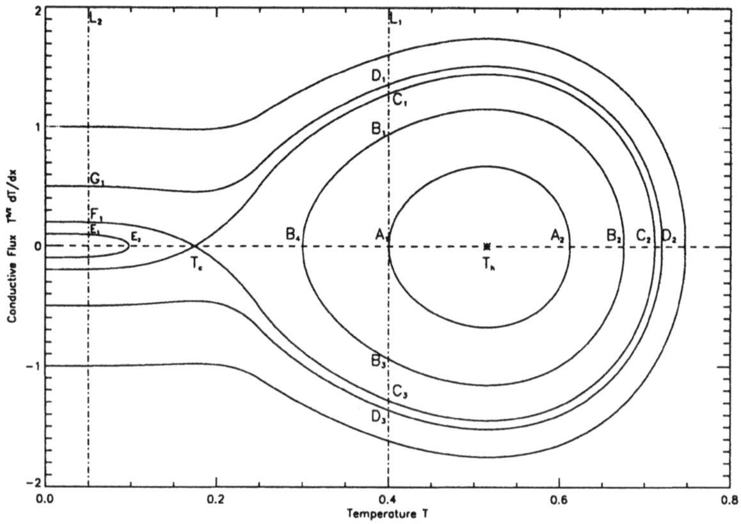

Figure 7. **The phase diagram of solutions to the energy energy equation when gravity is neglected and there is a constant heating along the loop (From [62]).**

in the third dimension. A full solution to the problem is then a curve in this three dimensional parameter space. Including gravity and loop area would further complicate the picture and a four or higher dimensional parameter space is required for such models.

In Fig. 8 it is seen how the static temperature profile along a loop changes significantly as the location of the heat deposition is changed. When the energy is mainly deposited close to the foot points then the temperature rises sharply over a short distance and is maintained nearly constant over the middle part of the loop. If, instead, the heat is deposited near the summit or more uniformly along the loop, then the temperature profile becomes parabolic. When the same amount of energy is deposited along the loop then energy released at the summit gives a much higher apex temperature than footpoint heating.

Priest et al. [50] inverted the problem, from observations of a large quiescent loop they determined the temperature profile along the loop. They then solved Eq. (9) - Eq. (10) with different heating profiles and used least square fitting to the temperature profile to determine the best fit. They found that uniform heating along the loop is most consistent with the data set. As pointed out by Mackay et al. [36] there are problems with this conclusion. First, the data only provides a temperature profile down to around 1.6 MK, and the fitting of those data gives a statistically insignificant result. Second, the chosen loop was directly on the solar limb

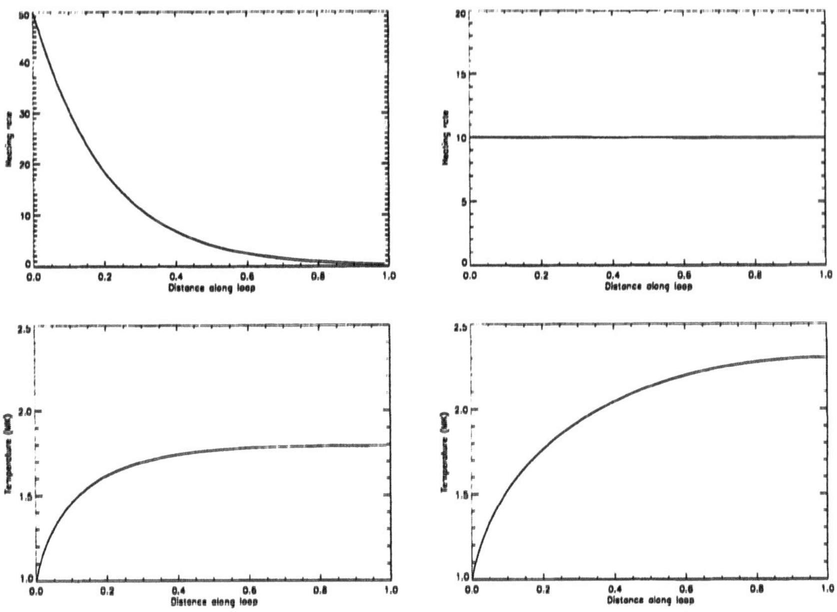

Figure 8. The two top panels show the shape of the energy release along the loop and the two bottom panels give the temperature profile corresponding to the energy deposition. The same amount of energy is released along the two loops. Heat released close to the base gives much lower loop temperatures then summit or uniform heating (image courtesy D. Mackay)

and the temperature data goes close to the limb. The loop temperature therefore has to decrease significantly over a very short distance to reach the photospheric temperature at it's footpoints. By trying to fit models down to chromospheric temperature values, Mackay et al. [36] showed that the increase in loop length is substantial for all models with smooth energy deposition. Only when foot point heating over a rather short length scale is used can the extra criteria about a very small increase in loop length be met. They conclude that requirements for using this method to determine the profile of the heat deposition along loops are; 1) that the measured temperature range goes down to at least one million Kelvin; 2) that both the accuracy of the measurements and the time cadence of the observations have to be increased before statistical significant results can be obtained.

The problem with comparing a steady state model with observations from the sun is naturally that the sun's atmosphere is a dynamically changing environment. Static models may therefore only provide believable results if the time variations of the observed loop is long compared with the observation time. Kano [29] showed that in active regions there are significant temperature variations on time scales down to order of minutes – a time scale that is probably limited by the time cadence of the instrument. A problem with observations of time dependent fluctuations in the temperature profile on time scales shorter than the integration time of the instrument, is that the signal is being time averaged in a non-linear way. The derived temperature profile may therefore not necessarily have a close relation with the actual time dependent energy release in the loop and deriving an average energy release may therefore give directly misleading answers. These problems have to be investigated in detail before it is known how strict the theoretical constraints are. The problem relates to the different time scales that are involved in determining the signature of the loop (time scales for the energy release, heat conduction, radiative loss, and dynamical time scale) and the problems that observations have with identifying a particular magnetic field line in time.

The theoretical approach to understand the heating of the corona is therefore to investigate how different time dependent heating models influence the temporal behaviour of the temperature profile. The hope is that different heating mechanisms produce different characteristic signatures that may easily be recognised. Work along these lines is in progress. Preliminary investigations by Walsh et al. [63] using heating input data arising from the flux braiding model discussed in Section 2 show that flux braiding gives a time and space dependent variation of the energy deposition with its average profile along the loop depending on the driving conditions. The time dependent fluctuations in the energy release, both in time and space, are clearly reflected in the temperature profile, where significant temperature fluctuations are found with time scales down to the order of one minute (with the assumed time scale for the braiding experiment).

6. Summary

The MHD equations are used in a range of different approaches towards understanding the properties of the solar atmosphere. The complexity of these equations and their coupled interaction means that analytical solutions cannot be found for most realistic cases. Much of the recent progress is therefore being increasingly guided by numerical models.

Investigations of the nature of magnetic reconnection have revealed that 3D reconnection is very complicated compared to 2D reconnection. The ability for magnetic reconnection to take place in regions where the magnetic field is non-zero provides a much richer environment than seen in 2D. The investigations of 3D reconnection are rather recent and only little is known yet about the true nature of the processes. Much work therefore has to be done in this field of research before a full understanding of the implications of 3D magnetic reconnection and its role in heating the solar corona is understood.

In the meanwhile, one should seek detailed knowledge of the magnetic field topology in the solar atmosphere. Significant progress is occurring in this field. Although the lack of simple comparisons with observations is a problem. How do we determine the magnetic structure with confidence when the means of comparisons with observations are rather poor?

Observations are starting to provide information about the temperature distribution along assumed magnetic field lines. The theoretical challenge now is to develop realistic models that can be directly compared with observations. For this to succeed the quality of observations has to improve further concerning both the time resolution and the accuracy of the data. From the theoretical point of view different heating models have to be expanded in the direction where both spatial and time dependent energy release have to be derived. Hopefully, these models will have different characteristics that can be used to identify the most important coronal heating mechanism.

Acknowledgement: The author would like to thank the organisers of the NATO ASI summer school for the invitation to participate in the school, PPARC for general financial support and D. Mackay in preparation of the manuscript.

References

1. Arber, T.D., Longbottom, A.W. and Van der Linden, R.A.M. (1999) Unstable coronal loops: Numerical simulations with predicted observational signatures, *ApJ*, **517**, 990
2. Aschwanden, M.J. (2000) Do EVU nanoflares account for coronal heating?, *Solar Phys.*, in press
3. Benz, A. (1994) Observations of fragmented energy release, in G.H.J. van den Oord (ed.), Procs. Utrecht Workshop "Fragmented Energy Release in Sun and Stars", *Space Sci. Rev.* **68**
4. Benz, A.O. and Aschwanden, M.J. (1992) Characteristics of the impulsive phase of flares, in B. Jackson, M. Machado, Z. Svestka (eds.), Eruptive Solar Flares, Lecture Notes in Physics, Springer, 106
5. Berger, M. (1985) Structure and stability of constant-α force-free fields, *ApJS* **59**, 433
6. Berger, M. and Field, G.B. (1984) The topological properties of magnetic helicity, *J. Fluid Mech.* **147**, 133
7. Biskamp, D. (1986) Magnetic reconnection via current sheets, *Phys. Fluids* **29**, 1520
8. Cook, J.W., Cheng, C.-C., Jacobs, V.L. and Antiochos, S.K. (1989) Effect of coronal elemental abundances on the radiative loss function, *ApJ* **338**, 1176
9. Craig, I. and Fabling, R.B. (1996) Exact solutions for steady state, spine, and fan magnetic reconnection, *ApJ* **462**, 969
10. Craig, I. and Fabling, R.B. (1998) Dynamical magnetic reconnection in three space dimensions: fan current solutions, *Phys of Plasma* **5**(3), 635
11. Craig, I., Fabling, R.B. and Watson, P. (1997) The power output of spine and fan magnetic reconnection solutions, *ApJ* **485**, 383

12. Crosby, N., Aschwanden, M.J. and Dennis, B.R. (1993) Frequency distribution and correlations of solar X-ray flare parameters, *Solar Phys.* **143**, 275
13. Dennis, B.R. (1985) Solar hard X-ray bursts, *Solar Phys.* **100**, 465
14. Galsgaard, K. and Nordlund, Å. (1996) Heating and activity of the solar corona: 1. Boundary shearing of an initially homogeneous magnetic field, *J. Geophys. Res.* **101**, 13445
15. Galsgaard, K. and Nordlund, Å. (1997) Heating and activity of the solar corona: 3. Dynamics of a low beta plasma with 3D null points, *J. Geophys. Res.* **102**, 231
16. Galsgaard, K., Priest, E.R. and Nordlund, Å. (2000) Three-dimensional separator reconnection - How does it occur?, *Solar Phys.*, in press
17. Galsgaard, K., Rickard, G. and Reddy, R. (1997) Energy release sites in magnetic fields containing single or multiple nulls, *Solar Phys.* **176**, 299
18. Glasgaard, K. (2000) Numerical investigation of quasi separator layers, acc. for *JGR*
19. Goossens, M. and Ruderman, M.S., (1995) Conservation laws and connection formulae for resonant MHD waves, *Physica Scripta* **T60**, 171
20. Green, J.M. (1989) Geometrical properties of 3D reconnecting magnetic fields with nulls, *J. Geophys. Res.* **93**, 8583
21. Hendrix, D.L. and Van Hoven, G. (1996) Magnetohydrodynamic turbulence and implications for solar coronal heating, *ApJ* **467**, 887
22. Hesse, H. and Schindler, K. (1988) A theoretical foundation of general magnetic reconnection, *J. Geophys. Res.* **93**, 5559
23. Heyvearts, J. and Priest, E.R. (1983) Coronal heating by phasemixed shear Alfvén waves, *A&A* **117**, 220
24. Hildner, E. (1974) The formation of solar quiescent prominences by condensation, *Solar Phys.* **35**, 123
25. Hornig, G. and Rastätter, L. (1997) The magnetic structure of B $\neq$ 0-reconnection, *Physica Scripta* **T74**, 34
26. Hudson, H.S. (1991) Solar flares, microflares, nanoflares, and coronal heating, *Solar Phys.* **133**, 357
27. Inverarity, G.W. and Titov, V.S. (1997) Formation of current layers in three-dimensional, inhomogeneous coronal magnetic fields by photospheric motions, *J. Geophys. Res.* **102**, 22285
28. Ionson, J. (1978) *ApJ* **226**, 650
29. Kano, R. (1997) Temperature structure of quasi-steady coronal loops revealed with YOHKOH, in The Coronal and Solar Wind near Minimum Activity, Vol. 404 of *ESA-SP*, ESA, Nordwijkerhout, Holland, p.17
30. Kerr, R. and Brandenburg, A. (1999) Evidence for a singularity in ideal magnetohydrodynamcis: implications for fast reconnection, *Phys. Rev. Lett.*, in press
31. Krucker, S. and Benz, A.O. (1998) Energy distribution of heating processes in the quiet solar corona, *ApJ* **501**, l213
32. Lau, Y.T. and Finn, J.M. (1990) 3D kinematic reconnection in the presence of field nulls and closed field lines, *ApJ* **350**, 672
33. Lee, J., White, S.M., Kundu, M.R., Mikić, Z. and Mcclymont, N.A. (1999) A test for coronal magnetic field extrapolations, *ApJ* **510**, 413
34. Longbottom, A.W., Rickard, G.J., Craig, L. J.D. and Sneyd, A.D. (1998) Magnetic flux braiding: force-free equilibria and current sheets, *ApJ*, **500**, 471
35. Longcope, D.W. and Sudan, R.N. (1994) Evolution and statistics of current sheets in coronal magnetic loops, *ApJ* **437**, 491
36. Mackay, D., Galsgaard, K. and Priest, E.R. (2000) How accurately can we determine the coronal heating mechanism in the large-scale solar corona ?, *Solar Phys.*, in press
37. Mackay, D.H., Gaizauskas, V., Rickard, G. and Priest, E. (1997) Force-free and potential models of a filament channel in which a filament forms, *ApJ* **486**, 534
38. Mackay, D.H., Longbottom, A.W. and Priest, E.R. (1999) Dipped magnetic field configurations associated with filaments and barbs, *Solar Phys.* **185**, 87

39. Mikić, Z., Linker, J.A., Schnack, D.D., Lionello, R and,Tarditi, A. (1999) Magnetohydrodynamic modeling of the global solar corona, *Phys. Plasma* **6, no 5**, 2217
40. Nakariakov, V., Ofman, L., DeLuca, E., Roberts, B. and Davila, J. (1999) TRACE observations of damped coronal loop oscillations: implications for coronal heating, *Science* **285**, 862
41. Ng, C.S. and Bhattachrjee, A. (1998) On non-equilibrium and current sheet formation in line-tied magnetic fields, *Physics of Plasma* **5**(11), 4028
42. Parker, E.N. (1957) Sweet's mechanism for merging magnetic fields in conducting fluids, *J. Geophys. Res.* **62**, 509
43. Parker, E.N. (1972) Topological dissipation and the small-scale fields in turbulent gases, *ApJ* **174**, 499
44. Parnell, C.E. and Jupp, P.E. (2000) Statistical analysis of the energy distribution of nanoflares in the quiet Sun, *ApJ* **529**, 554
45. Parnell, C.E., Smith, J.M., Neukirch, T. and Priest, E.R. (1996) The structure of three-dimensional magnetic null points, *Physics of Plasma* **3**, 759
46. Priest, E. and Forbes, T. (1989) Steady magnetic reconnection in three dimensions, *Solar Phys.* **119**, 211
47. Priest, E. and Forbes, T. (2000) Magnetic reconnection: MHD theory and applications, Cambridge University Press, in press
48. Priest, E.R. (1984) Solar magneto hydrodynamics, D. Reidel Publ. Comp.
49. Priest, E.R. and Démoulin, P. (1995) Three-dimensional magnetic reconnection without null points 1. Basic theory of magnetic flipping, *J. Geophys. Res.* **100**, 23443
50. Priest, E.R., Foley, C., Heyvaerts, J., Arber, T., Culhane, J. and Acton, L. (1998) The nature of the heating mechanism for the diffuse solar corona, **Nat.** *393*, 545
51. Priest, E.R., Lonie, D.P. and Titov, V.S. (1996) Bifurcations of magnetic topology by the creation or annihilation of null points, *J. Plasma Physics* **56**, 507
52. Priest, E.R. and Titov, V.S. (1996) Magnetic reconnection at three-dimensional null points, *Phil. Trans R. Soc. Lond. A* **354**, 2951
53. Rickard, G.J. and Titov, V.S. (1996) Current accumulation at a three-dimensional magnetic null, *ApJ* **472**, 840
54. Sakurai, T. (1989) Computational modeling of magnetic fields in solar active regions, *Space Science Reviews* **51**, 11
55. Spitzer, L. (1962) Physics of fully ionized gases, *Interscience, New York*
56. Sweet, P.A. (1958) The neutral point theory of solar flares, in B. Lehnert (ed.), Electromagnetic Phenomena in Cosmical Physics, Cambridge Univ. Press, p.123
57. Titov, V.S., Démolin, P. and Hornig, G. (1999) Quasi-separatrix layers: refined theory and its application to solar flares, in Magnetic fields and solar processes, 9-th European Meeting on Solar Physics, Firenze, Italy, September 12-18
58. Tsinganos, K.C. (1992) The MHD description of cosmic plasmas, in J.T. Schmelz, J.C. Brown (eds.), The Sun: A laboratory for Astrophysics, NATO ASI Series C vol 373, Kluwer Academic Publishers, p.139
59. Tsuneta, S. (1996) Magnetic reconnection: open issues, in B. Bentely, J. Mariska (eds.), Yohkoh Conference on Observations of Magnetic Reconnection in the Solar Atmosphere, Vol. 111, ASP Conference Proceedings, p.409
60. van Ballegooijen, A.A. (1985) Electric currents in the solar corona and the existence of magnetostatic equilibrium, *ApJ* **298**, 421
61. van Ballegooijen, A.A. (1998) Force free fields and coronal heating, *Geophys. Astrophys. Fluid Dyn.* **41**, 181
62. Walsh, R. (1999) Modeling the thermal structure of coronal loops, in Plasma dynamics and diagnostics in the solar transition region and corona, Vol. 446 of *ESA-SP*, ESA, Nordwijkerhout, Holland, p.687
63. Walsh, R.W., Galsgaard, K and, Arber, T. (2000) in preparation
64. Withbroe, G.L. and Noyes, R.W. (1977) Mass and energy flow in the solar chromosphere and corona, *ARA&A* **15**, 363

NON-LTE RADIATION PROCESSES: APPLICATION TO THE SOLAR CORONA

S. COLLIN
Observatoire de Paris
F-92195 Meudon, France

Abstract. These lectures are intended to present a simple but relatively complete description of the theory needed to understand the formation of lines in non-local thermodynamical equilibrium (NLTE), without appealing to any previous knowledge except a few basics of physics and spectroscopy. After recalling elementary notions of radiation transfer, the chapter is focussed on the computation of the level populations, the source function, the ionization state, and finally the line intensity. An application is made to forbidden coronal lines which were observed during eclipses since decades.

1. Introduction

Astrophysical media, at least for the fraction we can observe, are often very dilute compared to those we are used to on Earth. In the interstellar medium for instance, the number density is in average one atom per cm^3, in HII regions and planetary nebulae it is about 10^4 atoms per cm^3 and in the solar corona 10^8 atoms per cm^3, orders of magnitudes less than those one can get in laboratory experiments. Therefore the notion of "thermodynamical equilibrium" does not hold, and quite unusual phenomena are taking place, such as the emission of intense "forbidden lines" never observed on Earth. We will in particular show how to compute the intensity of visible forbidden coronal lines, and show how these lines, and others, can be used to get physical parameters of the solar corona.

J.-P. Zahn and M. Stavinschi (eds.),
Advances in Solar Research at Eclipses from Ground and from Space, 171–200.

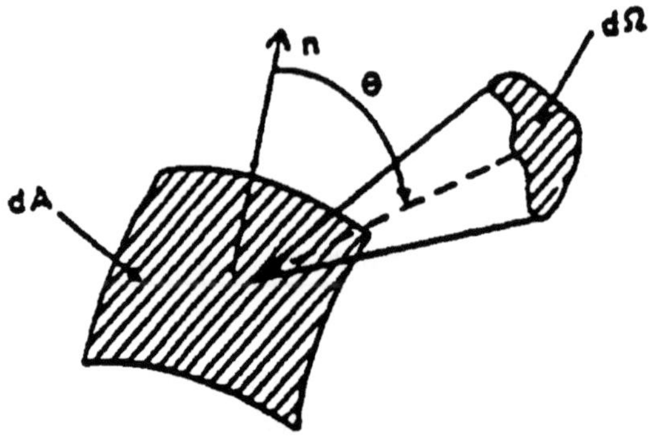

Figure 1. The specific intensity.

2. Basics of Transfer

2.1. PHOTOMETRIC QUANTITIES

We shall first define a few basic photometric quantities. Note that CGS units will generally be used.

- **The specific intensity** I_ν: The energy dE_ν crossing a surface of area dA in the direction of the normal, in a solid angle $d\Omega$, during a time dt, in a frequency interval $d\nu$, is (cf. Fig. 1):

$$dE_\nu = I_\nu \; d\nu \; d\Omega \; dt \; dA. \tag{1}$$

It can be expressed in Watt m^{-2} ster^{-1} Hz^{-1}, but other units are used in the visible or in the X-ray range. It can be also defined per interval of wavelength, and according to the relation $\lambda\nu = c$ (speed of light), one gets $I_\lambda = I_\nu c/\lambda^2$.

The specific intensity can be used both for the source of radiation or for the receptor, and it is constant along a path ray in the vacuum.

- **The mean intensity** J_ν:

$$J_\nu = \frac{1}{4\pi} \int I_\nu \; d\Omega \tag{2}$$

- **The flux** F_ν:

It is the power crossing a unit surface per unit frequency interval, in all directions (cf. Fig. 2):

$$F_\nu = \int I_\nu \; \cos\theta \; d\Omega \tag{3}$$

F_ν is generally expressed in Jansky in the radio and far infrared range: 1 J = 10^{-23} in CGS, or 10^{-26} in MKS (Watts m^{-2} Hz^{-1}).

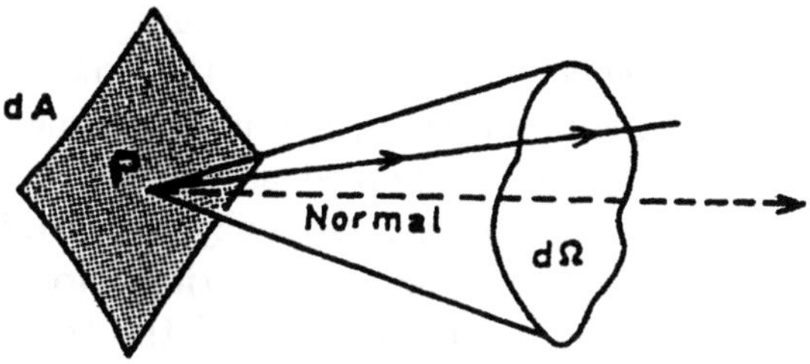

Figure 2. The flux.

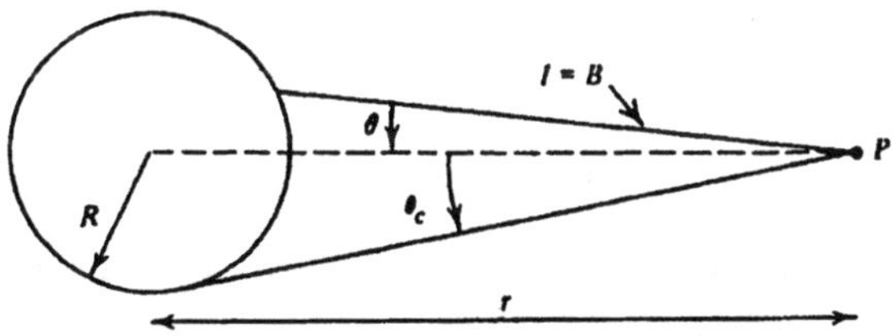

Figure 3. The flux received from a star.

As an application, we can compute the flux from a uniform sphere radiating isotropically (not a very good approximation for a star, actually), cf. Fig. 3):

$$F_\nu = \int I_\nu \; \cos\theta \; d\Omega = I_\nu \int_0^{2\pi} d\phi \int_0^{\theta_c} \sin\theta \; \cos\theta \; d\theta, \tag{4}$$

where θ_c is the angle under which the sphere is seen from the observer. If R is the radius of the star, and r its distance, one gets $\sin\theta_c = R/r$, and

$$F_\nu = \pi I_\nu (R/r)^2. \tag{5}$$

Note that the flux at the surface of the sphere is πI_ν.

2.2. TRANSFER EQUATION

When a light ray does not propagate in the vacuum, the specific intensity is not constant: *emission* adds energy, and *absorption* removes energy. There is also *diffusion*, in which the global luminous energy is not changed, but can be increased in one direction, and decreased in another. We shall not take into account the polarization of the radiation, and the possibility of non stationary phenomena.

2.2.1. *Transfer equation in a non diffusive medium*
The monochromatic emissivity η_ν is defined as the power emitted per unit solid angle, per unit frequency interval, per unit volume. The monochromatic absorption coefficient χ_ν is defined as follows: $\chi_\nu I_\nu$ is the power absorbed per unit solid angle, per unit frequency interval, by a slab of unit length normal to the direction of the propagation. Note that here χ_ν is the inverse of a length. Some people use instead the "opacity coefficient", defined per unit mass, and the emissivity per unit mass and not per unit volume. The variation of I_ν on a path length $d\ell$ *in the direction of the light ray* is therefore:

$$dI_\nu = (-\chi_\nu I_\nu + \eta_\nu)d\ell. \tag{6}$$

Let us define:

$$d\tau_\nu = -\chi_\nu d\ell \text{ and } \tau_\nu = \int -\chi_\nu d\ell, \tag{7}$$

τ_ν is the *optical depth*, which decreases towards the observer.

The transfer equation writes then:

$$\frac{dI_\nu}{d\tau_\nu} = I_\nu - S_\nu, \tag{8}$$

where $S_\nu = \eta_\nu/\chi_\nu$ is called the *source function.*

If there are several emission and absorption processes, they have to be all taken into account in the source function and in the absorption coefficient. For instance, if at a frequency ν a line is superposed onto a continuum, then $S_\nu = S_\nu^{line} + S_\nu^{cont}$, and $d\tau_\nu = d\tau_\nu^{line} + d\tau_\nu^{cont}$.

2.2.2. *Transfer equation in a diffusive medium*
If there is a diffusion process, it must also be taken into account in the transfer equation. The *diffusion coefficient* is defined like the absorption coefficient: $\sigma_\nu I_\nu$ is the power diffused per unit solid angle, per unit frequency interval, by a slab of unit length normal to the direction of the propagation. Let us assume that diffusion is coherent (i.e. without any change of frequency), and isotropic. As a consequence there is a corresponding emission coefficient, which is equal to $\sigma_\nu J_\nu$, and the transfer equation becomes:

$$dI_\nu = [(-\chi_\nu - \sigma_\nu)I_\nu + (\sigma_\nu J_\nu + \eta_\nu)]d\ell. \tag{9}$$

If one defines now an *extinction coefficient*, $\tau_\nu^{tot} = \int d\tau_\nu^{tot}$ with $d\tau_\nu^{tot} = -(\chi_\nu + \sigma_\nu)d\ell$, one gets for the transfer equation in presence of diffusion:

$$\frac{dI_\nu}{d\tau_\nu^{tot}} = I_\nu - S_\nu^{tot}, \tag{10}$$

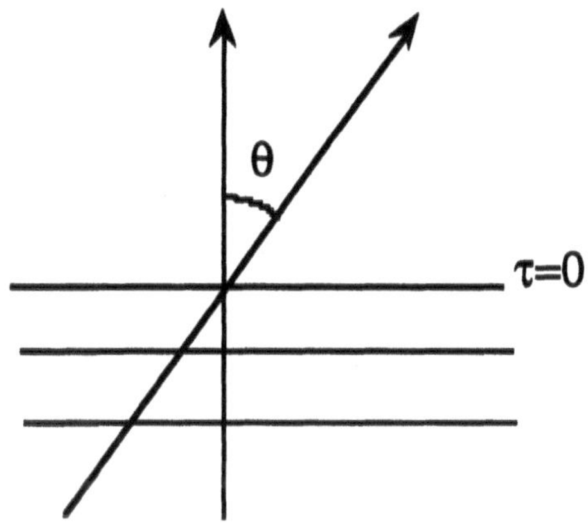

Figure 4. The plane-parallel approximation.

where $S_\nu^{tot} = (\sigma_\nu J_\nu + \eta_\nu)/(\chi_\nu + \sigma_\nu)$. Although it is formally similar to Eq. 8, it differs in that the intensity appears directly in the second term, so it is an integro-differential equation. Actually, the diffusion process is a probabilistic one, similar to a random walk, and one can show that the distance that a photon will travel before being absorbed is equal to $[\chi_\nu(\chi_\nu + \sigma_\nu)]^{-1/2}$, while in the case of pure absorption it is equal to χ_ν^{-1}.

In a purely diffusing medium, the transfer equation writes:

$$\frac{dI_\nu}{d\tau_\nu^{dif}} = I_\nu - J_\nu, \tag{11}$$

In the corona, the diffusion process is due to Thomson scattering by free electrons. As a first approximation it can be considered as a coherent and isotropic process. The diffusion coefficient σ_ν does not depend on frequency, and is equal to $\sigma_T N_e$, where N_e is the number of electrons per unit volume, and σ_T is the Thomson cross section, equal to 6.65 10^{-25} cm^2. This process is very important for the continuum in the visible range, as it is responsible of the emission of the K corona. As a consequence one can consider that the corona is a purely diffusive medium for the continuum in the visible range. On the other hand we will show that diffusion is negligible in the transfer of the coronal lines, on which we will focus later on, so we will not consider it in the following sections.

2.3. APPLICATION TO A PLANE-PARALLEL MEDIUM

A common utilisation of the transfer equation concerns a stratified medium, where all physical quantities are constant on infinite parallel planes. It is the usual approximation made for stellar atmospheres (cf. Fig. 4). Often interstellar clouds or diffuse nebulae are considered also as plane parallel media. The optical depth is then defined in the direction of the normal, and the transfer equation becomes, for a light ray which makes an angle θ with the normal to the planes:

$$\cos\theta \, \frac{dI_\nu}{d\tau_\nu} = I_\nu - S_\nu, \tag{12}$$

The formal solution of this equation is:

$$\left[I_\nu \exp\left(-\frac{\tau_\nu}{\mu}\right)\right]_{\tau_{\nu 1}}^{\tau_{\nu 2}} = -\int_{\tau_{\nu 1}}^{\tau_{\nu 2}} S_\nu \exp\left(-\frac{\tau_\nu}{\mu}\right) \frac{d\tau_\nu}{\mu} \tag{13}$$

where $\mu = \cos\theta$.

We can apply this solution to different cases, according to the boundary conditions.

2.3.1. *for a stellar atmosphere*

In this case the boundary conditions are:

- no radiation incident on the surface, or $I_\nu(\mu < 0,\ \tau_\nu = 0) = 0$, i.e.:

$$I_\nu(\mu < 0, \tau_\nu) = \int_0^{\tau_\nu} \frac{S_\nu}{|\mu|} \exp\left(-\frac{(t-\tau_\nu)}{\mu}\right) dt, \tag{14}$$

- the radiation at $\tau_\nu = \infty$ remains finite, i.e.:

$$I_\nu(\mu > 0, \tau_\nu) = \int_{\tau_\nu}^{0} \frac{S_\nu}{|\mu|} \exp\left(-\frac{(t-\tau_\nu)}{\mu}\right) dt\ , \tag{15}$$

and the intensity at the surface is:

$$I_\nu(\tau_\nu = 0) = \int_0^{\infty} S_\nu \exp\left(-\frac{\tau_\nu}{\mu}\right) \frac{d\tau_\nu}{\mu}. \tag{16}$$

One gets also the flux emerging from the surface:

$$\begin{aligned} F_\nu(\mu > 0) &= 2\pi \int_0^1 \mu\, d\mu \int_0^{\infty} S_\nu \exp\left(-\frac{\tau_\nu}{\mu}\right) \frac{d\tau_\nu}{\mu} \\ &= 2\pi \int_0^{\infty} S_\nu\, \mathcal{E}_2(\tau_\nu) d\tau_\nu\,, \end{aligned} \tag{17}$$

where $\mathcal{E}_n(x)$ is the order n integro-exponential: $\mathcal{E}_n(x) = \int_0^\infty exp(-ux)u^{-n}du$.

Although Eqs. 16 and 17 are only formal solutions which require to know the variation of S_ν as a function of depth at all frequencies, they can be of some help to understand intuitively two observations.

First we see that the intensity is approximately equal to the source function at $\tau_\nu = \mu$, and that the internal layers below do not contribute to the radiation. It explains the limb darkening effect at the surface of the Sun. According to the Eddington-Barbier relation, the source function in the photosphere is proportional to τ, so it decreases towards the surface (but not necessarily in the chromosphere, cf. P. Heinzel's lectures). Thus when we observe the limb ($\mu \to 0$), we are seeing layers close to the "surface" where S_ν is small, so I_ν is also small. When we observe the center of the Sun ($\mu = 1$), we are seeing deeper layers where S_ν is large, and I_ν is large too. So the center of the disk will appear brighter than the limb.

Second, in the case of stars, we do not observe the intensity, but the flux, which is approximately equal to the source function at $\tau_\nu = 1$. Assuming again that S_ν decreases with height in the photosphere, we can understand why we see absorption lines in the stellar spectrum. The absorption coefficient χ_ν in a line is larger than in the surrounding continuum, so we are seeing in a line the layers close to surface, and in the surrounding continuum the deeper layers: the lines are then in absorption. The effect is inverted if the source function increases with the height, as it is the case for lines formed in the chromosphere, or in extended envelopes of stars: the lines are then in emission.

2.3.2. *for a homogeneous slab of finite thickness*

One defines an optical thickness increasing towards the observer, $d\tau_\nu = \chi_\nu\, d\ell$. Since the medium is assumed homogeneous, the optical thickness of the slab is $T_\nu = \int \chi_\nu\, d\ell = \chi_\nu\, H$ in the direction of the normal, where H is the geometrical thickness.

The boundary conditions are now:

- incident intensity $I_\nu(\tau_\nu = 0) = I_{\nu 0}$,
- no incident intensity at $\tau_\nu = T_\nu$.

The solution of the transfer equation is thus (for $S_\nu =$ const.):

$$I_\nu(T_\nu, \mu) = I_{\nu 0}\ \exp\left(-\frac{T_\nu}{\mu}\right) + S_\nu \left[1 - \exp\left(-\frac{T_\nu}{\mu}\right)\right] . \qquad (18)$$

If $T_\nu < 1$, the slab is called "optically thin"; if $T_\nu > 1$ it is called "optically thick".

There are three interesting cases:

- 1. a non emissive layer (a cold cloud in front of an intense source):

$$I_\nu(T_\nu, \mu) = I_{\nu 0}\ \exp\left(-\frac{T_\nu}{\mu}\right) \qquad (19)$$

– 2. an optically thin layer with $T_\nu \ll 1$ (assuming that μ is not $\ll 1$):

$$\begin{aligned} I_\nu(T_\nu,\mu) &= I_{\nu 0}\ \exp\left(-\frac{T_\nu}{\mu}\right) + \frac{S_\nu\, T_\nu}{\mu} \\ &= I_{\nu 0}\ \exp\left(-\frac{T_\nu}{\mu}\right) + \frac{H\eta_\nu}{\mu} \end{aligned} \tag{20}$$

– 3. an optically thick layer with $T_\nu \gg 1$:

$$I_\nu(T_\nu,\mu) \sim S_\nu. \tag{21}$$

For a finite slab with no incident radiation one can show that $F_\nu = 2\pi S_\nu[0.5 - \mathcal{E}_3(T_\nu)]$, and deduce that:

- 1. in the optically thin case: $F_\nu = 2\pi S_\nu T_\nu = 2\pi\eta_\nu H$, or: $L_\nu = 4\pi\eta_\nu\times$ Volume (where L_ν is the total power emitted by the slab),
- 2. in the optically thick case: $F_\nu = \pi S_\nu$, or $L_\nu = \pi S_\nu\times$ Surface.

To summarize, in the optically thin case, one "sees" the emissivity, and the power is proportional to the volume. In the optically thick case, one "sees" the source function, and the power is proportional to the surface.

Finally note that in a purely diffusing medium, the same equations hold, replacing S_ν by J_ν, and τ_ν by the diffusion coefficient. So one gets in particular for an optically thin medium (i.e. $T_\nu^{dif} \ll 1$) with no incident radiation on the line of sight:

$$I_\nu = \frac{J_\nu T_\nu^{dif}}{\mu}\ . \tag{22}$$

This equation applies to the continuum of the solar corona in the visible range (the K corona). For a medium which is absorbing and diffusing, and optically thin both for diffusion and for absorption, but where the diffusion coefficient is negligible compared to the absorption coefficient, the solution of the transfer equation becomes:

$$I_\nu = \frac{J_\nu T_\nu^{dif}\ + H\eta_\nu}{\mu}\ . \tag{23}$$

This equation applies to the visible lines emitted by the solar corona (cf. later)

3. Local Thermodynamical Equilibrium (LTE)

3.1. RECALLING THE LAWS OF THERMODYNAMICAL EQUILIBRIUM (TE)

The *thermodynamical equilibrium* is the stationary state of an ensemble of interacting particles and photons which should be achieved in an infinitely thick medium (called a "Black Body") after an infinite time. Photons and particles have then the most probable energy distribution, which corresponds to microreversibility of all processes. For instance, there are as many radiative (resp. collisional) excitations from the level A to the level B of an atom, as radiative (resp. collisional) deexcitations from the level B to the level A per unit time.

3.1.1. *Energy distribution of photons: the Planck law*
It writes:

$$I_\nu \equiv B_\nu = \frac{2h\nu^3}{c^2}\left[\exp\left(\frac{h\nu}{kT}\right) - 1\right]^{-1}, \tag{24}$$

- h : Planck constant = 6.6262 10^{-27} erg s,
- k : Boltzmann constant = 1.3806 10^{-16} erg K^{-1}
- T : temperature.

Caution: it can also be expressed in units of wavelength, and it writes then:

$$B_\lambda = \frac{2hc^2}{\lambda^5}\left[\exp\left(\frac{hc}{\lambda kT}\right) - 1\right]^{-1}, \tag{25}$$

which has different shape and position of the maximum.

The integration over ν or λ gives:

$$B = \int B_\nu \, d\nu = \frac{\sigma T^4}{\pi}, \tag{26}$$

where σ is the Stefan constant = 5.6696 10^{-5} erg cm^{-2} s^{-1} K^{-4}.

There are two limiting cases :

- for $h\nu \ll kT$, the Rayleigh-Jeans law (used in radio-astronomy):

$$B_\nu = \frac{2h\nu^3}{c^2}\frac{kT}{h\nu} = \frac{2kT}{\lambda^2}; \tag{27}$$

- for $h\nu \gg kT$, the Wien law (used in the X-ray range):

$$B_\nu = \frac{2h\nu^3}{c^2}\exp\left(-\frac{h\nu}{kT}\right). \tag{28}$$

3.1.2. *Energy distribution of particles in non quantified levels: the Maxwell law*
It gives the number of particles per unit volume whose velocity projected on an axis z is between v_z and $v_z + dv_z$:

$$dN_z = N\left(\frac{M}{2\pi kT}\right)^{1/2}\exp\left(-\frac{Mv_z^2}{2kT}\right) dv_z\ , \tag{29}$$

where N is the number of particles per unit volume, and M is the mass of the particles. This expression is used e.g. to compute the profile of spectral lines broadened by thermal Doppler effect (cf. later).

Since at TE the velocity distribution is isotropic, the number of particles per unit volume whose absolute velocity is between $v = (v_x^2 + v_y^2 + v_z^2)^{1/2}$ and $v + dv$ is equal to:

$$\begin{aligned} dN &= N\left(\frac{M}{2\pi kT}\right)^{3/2} \exp\left(-\frac{M(v_x^2+v_y^2+v_z^2)}{2kT}\right) dv_x\, dv_y\, dv_z \\ &= N\left(\frac{M}{2\pi kT}\right)^{3/2} \exp\left(-\frac{Mv^2}{2kT}\right) 4\pi v^2\, dv\ , \end{aligned} \tag{30}$$

and the number of particles per unit volume with energy between $E = Mv^2/2$ and $E + dE$ is:

$$dN = N\left(\frac{1}{\pi kT}\right)^{3/2} \exp\left(-\frac{E}{kT}\right) 2\pi\, E^{1/2}\, dE\ . \tag{31}$$

This expression is used for instance to compute the rate of collisional excitations (cf. later).

Note that the Maxwellian distribution is normalized ($\int dN = N$).

3.1.3. *Energy distribution of particles in quantified levels: the Boltzmann law*

Consider two levels of energy E_m and E_n ($E_m > E_n$), with $E_m - E_n = E_{nm} = h\nu_{nm}$, where ν_{nm} is the frequency of the corresponding atomic transition. The number of particles per unit volume in each level (called the "level population") is given by:

$$\frac{N_m}{N_n} = \frac{g_m}{g_n} \exp\left(-\frac{E_{nm}}{kT}\right) \tag{32}$$

where g_m and g_n are the "statistical weights" of the levels m and n (they are given by spectroscopic tables, as well as the energies of the levels).

3.1.4. *Distribution of particles in different ionization states: the Saha law*

It gives the distribution of particles of a given species in a given ionization state:

$$\frac{N_{i+1}^1 N_e}{N_i^1} = \frac{2g_{i+1}^1}{g_i^1}\left(\frac{2\pi mkT}{h^2}\right)^{3/2} \exp\left(-\frac{\chi_i}{kT}\right)\ ; \tag{33}$$

– N_{i+1}^1 and N_i^1 are the numbers of ions on the fundamental level of states ionized $i+1$ and i times, per unit volume,

- N_e is number of electrons, per unit volume,
- g^1_{i+1} and g^1_i are the statistical weights of the fundamental level of states $i+1$ and i,
- χ_i is the ionization potential of state i (i.e. the energy needed to extract an electron from the fundamental level),
- m is the electron mass (9.1096 10^{-28} g).

Using Boltzmann law, one gets:

$$\frac{N_{i+1}N_e}{N_i} = \frac{2U_{i+1}}{U_i}\left(\frac{2\pi mkT}{h^2}\right)^{3/2} \exp\left(-\frac{\chi_i}{kT}\right) \; ; \tag{34}$$

- N_i and N_{i+1} are the total numbers of ions in states ionized i and $i+1$ times, per unit volume,
- U_i and U_{i+1} are called the "partition functions":

$$U_i(T) = \sum g_m \exp\left(-\frac{E_{1m}}{kT}\right) \; . \tag{35}$$

For small T ($E_{1m}/kT >> 1$), $U_i(T) = g_1$. For large T, the summation must be made on a large number of levels (depending on the density).

One can show (cf. books on statistical mechanics) that *microreversibility is equivalent to Maxwell + Boltzmann + Saha equations.*

At TE the optical thickness is infinite and the intensity is isotropic, so:

$$B_\nu(T) = I_\nu(\tau_\nu \to \infty) = S_\nu \tag{36}$$

which writes:

$$\eta_\nu \;=\; \chi_\nu \; B_\nu(T). \tag{37}$$

It is the well known *Kirchhoff law*, which is used to compute the emissivity when the absorption coefficient is known (or inversely).

3.2. FROM TE TO LTE

At TE, I_ν is equal to $B_\nu(T)$ and since T is constant, $dI_\nu/d\tau_\nu = 0$ and there is no transfer of radiation.

Let us now consider a state where T *varies in the medium, but Maxwell, Boltzmann and Saha equations are satisfied locally* (cf. Fig. 5). In such a state *microreversibility is achieved.* It is called *Local Thermodynamical Equilibrium (LTE)* . We will show later that this state is reached if the density and/or the optical thickness are large. The source function is then given by the Planck law (but not I_ν, which varies in the medium and is the solution of the transfer equation).

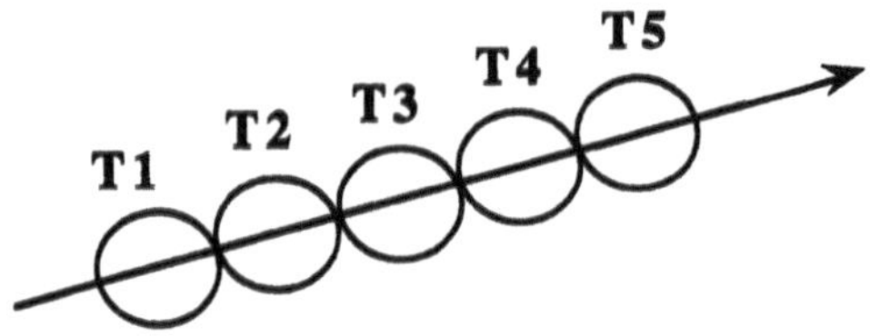

Figure 5. Local Thermodynamical Equilibrium.

4. Source function of a spectral line in non-LTE (NLTE)

A spectral line corresponds to the radiative transition between two bound levels of a given atom. If LTE is not reached, there is no microreversibility between the transitions but simply a stationary equilibrium in which the number of ALL processes populating a level is equal to the number of ALL processes depopulating this level. The main direct and inverse atomic processes populating and depopulating a bound level are:

- collisional excitations and deexcitations
- radiative excitations and deexcitations
- collisional ionizations and recombinations
- radiative ionizations and recombinations.

Several other processes can be important, such as dielectronic recombinations and auto ionizations, Auger processes, charge exchanges, but we will not consider them here.

4.1. RADIATIVE EXCITATIONS AND DEEXCITATIONS

Consider again the two bound levels m and n with populations N_m and N_n, energies $E_m > E_n$, $E_{nm} = E_m - E_n$.

4.1.1. *Einstein probabilities*

The Einstein probabilities A_{mn}, B_{mn} and B_{nm} are defined as follows:

- the number of radiative spontaneous transitions from level m to level n per unit volume per second is $N_m A_{mn}$. Each corresponds to the *emission* of a photon $h\nu = E_{nm}$.
- the number of radiative transitions from level n to level m per unit volume per second, corresponding to the *absorption* of a photon $h\nu = E_{nm}$, is $N_n J_\nu B_{nm}$.
- the number of radiative transitions from level m to level n, induced by a photon $h\nu = E_{nm}$, per unit volume per second and leading to the emission of a second photon $h\nu = E_{nm}$, is $N_m J_\nu B_{mn}$. It is called

"induced emission" (discovered by Einstein). Note that J_ν is the mean intensity at frequency $\nu = E_{nm}/h$.

One can determine the relations between these coefficients in TE. Microreversibility between radiative excitation and deexcitation gives:

$$N_m(A_{mn} + B_{mn}J_\nu) = N_n B_{nm} J_\nu \tag{38}$$

Boltzmann equation between levels m and n, and the Planck law $J_\nu \equiv B_\nu$ lead to:

$$J_\nu = \left(\frac{A_{mn}}{B_{mn}}\right) \frac{g_m B_{mn}}{g_n B_{nm}} \left[\exp\left(\frac{E_{nm}}{kT}\right) - 1\right]^{-1} \tag{39}$$

Identifying the terms depending and not depending on T (as these relations should be verified for any T), one gets:

$$g_n B_{nm} = g_m B_{mn}, \quad \text{and} \quad \frac{A_{mn}}{B_{mn}} = \frac{2h\nu^3}{c^2}. \tag{40}$$

These relations are equally valid in non-LTE, since A_{mn}, B_{mn} and B_{nm} are atomic parameters and do not depend on the physical conditions of the medium. The A_{mn} are given in spectroscopic tables (or equivalently the oscillator strength f_{nm}, cf. later). Note that a different definition of the Einstein probabilities is sometimes used, with the density of radiation instead of the intensity in Eq. 38.

4.1.2. *Relations of the Einstein probabilities with the absorption coefficient, the emissivity, and the source function*

Spectral lines are broadened by several processes, corresponding to an absorption profile $\Phi(\nu)$ and an emission profile $\Psi(\nu)$, normalized so that $\int \Phi(\nu)d\nu = 1$ and $\int \Psi(\nu)d\nu = 1$. The emissivity then writes:

$$\eta_\nu = \frac{h\nu}{4\pi} N_m A_{mn} \Psi(\nu), \tag{41}$$

and the absorption coefficient:

$$\chi_\nu = \frac{h\nu}{4\pi} N_n B_{nm} \Phi(\nu) \left(1 - \frac{N_m B_{mn}}{N_n B_{nm}}\right). \tag{42}$$

It contains a negative term of induced emission, $(N_m B_{mn})/(N_n B_{nm}) = (N_m g_n)/(N_n g_m)$, which can be larger than unity in the radio range (because the exponential term in the Boltzmann equation is close to unity). In this case the medium becomes an amplifier instead of an absorbant (MASER effect).

If $\Psi(\nu) = \Phi(\nu)$, one gets:

$$S = \frac{\eta_\nu}{\chi_\nu} = \left(\frac{N_m A_{mn}}{N_n B_{nm}}\right)\left(1 - \frac{N_m B_{mn}}{N_n B_{nm}}\right)^{-1} \tag{43}$$

and the source function is constant along the line profile.

The absorption and emission profiles are generally the convolution of a Gaussian and a Lorentzian function, the first being due to thermal and turbulent broadening, the latter due to radiative and collisional damping. Such a convolution is called "Voigt profile". The emission and absorption profiles are identical in the case of coherent scattering, or if on the contrary there is complete redistribution of frequencies in the absorption-reemission process. It is the case for a majority of lines, in particular for the coronal lines considered later in this chapter. Since they correspond to forbidden transitions, radiative damping is negligible, and since they are formed in a dilute medium, collisional damping is also negligible. So the emission/absorption profile reduces to a gaussian function. Let us rapidly recall the theory.

If an atom with a transition $h\nu_0 = E_{nm}$ has a velocity projected on the line of sight equal to v_z, it absorbs or emits photons propagating towards the observer, shifted by Doppler effect to $\nu = \nu_0(1 + v_z/c)$. For an assembly of atoms with a Maxwellian distribution of velocities, the number of atoms able to absorb or emit at a frequency ν, in a unit frequency interval, is therefore, according to Eq. 29:

$$\frac{dN_z}{dv_z}\frac{dv_z}{d\nu} = \frac{c}{\nu_0} N_n \left(\frac{M}{2\pi kT}\right)^{1/2} \exp\left(-\frac{Mv_z^2}{2kT}\right). \tag{44}$$

and one deduces for the absorption/emission profile:

$$\Phi(\nu) = \frac{1}{\sqrt{\pi}\Delta\nu_D} \exp\left[-\left(\frac{\nu - \nu_0}{\Delta\nu_D}\right)^2\right], \tag{45}$$

where

$$\Delta\nu_D = \frac{\nu_0}{c}\left(\frac{2kT}{M}\right)^{1/2} \tag{46}$$

is called the "Doppler width" of the line. If the gas has also a microturbulent velocity $V_{\rm turb}$, it should be taken into account in the Doppler width:

$$\Delta\nu_D = \frac{\nu_0}{c}\left(\frac{2kT}{M} + V_{\rm turb}^2\right)^{1/2} \tag{47}$$

as turbulence corresponds also to a Gaussian distribution of velocities.

The absorption coefficient at a frequency ν thus writes:

$$\chi_\nu = \frac{h\nu_0}{4\pi} N_n B_{nm} \frac{1}{\sqrt{\pi}\Delta\nu_D} \exp\left[-\left(\frac{\nu-\nu_0}{\Delta\nu_D}\right)^2\right]\left(1-\frac{N_m}{N_n}\frac{B_{mn}}{B_{nm}}\right) \tag{48}$$

and the absorption coefficient at the line center is:

$$\chi_{\nu_0} = \frac{h\nu_0}{4\pi} N_n B_{nm} \frac{1}{\sqrt{\pi}\Delta\nu_D}\left(1-\frac{N_m}{N_n}\frac{B_{mn}}{B_{nm}}\right) . \tag{49}$$

Instead of the Einstein probability, one often uses *the oscillator strength* f_{nm} which is related to B_{nm} by:

$$B_{nm} = \frac{\pi e^2}{mc} f_{nm} \frac{4\pi}{h\nu} \tag{50}$$

leading to:

$$\chi_{\nu_0} = N_n \frac{\pi e^2}{mc} f_{nm} \frac{1}{\sqrt{\pi}\Delta\nu_D}\left(1-\frac{N_m}{N_n}\frac{B_{mn}}{B_{nm}}\right) . \tag{51}$$

Numerically, $\pi e^2/mc = 0.027$ in CGS, $g_n f_{nm} = 1.5 \times 10^{-8} g_m A_{mn} \lambda^2$ (λ in micron, A_{mn} in s^{-1}), so f is of the order of unity for permitted lines, and $\ll 1$ for forbidden lines. As an example, for hydrogen:

$$f_{nm} = \frac{2^6}{3\sqrt{3}\pi}\frac{1}{g_n}\left(\frac{1}{n^2}-\frac{1}{m^2}\right)^{-3}\frac{1}{m^3}\frac{1}{n^3}\, g \tag{52}$$

where g is the so-called "Gaunt factor", of the order of unity.

4.2. COLLISIONAL EXCITATIONS AND DEEXCITATIONS

In passing close to an atom, a perturber P, having a kinetic energy $E_{\rm kin}$, can loose kinetic energy in exciting an atom from the bound level n to the bound level m. The kinetic energy $E'_{\rm kin}$ of the perturber after the collision is equal to $E_{\rm kin} - E_{nm}$. In the inverse process, the perturber gains the energy E_{nm} in deexciting the atom from the level m to the level n.

One defines the "collisional excitation rate" C_{nm} and the "collisional deexcitation rate" C_{mn} as follows:

- the number of collisional excitations per second per unit volume is $N_P\, N_n\, C_{nm}$
- the number of collisional deexcitations per second per unit volume is $N_P\, N_m\, C_{mn}$,

where N_P is the number density of the perturbers.

The perturbers are most often electrons, except in cold interstellar medium, where they can be hydrogen atoms or molecules. Let us call $\sigma_{mn}(v)$

(resp.. $\sigma_{nm}(v)$) the cross section for a collisional deexcitation (resp.. a collisional excitation), v being the relative velocity of the interacting particles. Generally the mean kinetic energy of different types of particles (in other words their temperature) is the same. The velocity of the electrons is thus much larger than the velocity of the (more massive) ions. If the perturbers are electrons, the relative velocity v reduces therefore to the velocity of the electrons. According to the definition of C_{mn} and C_{nm}, one gets:

$$C_{mn}\ (\text{respt.}C_{nm}) = < \sigma_{mn} v > \ (\text{respt.} < \sigma_{nm} v >)\ \text{s}^{-1}\,\text{cm}^3\,,$$

where the average is taken over the velocity distribution of the perturbers, i.e. the Maxwellian distribution of the electrons:

$$C_{mn}(T_e) = \left(\frac{m}{2\pi kT_e}\right)^{3/2} \int_0^\infty \sigma_{mn}(v) \exp\left(-\frac{mv^2}{2kT_e}\right) 4\pi v^3\, dv\ , \qquad (53)$$

$$C_{nm}(T_e) = \left(\frac{m}{2\pi kT_e}\right)^{3/2} \int_{E_{mn}}^\infty \sigma_{nm}(v) \exp\left(-\frac{mv^2}{2kT_e}\right) 4\pi v^3\, dv\ . \qquad (54)$$

where T_e is the electron temperature (generally equal to the ion temperature). One should note that for $C_{nm}(T_e)$ the integration is performed from E_{mn} to ∞, since the perturber must have a kinetic energy larger than E_{nm} to be able to excite the transition. The cross sections are determined experimentally and/or theoretically as functions of v, then integrated over the Maxwellian distribution, and $C_{mn}(T_e)$ is tabulated.

As for the radiative rates, one can determine the relations between C_{nm} and C_{mn} in TE:

- 1. microreversibility between collisional excitation and deexcitation gives:

$$N_n\ N_e\ C_{nm} = N_m\ N_e\ C_{mn}\ , \qquad (55)$$

- 2. Boltzmann equation between levels m and n gives therefore:

$$C_{nm} = \frac{g_m}{g_n}\ C_{mn} \exp\left(-\frac{E_{nm}}{kT_e}\right) \qquad (56)$$

where T_e is the electron temperature and N_e the electron number density. This relation is also valid in non-LTE.

Numerically one has:

$$C_{mn} = \frac{8.63\,10^{-6}}{T_e^{1/2}}\ \frac{\Omega_{mn}}{g_m}\ \ \text{s}^{-1}\,\text{cm}^3 \qquad (57)$$

where Ω_{mn} is a dimensionless "collision strength", given in different tables and articles.

4.3. PHOTOIONIZATIONS AND RADIATIVE RECOMBINATIONS

A photon with an energy $h\nu$ larger than the energy required to extract an electron from a bound level n, χ_n, can ionize an atom in the state n, ejecting an electron with a kinetic energy $E_{\rm kin}$. This is a "bound-free" process, because the energy of the ejected electron is not quantified, while the excitation between two bound levels is a "bound-bound" process. It is called *photoionization* and corresponds to the *absorption* of the photon $h\nu$. Photoionization can take place either from outer valence shells, or from inner shells, when the ionizing photons have X-ray energies. The inverse process, corresponding to the *emission* of a photon $h\nu$, is the *radiative recombination.* Finally, as for bound-bound processes, there is also stimulated recombination induced by an incident photon.

Note: we do not discuss here *dielectronic recombinations* and the inverse process, *autoionizations.* Autoionizations are generally unimportant.

One defines the rate of photoionization per unit volume per second from level n as $N_n\ P_n$, and the rate of radiative recombinations onto level n (which can include dielectronic recombinations) as $N_e N_{i+1}\alpha_n$. α_n is called the *recombination coefficient.*

The absorption coefficient corresponding to a photoionization from the level n can be computed from quantum mechanics, or determined experimentally. For hydrogenic ions it writes:

$$\begin{aligned} \chi_{\nu n} &= N_n \frac{2^4\, e^2 c^2 R^2 Z^4 g}{3^{3/2}\, mc} \frac{1}{n^5}\frac{1}{\nu^3}\left[1-\exp\left(-\frac{h\nu}{kT_e}\right)\right] \\ &= 2.815\, 10^{29}\, N_n Z^4\ g\frac{1}{n^5}\frac{1}{\nu^3}\left[1-\exp\left(-\frac{h\nu}{kT_e}\right)\right]\ \ {\rm cm}^{-1}\ , \qquad (58) \end{aligned}$$

where g is a mean "Gaunt factor", of order unity (for example it is equal 0.9 for photoionization from level 2). Note that the absorption coefficient is proportional to ν^{-3}. The dependence in ν is generally different for non hydrogenic ions. The factor $\left[1-\exp\left(-\frac{h\nu}{kT_e}\right)\right]$ takes into account induced recombinations (for the demonstration, see Mihalas, 1978).

$\chi_{\nu n}$ is equal to zero for $h\nu < \chi_n$, which is called the *photoionization edge.* One uses currently also $a_{\nu n} = \chi_{\nu n}/N_n$, called the *photoionization cross section.* Its value at the edge, $a_{\nu_0\, n}$, writes for hydrogenic ions (not taking into account induced recombinations):

$$a_{\nu_0\, n} = 7.906\, 10^{-18}\, \frac{ng}{Z^2}\ {\rm cm}^2\ . \qquad (59)$$

This coefficient is sometimes extended to non hydrogenic ions by defining an "effective" Z, $Z+s$, s being of the order of unity, but it is better to use results from quantum-mechanical computations now available in several databases, such as obtained by the OPACITY project or the CHIANTI database.

Using the Kirchhoff law $\eta_{\nu n} = \chi_{\nu n} B_\nu(T)$ and the laws of TE one can compute the emissivity corresponding to the recombination process. For hydrogen, it gives:

$$\eta_\nu = N_e N_{H^+} \frac{C}{T^{3/2}} \frac{g}{n^3} \frac{h}{4\pi} \exp\left(\frac{\chi_n - h\nu}{kT}\right) \tag{60}$$

where $C = 3.26\ 10^{-6}$ in CGS.

We can now write the photoionization rate from a level n

$$P_n = \int_{\nu_n}^{\infty} \chi_{\nu n} \left[1 - \exp\left(-\frac{h\nu}{kT_e}\right)\right] \frac{4\pi}{h\nu} J_\nu \, d\nu \tag{61}$$

which gives, for hydrogen,

$$P_n = 2.815\, 10^{29}\, N_n \frac{4\pi}{h} \frac{g}{n^5} \int_{\nu_n}^{\infty} \left[1 - \exp\left(-\frac{h\nu}{kT_e}\right)\right] \frac{J_\nu}{\nu^4} d\nu \ .\mathrm{cm}^{-3}\,\mathrm{s}^{-1}\ , \tag{62}$$

The recombination rate onto level n is:

$$N_e\ N_{i+1}\ \alpha_n \ = \frac{4\pi}{h} \int_{\nu_n}^{\infty} \frac{\eta_{\nu n}}{\nu} d\nu\ \mathrm{cm}^{-3}\,\mathrm{s}^{-1}\ , \tag{63}$$

which gives for hydrogen:

$$\alpha_n \ = \ \frac{C}{T^{3/2}} \frac{g}{n^3} \exp\left(\frac{\chi_n}{kT}\right)\ \mathcal{E}_1\left(\frac{\chi_n}{kT}\right) \tag{64}$$

where $\mathcal{E}_1$ is the order 1 integro-exponential. The total recombination coefficient is $\alpha_H = \sum \alpha_n$. As shown by Eq. 64 it depends only weakly on the temperature. It is roughly given by $\alpha_H = 2\,10^{-11} T^{-1/2}\, \Phi\, \mathrm{cm}^{-3}\mathrm{s}^{-1}$, where Φ varies from 1 to 3 for T varying from 10^5 to 10^3 K. These expressions are valid for hydrogenic ions with a multiplication by Z^2. For other ions, published tables give recombination coefficients as functions of the temperature, or absorption coefficients as functions of the photon energy.

4.4. COLLISIONAL IONIZATIONS AND RECOMBINATIONS

An electron can also be ejected from an atom in the energy state n, when another free electron is passing close to it (or another perturber, but electrons are generally more important, as for collisional excitations): this is called a *collisional ionization.* After the collision the first electron has lost an amount of kinetic energy equal to the kinetic energy of the second electron plus the potential energy χ_n. The conservation of energy writes therefore:

$$E_{\mathrm{kin},1} \ = \ E'_{\mathrm{kin},1} + E_{\mathrm{kin},2} + \chi_n, \tag{65}$$

where $E_{\text{kin},1}$ and $E'_{\text{kin},1}$ are the kinetic energies of the perturber before and after the collision, and $E_{\text{kin},2}$ is the kinetic energy of the ejected electron. In the inverse process, a free electron induces the recombination of a second electron on the state n: this process is called *collisional recombination*, or *three body recombination* (because three particles are interacting).

The rate of collisional ionizations from a level n, per unit volume and time, and the rate of collisional recombinations on level n, are written respectively $N_n N_e C_{nc}$ and $N_e^2 N_{i+1} C_{cn}$. Note that with this definition C_{nc} and C_{cn} do not have the same dimension! Like for collisional excitations, the ionization rate is equal to $< \sigma_{nc} v >$, where the average is taken on the velocity distribution integrated from the ionization edge, and $\sigma_{nc}(v)$ is a cross section either measured or computed. A general expression valid for a Maxwellian distribution and for $kT_e < \chi_i$ is:

$$C_{1n} \approx 10^{-8}\, \mathbf{o}\, T_e^{1/2}\, \chi_{i(eV)}^{-2} \exp\left(-\frac{\chi_i}{kT_e}\right) \tag{66}$$

where $\mathbf{o}$ is here the number of optical electrons and $\chi_{i(eV)}$ is expressed in eV. As usual one finds the relation between the direct and inverse coefficients using the TE laws:

$$C_{n1} = C_{1n} \left[\frac{N_i}{N_e N_{i+1}}\right]_{\text{Saha}} . \tag{67}$$

4.5. STATIONARY EQUILIBRIUM OF AN ATOM (STATISTICAL EQUILIBRIUM EQUATIONS)

For each level n, one can write an equation corresponding to the equality of all processes depopulating and populating the level n:

$$\begin{aligned} \textstyle\sum_{m\neq n} \quad & N_m \left(B_{mn} \int J_\nu \Phi_\nu d\nu + A_{mn} + N_e C_{mn}(T_e)\right) \\ & \qquad + N_e N_{i+1} \alpha_n(T_e) + N_e^2 N_{i+1} C_{cn}(T_e) \\ = \quad \textstyle\sum_{m\neq n} \quad & N_n \left(B_{nm} \int J_\nu \Phi_\nu d\nu + A_{nm} + N_e C_{nm}(T_e) + P_n + N_e C_{nc}(T_e)\right) \end{aligned} \tag{68}$$

where $A_{mn} = 0$ for $m < n$.

These equations can be simplified using the relations between the inverse processes shown above. This is generally done replacing also the level populations by their ratios to the LTE populations, b_n, defined as:

$$N_n = b_n N_{i+1} N_e \frac{g_n}{2} \left(\frac{2\pi m k T}{h^2}\right)^{-3/2} \exp\left(\frac{\chi_n}{kT}\right) , \tag{69}$$

where χ_n is the ionization potential from level n.

This set of equations forms a linear system of n equations and n unknown (the level populations), when T_e, N_e, N_{i+1} and $\int J_\nu \Phi_\nu \, d\nu$ are known. If the medium is relatively highly ionized, and since it is electrically neutral and made mainly of hydrogen, the electrons are provided mainly by hydrogen, and N_e is equal roughly to the number of hydrogen nuclei, so $N_e \approx 1.2 N_{H^+} \approx 1.2 N_H$ (the correction of 20% being due to the contribution from helium and heavy elements). If the medium is weakly ionized, the electrons are provided by elements having a low ionization potential, like C^+ and metals, and one cannot estimate a priori the value of N_e. We will see later how the N_{i+1} are determined. T_e is known through an energy balance equilibrium, and $\int J_\nu \Phi_\nu \, d\nu$ requires the solution of the transfer equation. So the problem is very complex, and involves the sophisticated numerical methods described in P. Heinzel's lectures.

To simplify the discussion we will assume that the atom is reduced to two bound levels n and m. This picture is correct for some forbidden lines, such as the HI 21 cm transition, for the forbidden coronal lines considered later, and generally when two levels in the ground configuration can be decoupled from the resonant lines and from the continuum, due to a large difference in potential energies. A most frequent case is also that of 3 levels in the ground configuration, which are decoupled from the other levels and from the continuum. For a two level atom the equilibrium equation reduces then to:

$$N_m \left(B_{mn} \int J_\nu \Phi_\nu \, d\nu + A_{mn} + N_e \, C_{mn}(T_e) \right) = N_n \left(B_{nm} \int J_\nu \Phi_\nu \, d\nu + N_e \, C_{nm}(T_e) \right) . \tag{70}$$

Using Eq. (43) for the source function, one gets:

$$S(\text{line}) = \frac{\int J_\nu \Phi_\nu \, d\nu + \epsilon \, B_\nu(T_e)}{1+\epsilon} \quad \text{with } \epsilon = \frac{N_e C_{nm}}{A_{mn}} \left[1 - \exp\left(-\frac{E_{nm}}{kT_e} \right) \right] . \tag{71}$$

$\int J_\nu \Phi_\nu \, d\nu$ is a *diffusion term*, as it does not correspond to any emission or absorption of radiation. From this expression we immediately see that *if the density is high, ϵ dominates and $S(line)=B_\nu(T_e)$. In other words, the transition is at LTE.* We can also note that LTE is reached more easily when $\int J_\nu \Phi_\nu \, d\nu$ is small (if the medium is optically thin, for instance). The expression of S is actually more complicated in general because $\Phi(\nu) \neq \Psi(\nu)$.

In the case of a multi-level atom, it is possible to write the source function as:

$$S(\text{line}) = \frac{\int J_\nu \Phi_\nu \, d\nu + \epsilon \, B_\nu(T_e) + \epsilon'}{1 + \epsilon + \epsilon''}, \tag{72}$$

where ϵ' and ϵ'' contain the other transitions, and to solve the system by iteration. This method was used in the past for a limited number of levels, but more sophisticated methods converging much more rapidly are used presently (cf. P. Heinzel's lectures).

4.6. REMARKS CONCERNING THE NOTION OF TEMPERATURES IN A NON-LTE MEDIUM

In LTE a unique temperature is sufficient to describe the medium. This is not the case in non-LTE, so it is common to use several temperatures, according to the process considered. The most important temperatures which are introduced are:

- The *"electron temperature"* T_e. It is the temperature given by the mean value of the kinetic energy, assuming a Maxwellian distribution; Eq. 31 gives thus:

$$T_e = \frac{m < v^2 >}{3k} \tag{73}$$

  where $< v^2 >$ is the root mean square velocity. This is the temperature intervening in the collisional rates C_{mn} and C_{nm}, C_{cn} and C_{nc}.
- The *"excitation temperature"* $T_{\rm exc}$, defined by:

$$\frac{N_m}{N_n} = \frac{g_m}{g_n} \exp\left(-\frac{E_{mn}}{kT_{\rm exc}}\right), \tag{74}$$

  which gives:

$$S_\nu = \left(\frac{N_m A_{mn}}{N_n B_{nm}}\right)\left(1 - \frac{N_m B_{mn}}{N_n B_{nm}}\right)^{-1} = B_\nu(T_{\rm exc}). \tag{75}$$

  According to this equation, the absorption coefficient becomes:

$$\chi_\nu = \frac{h\nu}{4\pi} N_n B_{nm} \Phi(\nu) \left[1 - \exp\left(-\frac{E_{mn}}{kT_{\rm exc}}\right)\right] . \tag{76}$$

  A negative excitation temperature correspond to a negative absorption coefficient, i.e. to a maser or a laser effect.
- The *"radiation temperature"* T_r, defined by:

$$I_\nu = B_\nu(T_r) . \tag{77}$$

  In the Rayleigh-Jeans regime, it leads to:

$$T_r = \frac{\lambda^2}{2k} I_\nu \tag{78}$$

and the solution of the transfer equation for a finite layer becomes:

$$T_r = T_r(0) \exp\left(-\frac{T_\nu}{\mu}\right) + T_{\text{exc}}\left[1 - \exp\left(-\frac{T_\nu}{\mu}\right)\right]. \qquad (79)$$

This equation is commonly used in the radio range.

- The *"effective temperature"* T_{eff}, defined by:

$$F = \int F_\nu \, d\nu = \pi B(T_{\text{eff}}) = \sigma T_{\text{eff}}^4 \,. \qquad (80)$$

- The *"color temperature"* T_c, defined by:

$$\frac{dI_\nu}{d\nu} = \frac{dB_\nu(T_c)}{d\nu}. \qquad (81)$$

For instance in a nebula ionized by a hot star, T_c is constant in the visible and UV range, and equal to the surface temperature of the star. Also the color temperature of the solar corona is that of the photosphere in the visible range (owing to diffusion).

5. Application to the solar corona

5.1. IONIZATION EQUILIBRIUM IN NON-LTE

To solve the statistical equilibrium equations we need to know the number of ions in a given ionization state. If for each level the rate of collisional ionizations is equal to the rate of collisional recombinations, Saha's law applies. One can show that it occurs for large values of the density ($N_e \geq 10^{16}$ cm^{-3}) which are not reached in the corona and in nebular media. These media are therefore in non-LTE for ionization equilibrium, which is given by the stationarity equation:

rate of (photoionizations + collisional ionizations) from all levels
=rate of (radiative + collisional recombinations) on all levels.

Actually the problem is simplified by the fact that in these media the populations of excited levels are very small, so only ionizations from the ground level should be taken into account.

Other simplifications take place in nebular media and in the corona.

a. Nebular media:

HII regions and planetary nebulae are dilute media photoionized by hot stars. Their kinetic or electron temperature is relatively low (10^4 K). One can show that in these conditions radiative ionizations dominate over collisional ionizations, and radiative recombinations dominate over collisional recombinations in the ionization equilibrium, which reduces to:

photoionizations from the ground level
= radiative recombinations on all levels.

The photoionization rate depends on the distance, on the temperature, and on the luminosity of the hot star. For more details on this subject, one can refer to Osterbrock's book (1989).

b. Solar corona:

In the corona T_e is of the order of a few 10^6 K, while T_r is only 5800 K. Therefore photoionizations are negligible with respect to collisional ionizations. On the other hand the density is relatively low ($N_e \leq 10^{10}$ cm^{-3}), and collisional recombinations are negligible with respect to radiative recombinations. The ionization equilibrium thus writes:

collisional ionizations from the ground level
= radiative recombinations on all levels

or:

$$N_e \ N_i \ C_{ic}(T_e) \ = N_e N_{i+1} \ \alpha_i(T) \tag{82}$$

where $\alpha_i(T_e) = \sum_{n=1}^{\infty} \alpha_n(T_e)$, and $C_{ic}(T_e)$ is the rate of ionizations from the ground level.

It can be written approximately:

$$\frac{N_{i+1}}{N_i} = 2 \ 10^4 \frac{P_i(T_e)}{n} \left(\frac{\chi_H}{\chi_i}\right)^2 \exp\left(-\frac{\chi_i}{kT_e}\right) \left(\frac{\chi_i}{kT_e}\right)^{-1} \tag{83}$$

where $P_i(T_e)$ is a tabulated function, of the order of unity, n is the principal quantum number, and χ_H, χ_i are the ionization potentials of hydrogen and of the ion i.

This equilibrium is called *coronal equilibrium* and holds not only in the solar corona, but also in the hot interstellar medium, in the hot gas of galaxy clusters, in supernovae remnants, etc. Its particularity is that $N_i / \sum N_i$ *is a function of only* T_e. As an example we show the ionization equilibrium of iron on Fig. 6. We see that for the kinetic temperature of the corona, $T_e \sim 10^6$ K, iron should be in the form of highly ionized species, Fe^{+9} to Fe^{+13} in particular. This figure shows also that the ionic fraction is a rapidly varying function of the temperature, implying that if it can be obtained from the observations, the corresponding temperature will be accurately determined.

5.2. THE VISIBLE CORONAL LINES

Since the beginning of the century many lines of unknown origin have been observed in the visible during eclipses. They were attributed to a new element "the coronium" (remember the "nebulium" of planetary nebulae!). Grotrian (1939) understood that *the intense red line at 6374Å is due to a forbidden transition in the ground configuration of FeX* (cf. Fig. 7).

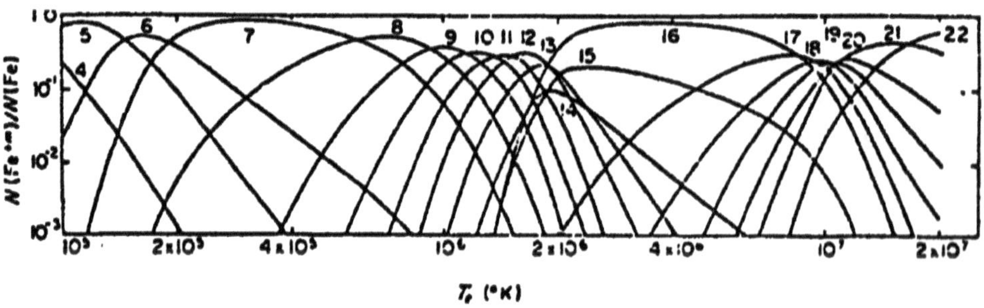

Figure 6. Coronal equilibrium of iron.

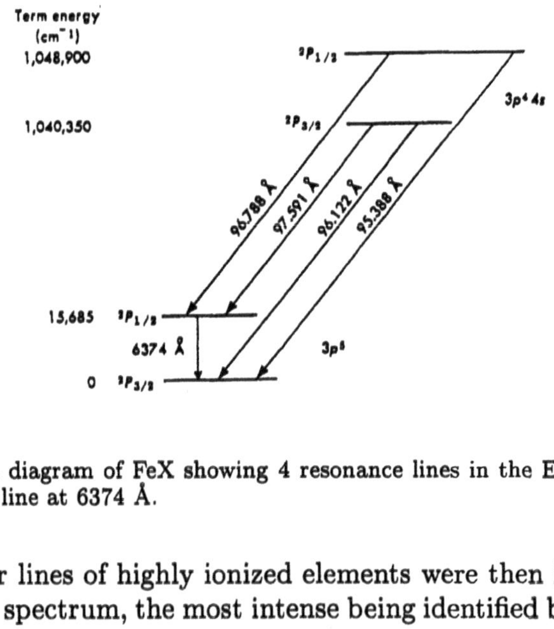

Figure 7. Term diagram of FeX showing 4 resonance lines in the EUV range and the forbidden visible line at 6374 Å.

Many other lines of highly ionized elements were then identified in the visible coronal spectrum, the most intense being identified by Edlen in 1942 to the forbidden transition in the 3s fundamental configuration of Fe^{+13}. These observations lead to the discovery of the high temperature of the corona. These lines are only visible on the limb, during eclipses, or using a coronagraph, as the disk is much too bright in the visible range (cf. Fig. 8).

It is not the same for extreme UV and X-ray lines of highly ionized atoms, which are emitted only by the corona, and can therefore be seen in emission on the disk itself because there is practically no underlying continuum in these bands. When the extreme ultraviolet and the X-ray bands were opened to observation with the launch of rockets and satellites, a number of permitted coronal lines due to resonant transitions were discovered. As an example Fig. 9 displays an image of the Sun obtained with EIT on board of SOHO in the FeX/XI lines at 171Å. It shows clearly strong vari-

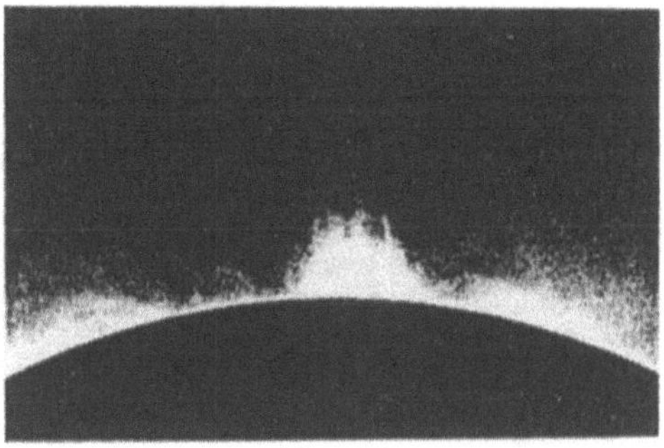

Figure 8. An eclipse seen in the forbidden FeXIV 5303Å line

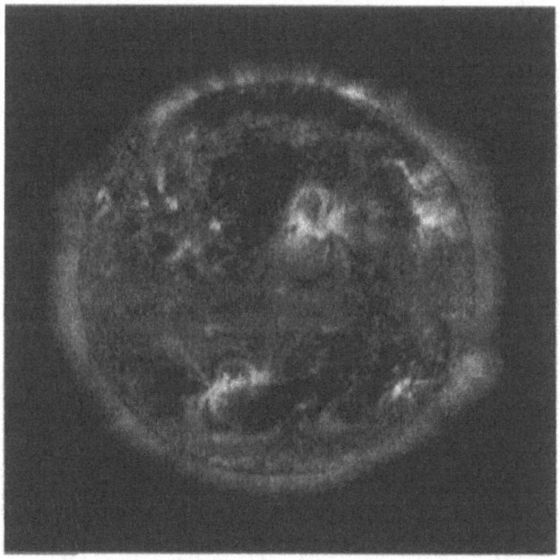

Figure 9. An image of the Sun obtained with EIT on board of SOHO in the FeX/XI lines at 171Å.

ations of the line intensity, due to variations of the temperature and of the density in the corona (cf. below).

But here we shall focus on forbidden visible lines, because they are more simple to handle than resonant lines, and as an example we will determine the intensity of the FeXIV 5303Å line emitted by the corona. The theory of these lines is established since decades, in particular with the pioneering work of Pottasch (1963).

The transition between levels 2 and 1 in the ground configuration of Fe^{+13} producing this line is represented on Fig. 10. One has first to check that it can be decoupled from the rest of the atom and from the continuum. This requires to show that the rates of radiative and collisional deexcitations from all upper levels onto level 2, and the rates of radiative and collisional

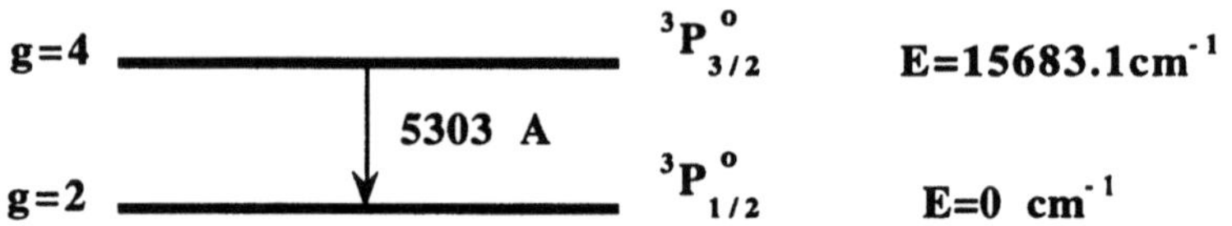

Figure 10. The transition FeXIV 5303 Å.

recombinations onto level 2, are negligible compared to $N_1 B_{12} \int J_\nu \Phi_\nu d\nu$ and $N_1 N_e C_{12}$ (populating level 2), and that the rates of photoionization and collisional ionization from level 2, and the rates of induced radiative and collisional excitations to all upper levels from level 2, are negligible compared to $N_2 B_{21} \int J_\nu \Phi_\nu d\nu$, $N_2 A_{21}$ and $N_2 N_e C_{21}$ (depopulating level 2). For instance collisional and induced radiative deexcitations from excited levels are not important because these levels are not highly populated, owing to their large energies, and collisional and radiative excitations to excited levels are negligible compared to excitations between the levels in the ground configuration, owing to the exponential term in the excitation rates. Collisional recombinations are negligible because of the low density. Actually this is not true for all visible coronal lines, and in several cases cascades from highly excited levels, are important in populating the upper level of the forbidden transition.

Let us consider thus a two-level atom in an ionized medium (i.e. perturbers are electrons), and compare the different terms of Eq. 70. The atomic data we need are: the Einstein coefficient A_{21} of the transition, equal to 60 s^{-1}, the statistical weights g_1 and g_2, respectively 2 and 4, and the "collision strength" Ω_{21} (cf. Eq. 57 and Fig. 10), of the order of unity (it decreases from 3 at $T = 10^5$ K to 0.3 at $T = 10^7$ K).

To compute the rate of radiative excitation, we need J_ν. We will not determine the optical thickness in the continuum at 5303 Å, as it would require to know the absorption coefficient in the continuum at this wavelength, due to free-free process that we have not discussed. But we know from observations that the corona is optically thin in the optical band, as we see the underlying disk without any absorption. We infer that the whole corona is illuminated by the photospheric radiation field, close to a black body of temperature $T_0 = 5800$K. Since the line is narrow ($\Delta\nu_D \ll \nu_0$), J_ν is approximately constant over the line profile. The corona being illuminated only from one side, one gets:

$$\int J_\nu \Phi_\nu d\nu = J_\nu = B_\nu(T_r) = \frac{B_\nu(T_0)}{2} \left(\frac{R_{\rm sun}}{z + R_{\rm sun}} \right)^2 \tag{84}$$

where $R_{\rm sun}$ is the radius of the sun and z the height in the corona, and one deduces the radiative excitation rate (using the relations between the

Einstein coefficients):

$$N_1 B_{12} J_\nu = N_1\ A_{21} \left[\exp\left(\frac{E_{12}}{kT_0}\right) - 1\right]^{-1} \left(\frac{R_{\rm sun}}{z + R_{\rm sun}}\right)^2 . \qquad (85)$$

For the basis of the corona it is equal to $0.9N_1$ s^{-1} cm^{-3}, and the induced deexcitation rate, $N_2 \frac{B_{12}}{2} J_\nu$, is equal to $0.45N_2$ s^{-1} cm^{-3}, therefore negligible with respect to the spontaneous deexcitation rate.

The exponential term in the collisional excitation rate $\exp\left(-\frac{E_{12}}{kT_e}\right)$ is roughly equal to unity, as $E_{12} \ll kT_e$, so $g_1 C_{12} = g_2 C_{21}$. According to Eq. 57 one gets thus for the rate of collisional deexcitation $N_2 N_e C_{21} \sim 2 \times 10^{-9} N_2 N_e$ s^{-1} cm^{-3} for $T_e = 10^6$ K, and the dependence on T_e is small. Comparing this with the radiative deexcitation rate $N_2 A_{21}$, we see that above a critical density $N_e \sim 10^{11}$ cm^{-3}, deexcitation would be dominated by collisional processes. It is a very high density, and one can admit that *deexcitation is always dominated by radiative spontaneous transitions.* For the excitation the critical density is much smaller, $\sim 10^9$ cm^{-3} at the basis of the corona, and it decreases with height, as the radiative rate is proportional to $R_{\rm sun}^2/(z + R_{\rm sun})^2$. So *excitation is sometimes dominated by radiative transitions, sometimes by collisional transitions.*

It is important to note here that we have used 3 different temperatures, namely the temperature of the underlying blackbody T_0, the radiation temperature T_r, and the electron temperature T_e. All are different from the excitation temperature T_{exc} (cf. below).

As a rough approximation we can consider the corona outside the disk as a finite slab of thickness $2R_{\rm sun}\sqrt{1 - \left(1 - \frac{H}{R_{\rm sun}}\right)^2} \sim 2\sqrt{HR_{\rm sun}}$, where H is the scale height of the corona ($< R_{\rm sun}$), with no incident intensity on the line of sight. The source function in the line is given by Eq. 71. But we do not need this source function to determine the line flux, as we can check that the corona is optically thin at the center of the line (and a fortiori in the wings). Eqs. 46, 50, 51 and the relations between the Einstein coefficients, give: $\chi_0 = 2.310^{-19} N_1$. N_1 can be written: $\frac{N_1}{N({\rm Fe}^{+13})} \frac{N({\rm Fe}^{+13})}{N({\rm Fe})} \frac{N({\rm Fe})}{N_H} \frac{N_H}{N_e} N_e$, where N_H is the number of hydrogen nuclei per unit volume. The iron abundance, $\frac{N({\rm Fe})}{N_H}$, is equal to $4\ 10^{-5}$, and the ratio $\frac{N_H}{N_e}$ is equal to 0.83 in the corona (owing to electrons from helium and heavy elements). Let us make the conservative assumption that iron is entirely in the form of $\rm Fe^{+13}$, and that N_1 is much larger than the populations of all the other levels, i.e. $N_1 = N({\rm Fe}^{+13})$. Both assumptions are reasonable in the region emitting the FeXIV line, actually. One finds then $\chi_0 \sim 10^{-23} N_e$, or $\tau_0 \sim 10^{-23} \mathcal{N}_e$, where $\mathcal{N}_e$ is the number of electrons on the line of sight per unit surface, in the emission region. Assuming that the whole corona is emitting the line (which is clearly an overestimation) one gets $\mathcal{N}_e =< N_e > 2\sqrt{HR_{sun}}$, where $< N_e >$ is the density averaged on the scale height, $N_e \sim 5\ 10^8$ cm^{-3}, and finally one gets for $H \sim 5\ 10^9$ cm: $\tau_0 \leq 10^{-4} \ll 1$.

At this stage it is also possible to show that the diffusion coefficient neglected in all this chapter is indeed negligible compared to the absorption coefficient in the line. Since diffusion is due to Thomson scattering, the diffusion coefficient is equal to $\sigma_T N_e = 0.66\ 10^{-24} N_e$, and we immediately see that it is smaller than χ_0. However, it is not possible to neglect the emissivity due to diffusion in the continuum, as it is the only emission process for the continuum underlying and surrounding the line. According to Eq. 22 for an optically thin purely diffusing medium, $I_\nu^{dif} = \tau_T J_\nu \sim \tau_T B_\nu(T_0)/2 \sim 0.3\ 10^{-24} \mathcal{N}_e B_\nu(T_0) \sim 3 10^{-6} B_\nu(T_0)$. This is indeed the intensity of the visible continuum observed at the basis of the corona. And according to Eq. 23, the intensity in the line should simply be added to the continuum intensity, since the absorption coefficient dominates on the diffusion one. So finally we see that the computation of the line intensity can be completely decoupled from the diffusion process.

The intensity in the line is therefore equal to $\int \eta_\nu(z)dz$ (as the emissivity is strongly dependent on the position, one cannot use the homogeneous approximation). It is proportional to the emissivity at the line center and to the emissivity integrated on the line profile (cf. Eqs. 41 and 45), which writes, using Eq. 43:

$$\eta_{line} = \frac{E_{12}}{4\pi} N_2 A_{21} = \frac{E_{12}}{4\pi} N_1 (N_e C_{12} + B_{12} J_\nu) \tag{86}$$

N_2 is always smaller than N_1, since when excitation is dominated by collision, $N_2/N_1 = N_e C_{12}/A_{21} < 1$, and when it is dominated by radiation $N_2/N_1 = B_{12} J_\nu / A_{21} \sim \exp(-E_{12}/kT_0) < 1$. (Note that T_{exc} which is defined either by $N_e C_{12}/A_{21} = 2\exp(-E_{12}/kT_{exc})$, or by $B_{12} J_\nu / A_{21} = 2\exp(-E_{12}/kT_{exc})$, is different from the previously used temperatures.) Thus $N_1 \sim N(\mathrm{Fe}^{+13}) \sim 4\ 10^{-5} N_e [N(\mathrm{Fe}^{+13})/N(\mathrm{Fe})]$, as assumed before.

Eq. 86 gives thus, in the two extreme cases of N_e much larger and much smaller than the critical density of 10^9 cm^{-3}:

- $N_e > 10^9$ cm^{-3}: $\eta_\nu \sim \frac{E_{12}}{4\pi} C_{12}\ N_e^2 [N(\mathrm{Fe}^{+13})/N(\mathrm{Fe})]$
- $N_e < 10^9$ cm^{-3}: $\eta_\nu \sim \frac{E_{12}}{4\pi} B_{12} J_\nu\ N_e [N(\mathrm{Fe}^{+13})/N(\mathrm{Fe})]$.

The first case corresponds to dense coronal loops, and the second one is close to the quiet corona.

Let us assume that the density is larger than the critical one (actually it is often the case for other forbidden coronal lines). The intensity writes thus:

$$I_{line} \sim 4\ 10^{-5} \frac{E_{12}}{4\pi} \int N_e^2\ C_{12}(T_e) \frac{N(\mathrm{Fe}^{+13})}{N(\mathrm{Fe})} dz. \tag{87}$$

The intensity, and consequently the flux measured at the Earth, is therefore a function of the density and of the temperature in the region where iron is in the form of Fe^{+13}. But $N(Fe^{+13})/N(Fe)$ depends so strongly on the temperature, that one considers it equal to unity in a small range of temperature, and zero outside. Moreover $C_{12}(T_e)$ is not strongly dependent on T_e. One can thus approximate the intensity by

$$I_{line} \sim 4\ 10^{-5}\ \frac{E_{12}}{4\pi}\ C_{12} \int E(T_e)dT_e \tag{88}$$

where $\int E(T_e)dT_e = \int N_e^2 dz$ is called the *emission measure.*

Therefore, measuring other visible forbidden coronal lines corresponding to different ions should allow to map the emission measure as a function of the temperature. More sophisticated methods involving inversion procedures can also be used to avoid the rough approximations made above.

We have focussed here on a forbidden line, but it is clear that the same kind of computation can be performed for other types of lines, such as resonance UV or X-ray lines. The equations giving the population of the upper level of the transition would not necessarily have been as simple as in the particular case considered here, and consequently the dependence of the emissivity on the temperature and on the density would have been more complicated, but the whole procedure would have been similar. Then, by comparing forbidden lines with resonance lines of the same ion in the extreme UV and in the X-ray, or more generally pairs of lines of the same ion, one can get powerful diagnostics of the temperature and/or of the density in the region where this ion is abundant. Besides, the EUV and X-ray lines are observed also on the disk, where the forbidden visible lines cannot be observed. For instance SiIX and SiX lines which are density sensitive have been observed with the coronal diagnostic spectrometer (CDS) on board SOHO and have allowed to obtain the density and the temperature in active regions on the limb and on the disk as a function of the radius and the position angle. One can guess that many other important results will be obtained from the data gathered in this mission.

Finally, one should not forget that the simple study of the line profiles using lines of different elements allows to determine both the kinetic temperature and the turbulent velocity, as shown also in this chapter.

Acknowledgements

I wish to thank Petr Heinzel for kindly reading the manuscript, which benefitted from the improvements he suggested.

References

1. Mihalas, D. (1978) "Stellar Atmospheres", Ed. Freeman and Co.
2. Osterbrock, D.E. (1989) "Astrophysics of gaseous nebulae and Active Galactic Nuclei", Ed. University Science Book.
3. Pottasch, S.R. (1963) *ApJ* **137**, 945.
4. Zirin, H. (1988) "Astrophysics of the Sun", Ed. Cambridge University Press.

MODELS OF THE SOLAR ATMOSPHERE

P. HEINZEL
Astronomical Institute
Academy of Sciences of the Czech Republic
CZ-25165 Ondřejov, Czech Republic

Motto:

My views have been colored by my long period of residence at the High Altitude Observatory in Boulder, where I was confronted daily by the ghastly reality of the Sun as seen at high time, spatial, and spectral resolution in a wide range of spectral bands. There is probably no more sobering an experience for a stellar atmospheres modeller than a detailed inspection of solar data !

(D. Mihalas, in 'Stellar Atmospheres: Beyond Classical Models' [1])

1. Introduction

Among the atmospheres of different stars, the *solar atmosphere* has a privilege to provide us an insight into various structural patterns and their dynamical behaviour. While most of stellar atmospheres are simply modelled as one-dimensional (1D) plane-parallel or spherically-symmetrical layer, the solar atmosphere must generally be resolved into various brightness components which also exhibit quite different behaviour. Therefore, a single solar atmospheric model cannot in general be treated as horizontally homogeneous, although historically this was the first step in modelling. While many stars have just the photospheres, the Sun - like some other stars - possesses also the chromosphere and the corona. The coronal temperatures reach millions degrees within very narrow chromosphere-corona transition region (CCTR) where steep temperature gradients allow an efficient heat conduction down to the lower layers of CCTR. The presence of these external layers largely complicates the solar atmospheric modelling at least for two reasons. First, due to their low density, strong departures from local thermodynamic equilibrium (LTE) exist. This is the principal problem which requires much more complicated so-called NLTE (or non-LTE) treatment of the radiative transfer. In fact, the solar atmosphere served

J.-P. Zahn and M. Stavinschi (eds.),
Advances in Solar Research at Eclipses from Ground and from Space, 201–220.

as a 'test laboratory' for a development of basic concepts of NLTE theory in late fifties and early sixties (see [2]). A second difficulty lies in the fact that the presence of the chromosphere and corona implies, by definition, existence of an extra heating (standard photospheres are supposed to be in radiative equilibrium). The amount of this nonradiative heating can be estimated, but the mechanisms responsible for it are still subject of an extensive debate.

In this review we will be concerned only with the so-called 'quiet solar atmosphere', i.e. we ignore all active structures and phenomena (sunspots, flares, prominences etc.). However, we shall mention the fine structure of the atmosphere and show that on such spatial scales it is in fact highly dynamical. This was clearly demonstrated namely within last few years during which the SOHO satellite has provided us with a large amount of high-resolution spectroscopic data in the ultraviolet (UV).

There are two basic approaches to solar atmospheric modelling, or more precisely to chromospheric and CCTR modelling. One assumes that the whole atmosphere or its particular brightness components can be represented by 1D plane-parallel layers extending from the base of the photosphere up to the CCTR (note that various 2D *periodic media* were also considered [3]). Another way of studying chromospheric fine structure is to treat individual *isolated structures* (loops, blobs, mottles etc.) as so-called *clouds* and to apply cloud-model methods as described by P. Mein in this volume. In this review we shall discuss the first approach, but one has to keep in mind that both these techniques are complementary and ultimately, in the future, they should be unified in very complex 3D models.

A vast literature exists on various aspects of the solar-atmospheric modelling and we provide the reader with some most important references. It was not our intention here to give an exhaustive review, but rather to formulate the basic principles of the modelling and to show the main results of both the classical as well as most advanced studies. Since we are primarily interested in a standard spectroscopic diagnostics and energy-balance aspects, we completely ignore the 'second solar spectrum' (O. Stenflo), i.e. the polarized light [4].

2. Basic Equations of Radiation Hydrodynamics (RHD)

A standard set of equations describing the structure, dynamics and radiation properties of the solar atmosphere consists of two parts: hydrodynamics (including the energy balance) plus the NLTE radiative transfer. These equations are fully time-dependent and thus capable of describing dynamical phenomena like atmospheric oscillations or flows (the reader should consult the comprehensive textbook of Mihalas and Mihalas [5]). However, most 'standard' solar atmospheric models are largely simplified by assuming static or stationary, 1D plane-parallel layers which are just graviationally stratified. We will see this later on as special cases of a general formulation.

2.1. EQUATIONS OF HYDRODYNAMICS

The following three equations describe the hydrodynamics and energy balance of a 1D plane-parallel atmosphere (in Eulerian coordinates):

Equation of continuity

$$\frac{\partial \rho}{\partial t} + \frac{\partial (\rho v)}{\partial z} = 0, \tag{1}$$

equation of momentum balance

$$\frac{\partial (\rho v)}{\partial t} + \frac{\partial (\rho v^2)}{\partial z} = -\frac{\partial p}{\partial z} - \rho g, \tag{2}$$

and equation of energy balance

$$\frac{\partial (\rho e)}{\partial t} + \frac{\partial (\rho v e)}{\partial z} + p\frac{\partial v}{\partial z} = -\frac{\partial}{\partial z}(F_c + F_r + F_h). \tag{3}$$

In these equations, v is the flow velocity, ρ is the gas density, p the total pressure, g the gravitational acceleration (on surface), e is the internal energy (thermal and ionization plus excitation energy). The fluxes F_c, F_r and F_h correspond to the conductive flux, radiative flux and any other heating flux, respectively. Density ρ, gas pressure p_g and temperature T are related through the equation of state

$$p_g = p_g(\rho, T). \tag{4}$$

2.2. NLTE EQUATIONS

The following set of equations covers the interaction of plasma with the radiation field, which has typically a non-local character (the basic principles of NLTE physics are discussed by S. Collin-Zahn in this volume)

The equation of radiative transfer in a 1D plane-parallel atmosphere (with vertically increasing coordinate z) has the form

$$\mu \frac{dI_{\nu\mu}}{dz} = -\chi_{\nu\mu} I_{\nu\mu} + \eta_{\nu\mu}, \tag{5}$$

where ν is the frequency, μ the directional cosine, $\chi_{\nu\mu}$ and $\eta_{\nu\mu}$ are the absorption and emission coefficients, respectively. The source function is then defined as $S_{\nu\mu}$

$$S_{\nu\mu} \equiv \frac{\eta_{\nu\mu}}{\chi_{\nu\mu}}. \tag{6}$$

For an isotropic opacity the optical depth can be defined as $d\tau_\nu \equiv -\chi_\nu dz$ and the transfer equation written in a simple standard form

$$\mu \frac{dI_{\nu\mu}}{d\tau_\nu} = I_{\nu\mu} - S_{\nu\mu}. \tag{7}$$

The equations of statistical equilibrium are

$$\frac{\partial n_{ik}}{\partial t} + \frac{\partial(n_{ik}v)}{\partial z} = \sum_{j\neq i}(n_{jk}P_{jik} - n_{ik}P_{ijk}). \tag{8}$$

The transition rates consist of radiative and collisional terms, i.e. $P_{ij} = R_{ij} + C_{ij}$. $C_{ij} = n_e\Omega_{ij}(T)$, where n_e is the electron density and $\Omega_{ij}(T)$ is a temperature-dependent collision rate. The radiative rates are discussed in more detail in section 3. Electrical neutrality of the gas is governed by the charge conservation equation

$$\sum_{i,k} n_{ik}Z_{ik} + n_e^{\mathrm{LTE}}(T, n_e) = n_e, \tag{9}$$

where the first term corresponds to ionization of all species k explicitly treated in NLTE (summation is over all such atoms and their energy levels, n_{ik} being their number density and Z_{ik} the ionization degree). Other species are for simplicity treated in LTE, using the Saha equations. Note a non-linearity of this equation in n_e, which is usually treated by the method of linearization (for details see [6]). Taking into account all bound states (atomic levels) and continuum states (ions) of all species, we can see that equations of statistical equilibrium are not linearly independent and we thus need one auxiliary equation to close the whole set. This can be the total particle-number conservation equation in the form

$$N = N(\rho). \tag{10}$$

All these NLTE equations describe the distribution of microscopic atomic states depending on the radiation and collisional processes and, conversely, the radiation field intensity which depends on the distribution of atomic states. These equations are strongly nonlinear. The transfer equation is generally non-local, which means that the intensity of the radiation field is not locally coupled to the plasma state - for a detailed discussion see [6].

2.3. RADIATION–HYDRODYNAMICAL COUPLING

Basic coupling between the plasma hydrodynamics and the radiation field comes via the gradient of the radiation flux F_r in Eq. (3) and via all terms which somehow depend on the ionization degree. The former corresponds to so-called radiation losses at a given depth of the atmosphere

$$\frac{dF_r}{dz} = \frac{1}{2}\int_0^\infty \int_{-1}^{1} (\eta_{\nu\mu} - \chi_{\nu\mu} I_{\nu\mu})\, d\nu d\mu. \tag{11}$$

We see that these losses are due to imbalance between the absorbed and emitted radiation, integrated over all wavelengths and over all directions. The problem of radiation losses is rather complex for the following reasons. In practice, we solve the transfer equation only for a rather limited number of atomic transitions in selected atomic species. This is a good approximation to obtain the ionization balance which tells us how the total particle density is distributed over atomic states. For such transitions one can evaluate the integral (11) explicitly. However, the contribution to total radiation losses from other species/transitions can be significant. In that case the usual procedure is to compute (11) for most the important optically-thick transitions which are known to be important for losses (e.g. hydrogen Lyman lines, lines of CaII and MgII) and finally to add all other contributions using an optically-thin approximation like in [7]. More sophisticated procedures take into account up to millions of lines (the so-called line blanketing problem).

Equations of RHD are summarized in Table 1.

TABLE 1. Summary of RHD equations and relevant quantities

Equation	Quantity	Notation
Continuity equation	Velocity	v
Momentum equation	Density	ρ
	or Particle density	N
Energy equation	Temperature	T
State equation	Gas pressure	p_g
Radiative transfer	Intensity	$I_{\nu\mu}$ or J_ν
Statistical equilibrium	Level populations	n_i
Charge conservation	Electron density	n_e

3. Solution of NLTE Problems

The most important ingredient of the NLTE physics is a proper treatment of the photon diffusion (e.g. the scattering in spectral lines). This problem is generally two-fold: (i) spatial diffusion and (ii) frequency and angular diffusion. In LTE, by definition, no diffusion is present because the collisional rates dominate the radiative ones and the source function is equal to the Planck function. In general, the line source function has the form (Mihalas [6])

$$S_\nu^L = \frac{n_j A_{ji}}{n_i B_{ij} - n_j B_{ji}} \rho_\nu, \tag{12}$$

where for the line transition $i \leftrightarrow j$, n_i and n_j are the respective level populations, A_{ji}, B_{ij} and B_{ji} are the Einstein coefficients and $\rho_\nu \equiv \psi_\nu / \phi_\nu$ is the ratio of emission to absorption profile. As far as the photon-frequency change during the scattering process is considered, we generally distinguish between two limiting cases: coherent scattering and so-called complete frequency redistribution (CRD). In reality, an intermediate case of the partial frequency redistribution (PRD) applies to strong resonance lines, while the subordinate lines are usually approximated by CRD (but see [8]). For CRD, the emission profile is supposed to be identical to the absorption one and thus $\rho_\nu = 1$. On the other hand, PRD treatment of the line scattering largely complicates the numerical solution of the transfer equation (for details see [9], [10]).

The spatial diffusion of photons is governed by the radiative transfer equations which are non-linearly coupled to the equations of statistical equilibrium for a multilevel atom. Classical numerical treatments like the so-called Equivalent-Two-Level-Atom (ETLA) method or the Complete Linearization (CL) technique are detailed in Mihalas' textbook [6]. Here we shall briefly outline the class of completely new methods based on the Accelerated Lambda Iteration (ALI) approach (for a review see [11]). The basic idea consists in a 'splitting' of the exact Λ-operator

$$\Lambda_{\nu\mu} = \Lambda_{\nu\mu}^* + (\Lambda_{\nu\mu} - \Lambda_{\nu\mu}^*), \tag{13}$$

where $I_{\nu\mu} = \Lambda_{\nu\mu}[S_{\nu\mu}]$ and Λ^* is the Approximate Lambda Operator (ALO). demonstrated by Rybicki and Hummer [12], the statistical equilibrium equations can be *preconditioned* using this type of operator splitting; here for the clarity of exposition we shall assume only a diagonal ALO. The ALI method is iterative, but contrary to standard Λ-iterations we write for the n-th iteration step

$$I_{\nu\mu}^n = \Lambda_{\nu\mu}^* \cdot S_{\nu\mu}^n + (\Lambda_{\nu\mu} - \Lambda_{\nu\mu}^*)[S_{\nu\mu}^{n-1}] = \Lambda_{\nu\mu}^* \cdot S_{\nu\mu}^n + \Delta I_{\nu\mu}^{n-1}. \tag{14}$$

Now taking the 'net radiative rate' for transition $i \leftrightarrow j$ in the form

$$R_{ij}^{\rm net} = n_j A_{ji} - (n_i B_{ij} - n_j B_{ji})\overline{J}_{ij} \quad (15)$$

and using the equations (12) and (14), we arrive directly at the preconditioned $R_{ij}^{\rm net}$ of the form

$$R_{ij}^{\rm net} = n_j A_{ji}(1 - \Lambda^*) - (n_i B_{ij} - n_j B_{ji})\Delta\overline{J}_{ij}^{n-1}. \quad (16)$$

In these expressions we have introduced the frequency-averaged mean intensity $\overline{J}_{ij} = \int_0^\infty \phi_\nu J_\nu d\nu$, the mean intensity being $J_\nu = 1/2 \int_{-1}^{1} I_{\nu\mu} d\mu$. Λ^* and $\Delta\overline{J}$ are the angle- and frequency-averaged quantities from Eq. (14). Note that in the case of PRD, Λ^* is averaged with the weight ρ_ν (see e.g. [13]). Comparing the two expressions for $R_{ij}^{\rm net}$ we see that while the first one depends non-linearly on level populations through the mean intensity, the second (preconditioned) one is now simply *linear* in level populations, provided that both Λ^* and $\Delta\overline{J}$ are computed using the quantities from previous $(n-1)$-th iteration. Thus the explicit dependence on unknown intensity was eliminated (note that the converged solution is exact, not approximate). With this formulation, the ALI techniques (called MALI for a multilevel case - [12]) have very good convergence properties and are much faster as compared to classical techniques which use for example the Feautrier method [6] to solve the transfer equation in a matrix form. They are particularly suitable for problems with very large number of atomic levels/line transitions, time-dependent problems related to RHD, multidimensional problems, etc. The example shown here can be easily generalized for the case of a continuum transition. However, when the electron density is not known as the model parameter, the statistical equations become non-linear in populations. In such a case, one can simply use the standard Newton-Raphson linearization of the statistical equilibrium equations, but contrary to CL we don't linearize the transfer equations since the statistical equations are already preconditioned (see Heinzel [14]).

It is important to note that both LTE as well as NLTE equations can be coupled to hydrodynamics. Time-dependent phenomena are easily governed by LTE where the ionization balance is described by the Saha equation in which the electron density closely follows the temperature variations. In the NLTE case, the situation is complicated by the presence of different kinds of relaxations.

4. Semiempirical Models of the Solar Atmosphere

In this section we shall describe so-called 'semiempirical' models of the solar atmosphere. These models have been developed to a high degree of sophistication and are still used as a kind of standard models. The basic idea behind these models is that the temperature structure of the atmosphere is determined *empirically*, which means that the energy-balance equation is not considered. The reason is that we don't understand very well the heating mechanisms, mainly in the chromosphere. The temperature at each atmospheric depth is thus adjusted empirically. This is done by comparing the theoretically computed spectrum (synthetic spectrum) for an estimated temperature structure with observed spectra in different continua and lines and making temperature adjustments to improve agreement. Such a comparison is iterated until a reasonable agreement is achieved. Semiempirical models are typically static.

4.1. STATIC MODELS

Because of the absence of any macroscopic velocities, hydrodynamical equations reduce simply to a momentum balance without flows, i.e. to the hydrostatic equation which describes the gravitational stratification of the whole atmosphere

$$\frac{dp}{dz} = -\rho g, \tag{17}$$

where $p = p_g + p_r + p_t$ is the total pressure composed of the gas pressure, radiation pressure and turbulent pressure. $p_g = NkT$ (N - total particle density), $p_t = 1/2\rho v_t^2$ (v_t is the mean microturbulent velocity). p_r can be neglected under solar atmospheric conditions. Introducing the column-mass scale $dm = -\rho dz$, one can solve Eq. (17) to get

$$p(m) = mg + p_0. \tag{18}$$

With this pressure and using a current estimate of the temperature structure, we solve the full NLTE transfer problem in all relevant lines and continua and iterate the temperature as described above.

The most complex semiempirical models so far developed are those of Vernazza, Avrett and Loeser [15], usually called VAL-models (an earlier widely used model was the Harvard-Smithsonian Reference Atmosphere, or so-called HSRA model [16]). To solve coupled equations of radiative transfer and statistical equilibrium, these authors use the ETLA method described in [17] and implemented in their code called PANDORA. PRD was used for hydrogen Lα and Lβ lines, but in the PCS approximation (see Appendix A in [15]). Several chemical species were treated in NLTE. As the basic observational constraint, the EUV spectra in the range 40 - 140 nm as obtained during the *Skylab* mission were used. In [15] the authors consider 6 plane-parallel models for various solar atmospheric structures having different brightness. These are: A - a dark point within a cell, B - the average cell center, C - the averaged quiet Sun, D - the average network, E - a bright network element and F - a very bright network element.

The model C is mostly referred to in the literature, we show its temperature structure in Fig. 1. In photospheric layers we see the temperature decreasing towards so-called *temperature minimum* $T = 4170$ K. Then T starts to increase again, which is a typical behaviour of the solar and stellar chromospheres. An extended temperature plateau is followed by extremely steep temperature increase in the CCTR. Note that T reaches almost coronal values within a few hundreds km. Such a steep temperature gradient is largely responsible for conductive heating of CCTR. In Fig. 1 one can also identify several spectral features (lines and continua) as they are formed at different atmospheric levels. Namely the continua indicate their sensitivity to kinetic temperature.

4.2. ENERGY BALANCE IN THE CHROMOSPHERE

Although the energy problem is not considered in semiempirical models of VAL, these authors evaluate the radiation losses at all depths of VAL-C model, using Eq. (11) (this formula can be used for any line or continuum, but for lines another expression is given by [15]). Resulting losses called *net radiative cooling rates* are shown in Fig. 2 for H and K lines of CaII, h and k lines of MgII, hydrogen lines, H^- continuum (an important opacity source in the lower atmosphere) and for other species, as calculated for model VAL-C. Note a strong peak at the base of CCTR which is due to most opaque hydrogen Lα line. Cooling rates or losses calculated in this way give us an idea how much of the heating we need to achieve the global energy balance. However, this doesn't tell us anything about the heating mechanisms.

4.3. ENERGY BALANCE IN CCTR

A small temperature plateau in CCTR around $T = 23000$ K is needed to match the intensity of hydrogen Lα line. However, to balance the Lα loss peak in Fig. 2, one needs an extra heating since dF_c/dz is insufficient at the base of CCTR. A solution to this problem was presented by Fontenla, Avrett and Loeser in a series of papers [18], [19] and [20] describing a new set of models referred to as FAL (instead of VAL). At lower depths, FAL models are almost identical to VAL ones (they are also semiempirical), but CCTR temperature structure is computed theoretically using the energy-balance equation with the ambipolar diffusion energy transport included. Ambipolar diffusion plays a role at atmospheric layers where the hydrogen is not still fully ionized, but the temperature gradient is high enough. Neutral atoms flow upwards towards higher temperatures where they are ionized. Simultaneously, hydrogen ions (protons) flow back to lower temperatures where they recombine thus releasing the ionization energy carried from

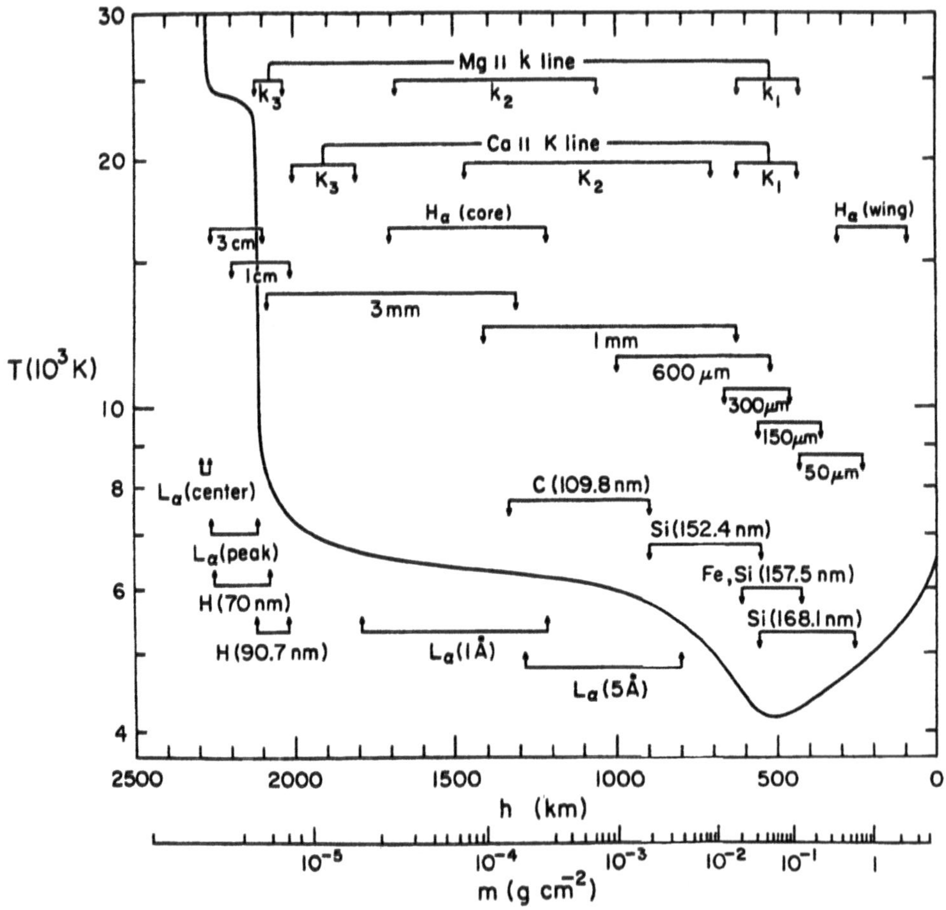

Figure 1. VAL model C of a mean atmosphere. Photospheric temperature decrease towards T_{min}, a chromospheric increase and strong CCTR temperature gradient are demonstrated. Note the temperature plateau around 23000 K where the Lα line is formed. This figure also shows the depths of formation of various spectral features used for empirical construction of the temperature curve. From Vernazza et al. [15].

higher levels. This transport of the ionization energy, called the ambipolar diffusion, seems to be important heating mechanism in the lower CCTR. It appears on the right side of Eq. (3) as an extra term proportional to the ambipolar-diffusion velocity v_{AD} ([18]). However, v in RHD equations is still zero, because the *mean* velocity is zero and thus the FAL models are

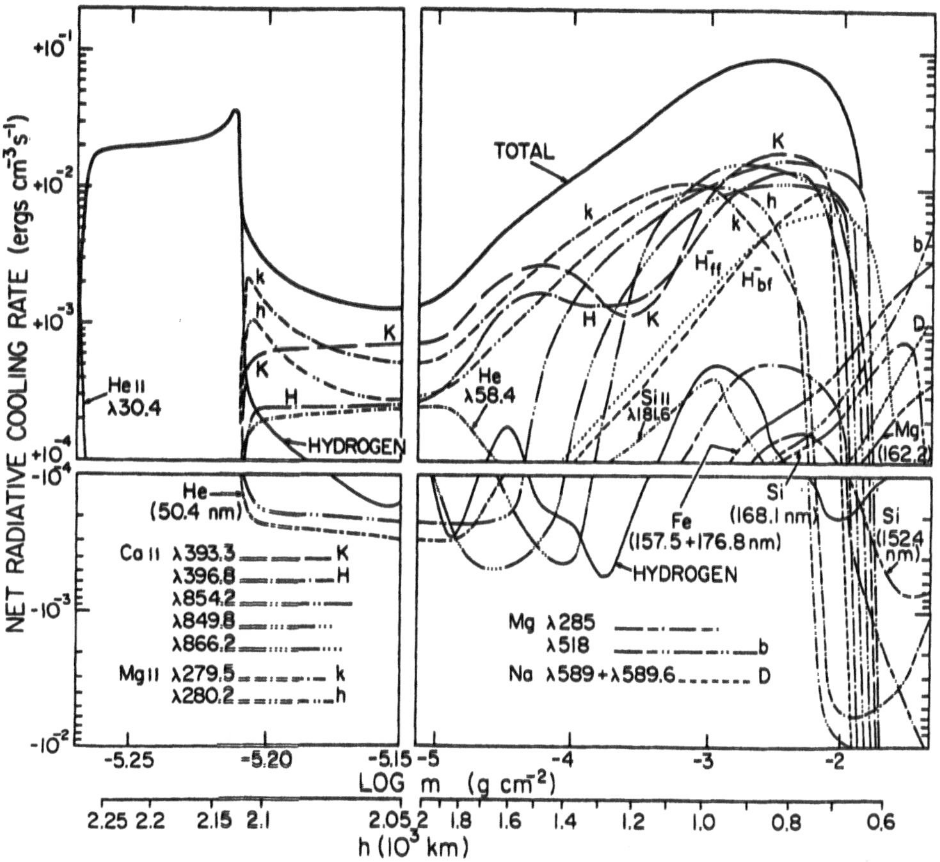

Figure 2. Net radiative cooling rates (losses) computed by VAL for model C. Contributions of different lines and continua of various species to the total loss function are shown. ¿From Vernazza et al. [15].

in hydrostatic equilibrium. Nevertheless, due to presence of flows of neutral hydrogen atoms and protons, one must use more general form of statistical-equilibrium equations than for a simple static case of VAL. Namely, the advection term - second term on lhs of Eq. (8), must be included, with $| v |= v_{AD}$. The new temperature structure as obtained by FAL for model C is shown in our Fig. 3, where one can note important modifications within the CCTR as compared with the earlier VAL-C model. Particularly the Lα

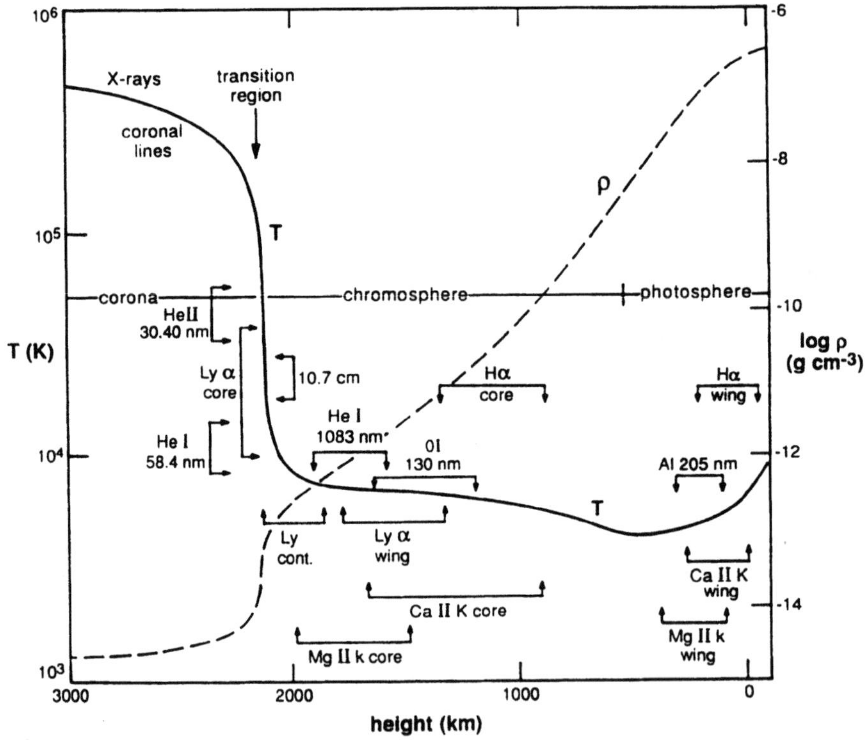

Figure 3. FAL model C of a mean atmosphere. Compare the temperature distribution with that shown in Fig. 1, namely in CCTR. Dashed line shows variations of the density ρ. ¿From Avrett [22].

line core is formed much higher, near T=40000 K. Quite recently, Avrett [21] has also considered the transport of ionization energy due to helium.

4.4. SPECTRAL LINE SYNTHESIS FROM VAL AND FAL MODELS

There is a considerable interest in evaluating the synthetic profiles of various lines of different species, using semiempirical models like VAL or FAL. Perhaps most popular for this purpose is the code MULTI written by Carlsson ([23]). In [15], the authors show a reasonable fit to existing data for hydrogen Lα line, while the computed Lβ has much lower wings as compared to OSO-8 measurements [24]. Heinzel and Hubeny [25] have suggested that strong Hα line wings could feed the Lβ ones via the Raman scattering process. Using highly sophisticated quantum formulation for such PRD multilevel interlocking (see [26]), Hubeny and Lites [27] have reevaluated the Lα and Lβ profiles for VAL-C model and obtained certain improvement for Lβ wings. Similar calculations but for FAL models with the advection terms and for the whole Lyman series are now under development by Avrett, Heinzel and Hubeny. These are to be compared with quite new data from SOHO/SUMER (Warren et al. [28]), obtained for the different brightness components used in the VAL and FAL models.

5. Dynamical Models

All emission profiles of the hydrogen Lyman series as observed by Warren et al. [28] correspond to time-averaged intensities; spatial averaging was done over typical brightness structures and also for the mean Sun. In this sense such data seem to be consistent with models VAL or FAL. However, even rough inspection of all line profiles displayed in [28] clearly shows a well pronounced asymmetry of the peak intensities, with enhanced red peak. This is an indication of a *dynamical behaviour* of the quiet solar atmosphere. Indeed, we have known for several years from optical observations that the solar atmosphere exhibits quasi-periodical oscillations, with periods around 5 minutes in the photosphere and around 3 minutes in the interiors of chromospheric supergranular cells (bright grains - for a review see [29]). In the magnetic network, these periods seem to be longer (Kalkofen [30]). In the UV, similar oscillations were detected by SOHO (SUMER or CDS), in different lines of various species (see review by Deubner [31]). Also hydrogen Lyman lines and continuum do exhibit oscillations with periods around 3 minutes in cell interiors ([32], [33]) - see Fig. 4. An interesting feature is a periodical variation of peak asymmetry of Lβ (Curdt et al. [34]), which, after time averaging gives the profile similar to that of Warren et al. [28]. How do all these oscillations affect the models of the solar atmosphere? This important question has been addressed in the past, but real numerical simulations of radiation hydrodynamics in the chromosphere have been performed only in this decade.

5.1. RADIATION HYDRODYNAMICS WITH SOUND WAVES

A principal goal of these RHD simulations is a study of the propagation of sound waves in the chromosphere to explain '3 minute' oscillations of bright

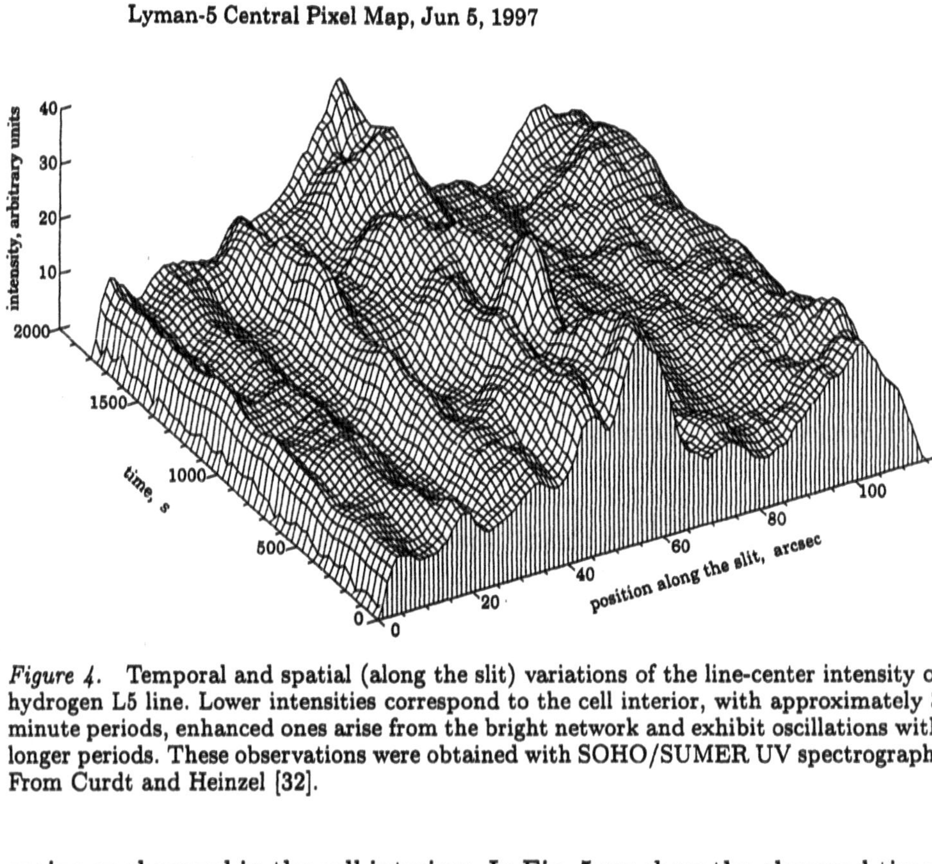

Figure 4. Temporal and spatial (along the slit) variations of the line-center intensity of hydrogen L5 line. Lower intensities correspond to the cell interior, with approximately 3 minute periods, enhanced ones arise from the bright network and exhibit oscillations with longer periods. These observations were obtained with SOHO/SUMER UV spectrograph. From Curdt and Heinzel [32].

grains as observed in the cell interiors. In Fig. 5 we show the observed time-sequence of the CaII H line intensities, showing typical brightenings in the H_{2v} peak.

To reproduce this, Carlsson and Stein [36], [37], [38] (hereafter referred to as CS) have suggested dynamical models schematically drawn in [37] (their Fig. 1). At time t=0 the whole atmosphere above the convection zone is in radiative equilibrium, so no extra heating is considered. The wave is then generated by a subphotospheric piston located inside the convective zone where combined radiative and convective equilibrium holds. The piston velocity variations were extrapolated down to h=–100 km,

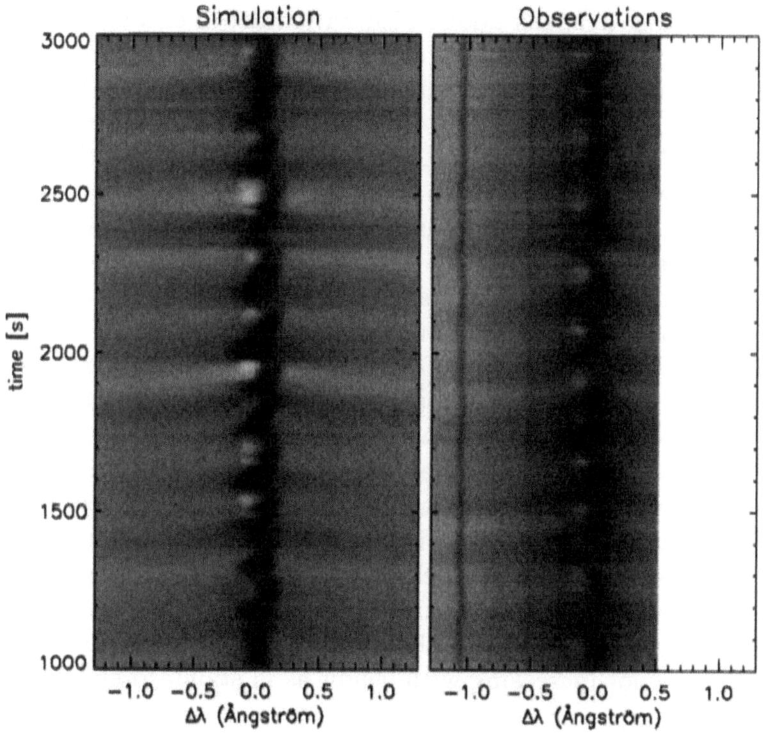

Figure 5. Comparison between observations and numerical simulations of temporal variations of CaII H line. 3 minute intensity enhancements in the blue wing are well reproduced. From Carlsson and Stein [35].

using the Doppler shifts of an Fe I line observed at the height 260 km. Perturbation caused by a piston, having certain spectral power, gradually propagates through the whole computational domain having a transparent upper boundary at about h=1800 km. A complete set of RHD equations as described in our section 2 is solved on an adaptive grid (governed by an extra grid-equation), HI and CaII NLTE transfer is fully time-dependent with flows (for further details see e.g. [37] and references therein). Propagating waves form shocks above 1000 km, which corresponds to regions where CaII line core and emission peaks are typically formed. The NLTE transfer problem is solved using the ALI technique, together with preconditioning ([39]).

There are several important messages that follow from this kind of simulation. First, the time-dependent synthetic CaII intensities match very well the observed ones, at least the qualitative behaviour of temporal variations. This is shown in Fig. 5. Second, due to presence of shocks, it is rather misleading to speak about a formation depth of a continuum. There are four important UV continua formed in the chromosphere: AlI (207 nm), SiI (152 nm), CI (110 nm) and HI Lyman continuum (91.2 nm). The contribution function (see Mein, this volume) for the continuum intensity may be bimodal with two peaks, one at $\tau_\nu \simeq 1$ and the other at smaller optical depth where a shock leads to an abrupt increase of the intensity (see Fig. 2 in [35]). Third and most important, the temporal variations of the continuum intensities at different depths give the *mean* synthetic intensity (i.e. averaged over time) which has the brightness temperature much higher than the mean kinetic temperature. This is because of the non-linear averaging of UV continuum intensity, which 'prefers' the highest intensities. Although the continuum source function in the chromosphere is not strongly coupled to the local Planck function (as would be, by definition, in LTE), the semiempirical models based on these continua and described above (VAL models) tend to overestimate the chromospheric temperature at heights below 1800 km. This is clearly illustrated in Fig. 6 taken from [36]. It was also shown that much of the mechanical energy of the shocks is directly converted into the radiation, and the kinetic temperature is enhanced only episodically during short time intervals. Thus the mean kinetic temperature is not increasing with height as in standard empirical models.

5.2. DOES THE SUN HAVE A FULL-TIME CHROMOSPHERE ?

This question was recently addressed by Kalkofen et al. [40] as a response to the abovementioned results of CS. These authors try to understand why observed UV emission lines formed below 1800 km show the emission *all of the time* (SOHO/SUMER observations of Carlsson et al. [41]), while CS simulations do predict a temperature decrease during a substantial part of the wave period, thus implying formation of absorption lines. They argue that CS simulations reproduce well the dynamical behaviour of the bright grains appearing in CaII H and K lines, but the chromospheric heating is not fully accounted for. Simplifications like 1D geometry, selected opacity sources, CRD instead PRD (crucial for quantitative reproduction of CaII line profiles), neglect of (NLTE) line blanketing and a restriction to heights below 1800 km seem to be of secondary importance. According to [40], the main problem comes from the lack of energy-supply by waves with frequencies above 10 mHz, which are not present in the power spectrum of the piston. They conclude that a full-time presence of UV emission lines

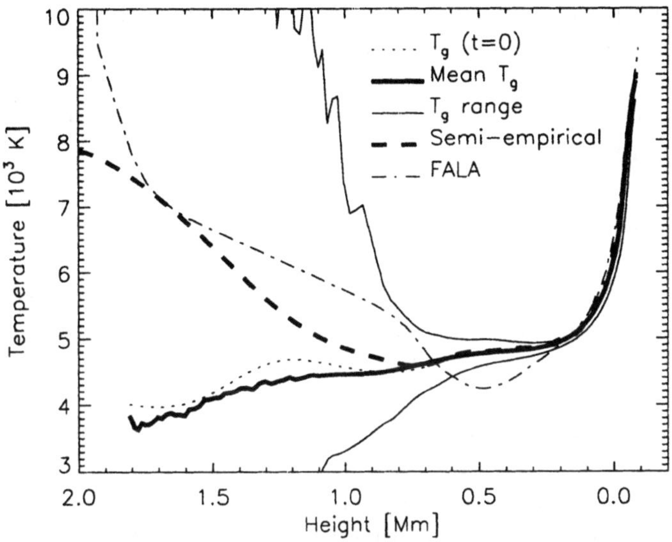

Figure 6. Dynamical models of the solar atmosphere. Temperature range in simulations is bounded by two thin lines denoted as T-range. The mean temperature (thick line) is then decreasing up to the height 1800 km. It surprisingly follows quite well the starting temperature structure at $t = 0$, which was computed using the radiative equilibrium condition (dotted line). Dashed curve shows a 'semiempirical' temperature which was calculated from these dynamical models; it is compared with the semiempirical model A of FAL (thin dot-dash). From Carlsson and Stein [36].

([41], [32]), requiring a full-time temperature rise, can be explained by the chromospheric heating with these high-frequency acoustic waves. A rather strong conclusion was made concerning the validity of VAL-type semiempirical models: 'they are valid as typical, instantaneous, thermal models of the chromosphere and not merely as time-averaged models'. Therefore, the Sun has the full-time chromosphere according to Kalkofen et al. [40].

Let us mention in this context that time and spatially well resolved SOHO (SUMER or CDS) observations of many UV lines never indicate a presence of any absorption. This is inconsistent with one of most critical CS assumptions, and namely that the atmosphere is in radiative equilibrium at the time $t = 0$ when the simulations start. We may thus speculate that the chromosphere is being *continuously* heated by e.g. above-suggested high-frequency waves (or other mechanisms) and stochastic wavetrains driven by a piston at lower frequency (3 minute oscillations) are just what we need to explain the calcium bright grains. The fact that we clearly see '3 minute' oscillations in all Lyman lines of hydrogen ([32], [33], [34]) indicates the penetration of waves to higher atmospheric levels - if such waves are not dissipated in shocks, they will heat the lower chromosphere even less than CS simulations show.

6. Conclusions

Currently the most important issue in solar atmospheric modelling seems to be the controversy between two rather different views on chromospheric and CCTR dynamics and heating. One of them, represented by Carlsson and Stein, considers the outer solar atmosphere so dynamical that mean semiempirical models have no reasonable physical meaning. On the contrary, Kalkofen et al. [40] give their arguments for the existence of a full-time chromosphere and, moreover, they claim that the semiempirical models represent not only time-averaged temperature structure, but also an instant situation. The latter statement could be justified if one would consider the semiempirical models as instantaneous 'snapshots' of a (slowly) evolving atmosphere. However, the highly dynamical nature of the CS simulations, consistent with time-resolved CaII observations, suggests that such snapshots will hardly represent a realistic atmosphere. The VAL and FAL models have not been aimed at describing an instantaneous situation.

Reaching higher atmospheric levels, we will certainly meet much more complicated CCTR structure than can be described by simple 1D models. Various kinds of magnetic 'canopies' will affect both the propagation and reflection of waves, as well as the thermodynamic properties of the models ([42]). These models will require 2D or fully 3D solution of the RHD problems. In CCTR, not only wave motions seem to be present, but also other kinds of flows have been considered like the ambipolar diffusion (FAL models) or velocity-redistribution effects on He II line formation (Andretta et al. [43]).

We certainly need new high-resolution data taken simultaneously in several spectral ranges, in order to study the chromospheric dynamics. Attempts were already made to observe simultaneously the optical lines like CaII or Hα (Tenerife) with UV lines accessible by SOHO/SUMER ([44], [34]). Their careful analysis can illuminate the problem of the wave propagation and heating. In parallel, we have to continue the highly sophisticated numerical RHD modelling as already mentioned in the text. For a review of 'next generation model atmospheres' see Avrett [45]. Several computer codes exist today which treat one or few specific problems (1D RHD, 2D NLTE line formation, cloud models for dynamical inhomogeneities, line blanketing, PRD interlocking etc.) and the basic question is how all these aspects affect the time-dependent as well as the time averaged synthetic spectrum which we compare to new observations. Finally, let us note that this entire effort is also extremely important for understanding the chromospheres of other solar-like stars.

Acknowledgements

This review was presented during the NATO-ASI Summer School 'Advances in Solar Research at Eclipses, from Ground and from Space' held in Bucharest (Romania), August 9-20, 1999. The author acknowledges the financial assistance of NATO. This work was also partly supported by the Grant No. A3003902 of the GA of Academy of Sciences of the Czech Republic. The author is indebted to Gene Avrett and Mats Carlsson for valuable comments and for allowing him to reproduce their figures.

References

1. Mihalas, D. (1991) in *Stellar Atmospheres: Beyond Classical Models*, Eds. L. Crivellari, I. Hubeny, D.G. Hummer, NATO ASI Ser., Vol. **341**, 127.
2. Thomas, R. and Athay, R. (1964) *Physics of the Solar Chromosphere*, Interscience, New York.
3. Auer, L., Fabiani Bendicho P. and Trujillo Bueno, J. (1994) *Astron. Astrophys.* **292**, 599.
4. Stenflo, J.O. (1994) *Solar Magnetic Fields (Polarized Radiation Diagnostics)*, Kluwer Academic Publishers, Dordrecht.
5. Mihalas, D. and Mihalas, B.W. (1984) *Foundations of Radiation Hydrodynamics*, Oxford University Press, New York (1999 - second edition).
6. Mihalas, D. (1978) *Stellar Atmospheres*, W.H. Freeman and Co., San Francisco.
7. Kuin, N.P.M. and Poland, A.I. (1991) *Astrophys J.* **370**, 763.
8. Hubeny, I. and Heinzel, P. (1984) *J. Quant. Spectrosc. Radiat. Transfer* **32**, 159.
9. Hubeny, I. (1985) in *Progress in Stellar Spectral Line Formation Theory*, Eds. J.E. Beckman and L. Crivellari, NATO ASI Series, Vol. **152**, 27.
10. Oxenius, J. (1986) *Kinetic Theory of Particles and Photons: Theoretical Foundations of Non-LTE Plasma Spectroscopy*, Springer-Verlag, Berlin.
11. Hubeny, I. (1992) in *The Atmospheres of Early-Type Stars*, Eds. U. Heber and C.J. Jeffery, Lecture Notes in Physics **401**, Springer-Verlag, Berlin.
12. Rybicki, G.B. and Hummer, D.G. (1991) *Astron. Astrophys.* **245**, 171.
13. Paletou, F. (1995) *Astron. Astrophys.* **302**, 587.
14. Heinzel, P. (1995) *Astron. Astrophys.* **299**, 563.
15. Vernazza, J.E., Avrett, E.H. and Loeser, R. (1981) *Astrophys. J. Suppl.* **45**, 635.
16. Gingerich, O., Noyes, R.W., Kalkofen, W. and Cuny, Y. (1971) *Solar Phys.* **18**, 347.
17. Avrett, E.H. and Loeser, R. (1987) in *Numerical Radiative Transfer*, Ed. W. Kalkofen, Cambridge University Press, Cambridge, 135.
18. Fontenla, J.M., Avrett, E.H. and Loeser, R. (1990) *Astrophys J.* **355**, 700.
19. Fontenla, J.M., Avrett, E.H. and Loeser, R. (1991) *Astrophys J.* **377**, 712.
20. Fontenla, J.M., Avrett, E.H. and Loeser, R. (1993) *Astrophys J.* **406**, 319.
21. Avrett, E.H. (1999) in *SOHO-8 Workshop*, ESA-SP 446.
22. Avrett, E.H. (1998) in *Solar Electromagnetic Radiation Study for Solar Cycle 22*, Eds. J.M. Pap, C. Frohlich, and R.K. Ulrich, Kluwer, Dordrecht, 449.
23. Carlsson, M. (1986) *A Computer Program for Solving Multi-Level Non-LTE Radia-*

tive Transfer Problems in Moving or Static Atmospheres, Report No. 33, Upsalla Astronomical Observatory.
24. Gouttebroze, P., Lemaire, P., Vial, J.C. and Artzner, G. (1978) *Astrophys J.* **225**, 655.
25. Heinzel, P. and Hubeny, I. (1985) in *Progress in Stellar Spectral Line Formation Theory*, Eds. J.E. Beckman and L. Crivellari, NATO ASI Series, Vol. **152**, 137.
26. Cooper, J., Ballagh, R.J. and Hubeny, I. (1989) *Astrophys J.* **344**, 949.
27. Hubeny, I. and Lites, B.W. (1995) *Astrophys J.* **455**, 376.
28. Warren, H.P., Mariska, J.T. and Wilhelm, K. (1998) *Astrophys. J.* **119**, 105.
29. Rutten, R.J. and Uitenbroek, H. (1991) *Solar Phys.* **134**, 15.
30. Kalkofen, W. (1999) in *Magnetic Fields and Oscillations*, Eds. B. Schmieder, A. Hofmann and J. Staude, ASP Confer. Series, Vol. **184**, 227.
31. Deubner, F.-L. and Steffens, S. (1999) in *Magnetic Fields and Solar Processes*, ESA-SP 448, 149.
32. Curdt, W. and Heinzel, P. (1998) *Astrophys J.* **503**, L95.
33. Heinzel, P. and Curdt, W. (1999) in *Magnetic Fields and Oscillations*, Eds. B. Schmieder, A. Hofmann and J. Staude, ASP Confer. Series, Vol. **184**, 201.
34. Curdt, W., Heinzel, P., Schmidt, W., Tarbell, T., von Uexkull, M. and Wilken, V. (1999), in *Magnetic Fields and Solar Processes*, ESA-SP 448, 177.
35. Carlsson, M. and Stein, R.F. (1994) in *Chromospheric Dynamics*, Ed. M. Carlsson, University of Oslo, Oslo, 47.
36. Carlsson, M. and Stein, R.F. (1995) *Astrophys. J.* **440**, L29.
37. Carlsson, M. and Stein, R.F. (1997) in *Solar and Heliospheric Plasma Physics*, Eds. G.M. Simnett, C.E. Alissandrakis and L. Vlahos, Springer-Verlag, Berlin, 159.
38. Carlsson, M. and Stein, R.F. (1998) in *New Eyes to See Inside the Sun and Stars*, Eds. F.-L. Deubner et al., IAU, 435.
39. Scharmer, G.B. and Carlsson, M. (1985) *J. Comput. Phys.* **59**, 56.
40. Kalkofen, W., Ulmschneider, P. and Avrett, E.H. (1999) *Astrophys J.* **521**, L141.
41. Carlsson, M., Judge, P.G. and Wilhelm, K. (1997) *Astrophys J.*, **486**, L63.
42. Rutten, R.J. (1999) in *Magnetic Fields and Solar Oscillations*, Eds. B. Schmieder, A. Hofmann and J. Satude, ASP Confer. Series, Vol. **184**, 181.
43. Andretta, V., Jordan, S.D., Brosius, J.W., Davila, J.M., Thomas, R.J., Behring, W.E. and Thompson, W.T. (1999) *Astrophys J.*, submitted.
44. Steffens, S., Deubner, F.-L., Fleck, B. and Wilhelm, K. (1997) in *Fifth SOHO Workshop*, ESA-SP 404, 685.
45. Avrett, E.H. (1996) in *Stellar Surface Structure*, Eds. K.G. Strassmeier and J.L. Linsky, IAU, 503.

INVERSION OF LINE PROFILES AND SOLAR FINE STRUCTURES

P. MEIN
Observatoire de Paris
F-92195 Meudon, France

Abstract. We present some methods to analyse data cubes produced by large solar telescopes. Applications to fine structures observed at the disc or at the limb are briefly reviewed.

1. Introduction

New large telescopes produce more and more data which must be interpreted in terms of physical parameters modelling the solar atmosphere. The capabilities of modern computers can be extended to processing of large data cubes, on the condition that fast algorithms for line profile inversions are available. To give a brief review of this topic, we shall discuss the following points:

- What can we deduce from solar line profiles?
- Intermediate quantities between profiles and model parameters:
  Center of Gravity, Moments, Bisector, Fourier coefficients...
- General inversion methods:
  Model fitting
  Small disturbances and Response functions
  Iterative methods
  Grids of models
- Examples of fine structures on the solar disc
- Isolated structures, Prominences and Filaments
- Conclusion

J.-P. Zahn and M. Stavinschi (eds.),
Advances in Solar Research at Eclipses from Ground and from Space, 221–244.

2. What can we deduce from solar line profiles?

The observed intensities of line profiles depend on the wavelength λ, the coordinates α, δ and the time t. They are functions of several parameters of the 3D solar atmosphere (abundance, temperature, density, velocity, magnetic field, microturbulence,...) that we would like to determine:

$$I(\lambda, \alpha, \delta, t)$$
$$\Downarrow$$
$$\text{Element Abundance}$$
$$T_e(x, y, z, t)$$
$$N_e(x, y, z, t)$$
$$\vec{V}(x, y, z, t)$$
$$\vec{B}(x, y, z, t)$$
$$\xi(x, y, z, t)$$
$$\ldots\ldots\ldots\ldots\ldots$$

Because of data noise, we must assume model atmospheres and restrict the unknown quantities to a small number of parameters. The inversion problem becomes:

$$I(\lambda, \alpha, \delta, t)$$
$$\Downarrow$$
$$\text{Model parameters}$$

In principle, we have only to invert the transfer equation

$$I(\lambda) = \int_0^{+\infty} S(\lambda, s) e^{-\tau(\lambda, s)} d\tau(\lambda, s)$$

with

s = geometrical depth along the line of sight (LOS)

S = source function (in LTE, Planck function which depends only on the local temperature and the wavelength)

τ = optical depth deduced from the absorption coefficient

If the number of available wavelengths is larger than the number of parameters, we can use a least-square method minimizing the departures between observed and theoretical intensities

$$\int (I^{obs}(\lambda) - I^{theor}(\lambda))^2 d\lambda$$

The above equations can be extended to several lines, and also to the case of Stokes parameters I, Q, U, V which characterize the polarized radiation [1],[2],[3],[4].

Let us give an *example* of a very simple model and straightforward inversion, namely the rough estimate of the temperature at τ=1.

Assuming LTE, optically thick case (disc), and B linear function of τ

$$B(\lambda, \tau) = a(\lambda) + b(\lambda)\tau$$

we can compute the line intensity (at disc center)

$$I(\lambda) = a(\lambda) + b(\lambda)$$

so that $I(\lambda) = B(\lambda, T_e(\tau(\lambda) = 1))$ provides an estimate of the temperature at the optical depth unity corresponding to the wavelength λ (Eddington-Barbier relation).

3. Intermediate quantities between line intensities and model parameters - some simple inversions

To connect profiles and parameters, it is often very useful to introduce quantities which can be easily derived from line intensities, and which are, at the same time, tightly connected to some model parameter, such as doppler velocities. As examples, we shall mention center of gravity, centered moments and Fourier coefficients:

$$I(\lambda, \alpha, \delta, t)$$

$$\Downarrow$$

Center of Gravity / Centered moments / Fourier coefficients / ...

$$\Downarrow$$

Model parameters

3.1. CENTER OF GRAVITY

In the case of emission lines (lines of prominences for example), we define the area of the profile and the wavelength of the center of gravity by

$$A = \int I(\lambda)d\lambda$$

$$\lambda_G = \frac{1}{A}\int I(\lambda)\lambda d\lambda$$

In the case of absorption lines (disc), we use

$$A = \int (I_c(\lambda) - I(\lambda))d\lambda$$

$$\lambda_G = \frac{1}{A}\int (I_c(\lambda) - I(\lambda))\lambda d\lambda$$

where $I_c(\lambda)$ is the continuum radiation.

Let us consider a simple model with **constant velocity v** along the line-of-sight. If the equation without velocity is written

$$I(\lambda) = \int_0^{+\infty} S(\lambda, s) e^{-\tau(\lambda,s)} d\tau(\lambda, s)$$

we must replace

$\tau(\lambda, s)$ by $\tau(\lambda - \lambda_0 v/c, s)$

$I(\lambda)$ by $I(\lambda - \lambda_0 v/c)$

λ_G by $\lambda_G + \lambda_0 v/c$ so that the shift of τ = shift of $I = \lambda_G - \lambda_{G(v=0)} = \lambda_0 v/c$.

In a similar way, in the case of **constant magnetic field** $B_{//}$ along the line-of-sight, the shift between profiles $I + V$ and $I - V$ is proportional to $B_{//}$ (Zeeman effect).

We conclude that, in simple cases, the center of gravity is a direct measure of velocity or magnetic field line-of-sight components.

3.2. CENTERED MOMENTS

Around the wavelength of the center of gravity, we can define the centered moments of order n of $I(\lambda)$ by

$$M_n = \frac{1}{A}\int I(\lambda)(\lambda - \lambda_G)^n d\lambda$$

Let us try to interpret the moments in terms of physical parameters in the simple case of **optically thin** ($\tau << 1$) **emission lines** (at the limb). Let us assume that the source function does not depend on the wavelength across the line profile:

$$I(\lambda) \simeq \int_{s=0}^{+\infty} S(s) d\tau(\lambda, s)$$

with

$$d\tau(\lambda, s) = \kappa(\lambda, s) ds$$

If we assume that the observed structure lies between s_1 and s_2, and that the absorption coefficient κ depends only on λ between s_1 and s_2:

$$I(\lambda) \simeq \kappa(\lambda) \int_{s_1}^{s_2} S(s)ds \sim \kappa(\lambda)$$

The line profile is proportional to the absorption coefficient profile, that we can write

$$\kappa(\lambda) \sim e^{-\frac{[(\lambda-\lambda_0)c/\lambda_0 - v]^2}{\xi^2+2kT/M}}$$

with

λ_0 = central wavelength of the line (without velocities)
v = LOS component of the velocity
ξ = microturbulence
T = temperature
M = atomic mass

The LOS velocity can be deduced from λ_G, and the sum $\xi^2 + 2kT/M$ from the dopplerwidth of the gaussian profile (proportional to $\sqrt{M_2}$).

3.3. BISECTOR

The bisector of a line is defined by the middle points of segments joining two points of equal intensities of the profile.

In the case of constant velocity across the atmosphere, the bisector is a straight line, at the wavelength shifted by $\lambda_0 V/c$. Any other shape reveals velocity gradients or unresolved structures. For example, the granulation pattern produces well-known "C-shapes", which can be explained by multi-column models including hot material going up and cool material going down [5], [6].

3.4. FOURIER COEFFICIENTS

Let us consider a wavelength interval $\lambda_0 - L/2, \lambda_0 + L/2$ around the line center. In this interval, the line profile can be expanded according to

$$I(\lambda) \simeq F(\lambda) = a_0/2 + \sum_{n=1}^{N} a_n cos[2\pi n(\lambda - \lambda_0)/L] + \sum_{n=1}^{N} b_n sin[2\pi n(\lambda - \lambda_0)/L]$$

The main advantages of Fourier coefficients a_n,b_n are:

3.4.1. *Least square fitting of theoretical profiles*
We have to minimize the departures between the observed profiles $I(\lambda)$ and the theoretical profiles $I'(\lambda)$. In the quantity

$$\frac{1}{L}\int_{-L}^{+L}(I'(\lambda)-I(\lambda))^2 d\lambda$$

we can replace I' by F' and I by F:

$$\frac{1}{L}\int_{-L}^{+L}(F'(\lambda)-F(\lambda))^2 d\lambda$$

that is (Parseval equality)

$$\frac{1}{2}(a_0'-a_0)^2+\sum_{n=1}^{N}((a_n'-a_n)^2+(b_n'-b_n)^2).$$

3.4.2. *Limited number of terms*
Since the expansions converge generally very fast, it is possible to restrict the expansions to a limited number of terms.

3.4.3. *Number of available model-parameters*
This number can be estimated by the number of terms which are larger than the data noise (degree of freedom).

3.4.4. *Physical interpretation*
In the case of small disturbances

- the **even** terms describe **symmetrical effects** on the profiles ($\Rightarrow$ temperature, density, macro- and micro-turbulence,...)
- the **odd** terms describe **asymmetries** ($\Rightarrow$ LOS velocity and magnetic field, gradients,...)

3.5. INVERSION USING SECOND DEGREE EXPANSION OF FOURIER COEFFICIENTS

Even in the case of NLTE chromosopheric lines, direct inversions can be performed in the following way [7]:

- Fourier expansion of "double profiles" (antisymmetry) for fast convergence
- Theoretical (NLTE) calculation of second degree expansions of each Fourier coefficient with respect to model parameters
- Inversion of second degree equations for each line of sight.

4. General inversion methods

4.1. MODEL FITTING

The model fitting uses generally trial and error methods. It is rather slow, and can be applied only to a small number of profiles. It is often used.

4.2. SMALL DISTURBANCES AND RESPONSE FUNCTIONS

4.2.1. *Definition*

Let us come back to the general equation

$$I(\lambda) = \int_0^{+\infty} S(\lambda, s) e^{-\tau(\lambda,s)} d\tau(\lambda, s)$$

In the **LTE case** let us assume an average model defined by B and κ:

$$I(\lambda) = \int_0^{+\infty} B(\lambda, s)\kappa(\lambda, s) exp[-\int_0^{s} \kappa(\lambda, s')ds']ds$$

Let us introduce small temperature disturbances $\delta T(s)$ which produce fluctuations $\delta B(\lambda, s)$ of the source function and $\delta\kappa(\lambda, s)$ of the absorption coefficient.

The intensity fluctuations can be written

$$\delta I(\lambda) = \int_0^{+\infty} R_T(\lambda, s)\delta T(s)ds.$$

R_T is named "Response (or Weighting) function" of temperatures at the wavelength λ.

There are **as many R as physical quantities** depending on s.

Similar definitions can be used in the NLTE case. NLTE Response functions are generally derived from numerical calculations.

The **Extension to Dopplershifts** d can be obtained by using the slope of the line profile at the wavelength λ

$d \simeq -\delta I/(\partial I/\partial\lambda)$

Definitions of response functions can be found in [8], [9], [10].

Extensions to Stokes parameters are detailed in [11]. The figure (1) shows examples of Response functions of Stokes parameters versus λ and $log(\tau)$.

4.2.2. *Response functions and formation depths; example of velocities*

The dopplershift can be written

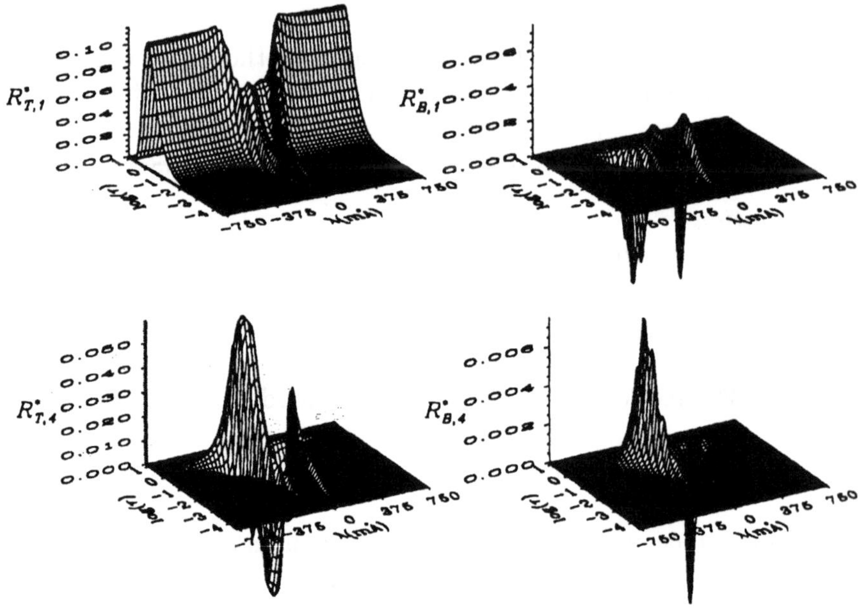

Figure 1. Examples of relative RFs to temperature (left) and to magnetic field strength (right) of Stokes I (top) and V (bottom) (from Ruiz Cobo et al., 1994).

$$d(\lambda) = \int_0^{+\infty} R_v(\lambda, s) v(s) ds$$

with the normalization

$$\int_0^{+\infty} R_v(\lambda, s) ds = \lambda_0 / c.$$

- **Velocity gradients:**
If the LOS velocity is a linear function of s

$$v(s) = a(s - s_0)$$

$$d(\lambda) = a \int_0^{+\infty} R_v(\lambda, s) s ds - a s_0 \int_0^{+\infty} R_v(\lambda, s) ds$$

$$d(\lambda) = (\lambda_0 / c) a (s_G - s_0) = (\lambda_0 / c) v(s_G)$$

where s_G is the depth of the barycenter of the response function.

The observed dopplershift corresponds to the velocity at $s = s_G$.

- Phase of waves:
Let us assume that a velocity wave is travelling along s across the atmosphere

$$v = exp[i(ks + \omega t + \phi)]$$

$$d(t) = \int_0^{+\infty} R_v(s) exp[i(ks + \omega t + \phi)] ds$$

$$d(t) \sim exp[i(\omega t + \phi')].$$

The phase of the dopplershift is equal to the phase of the wave at

$$s_\phi = (\phi' - \phi + 2n\pi)/k$$

which can be defined as a "depth of formation of the phase".

By measuring the time-delays $t_2 - t_1$ between the dopplershifts of 2 lines with different formation depths, it is possible to derive the wavenumber of the wave

$$k = \omega(t_1 - t_2)/(s_{\phi,2} - s_{\phi,1}).$$

4.3. ITERATIVE METHODS

4.3.1. *Integration of Radiative Transfer Equation + Least Square iteration*
We take the example of Stokes parameters and magnetic field observations. The data are the observed line profile(s) $I_0(\lambda)$, $Q_0(\lambda)$, $U_0(\lambda)$, $V_0(\lambda)$.

At each step, the iterative process consists in 2 parts:

1) The Radiative Transfer Equation

$$\frac{d\vec{I}}{d\tau} = K(\vec{I} - \vec{S})$$

is integrated with an approximate model defined by K and $\vec{S}$ (Milne-Eddington model for example)

2) a (non linear) least-square procedure provides corrections to minimize departures between computed and observed profiles

$$\int [(I(\lambda) - I_0(\lambda))^2 + (Q(\lambda) - Q_0(\lambda))^2 + (U(\lambda) - U_0(\lambda))^2 + (V(\lambda) - V_0(\lambda))^2] d\lambda$$

The new computed values of model parameters are used in the next step. Examples will be shown in the section (5.1.1).

4.3.2. *Use of Response Functions + Least Square iteration*
Examples will be mentioned in the section (5.1.3).

4.4. GRIDS OF MODELS + INTERPOLATION

4.4.1. *Grids and error bars*
Since the computers are able to store large numbers of profiles, the following process can be used :

1) A grid of profiles is computed inside a given range of model parameters

2) for each observed profile:

- the profile is compared successively to all the points of the grid
- The point with smallest departure is detected
- The best solution is determined by interpolation between the neighbouring points of the grid.

Let δI be the accuracy of intensity measurements. Grid profiles such that the rms of departures is smaller than δI are possible solutions. This condition specifies the error range for the observed profile, that is the multidimensional extension of error bars to the parameter space. Examples will be shown in the section devoted to prominences and filaments (6.3.8).

4.4.2. *Main advantages of grid methods:*
Many advantages can be mentionned:

- Any model geometry, temperature structure, NLTE case,... can be used
- The error ranges can be easily determined
- In case of ambiguities, the full set of solutions is obtained
- No iteration is needed
- The inversions are fast with modern computers
  if the number of parameters is not too large
  or if intermediate quantities (close to linear functions
  of model parameters) allow interpolations inside large-mesh grids.

4.4.3. *Grids using Principal Component Analysis (PCA)*
Let M be the dimension of the signal space, that is the number of wavelengths observed in the line profile. The vectors $I = (I(\lambda_1), I(\lambda_2), ...)$ defining the profiles of the grid in this space cluster generally around a smaller "feature space" of dimension m ($< M$).

Orthonormal eigenvectors e_j (eigen profiles) are determined so that

$$I = \overline{I} + \sum_{j=1}^{m} p_j e_j + I_\perp$$

with
$\overline{I}$ = mean signal profile
p_j = coefficients specifying a given profile
$I_\perp$ is small and can be neglected

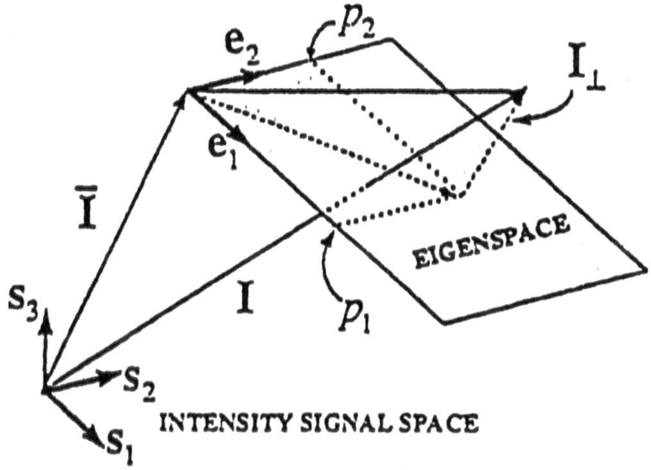

Figure 2. Relationship between PCA eigenspace and signalspace in the case $M = 3$ and $m = 2$ (from Rees et al., 1998).

The grid is built in the m-dimensional feature-space, instead of the M-dimensional profile-space ($m < M$). The method is especially efficient if the "features" are smooth functions of model parameters [12].

5. Examples of fine structures on the solar disc

5.1. INVERSION OF STOKES PROFILES

This concerns the important problem of magnetic fields observations. I already mentioned some books about polarized radiation ([1], [2]). A review of different inversion methods can be found in [13].

5.1.1. *Milne-Eddington atmosphere (HAO)*

The magnetic field, the dopplerwidth $\Delta\lambda_D$ and the ratio η_0 (line/continuum) are assumed to be constant along the line-of-sight.

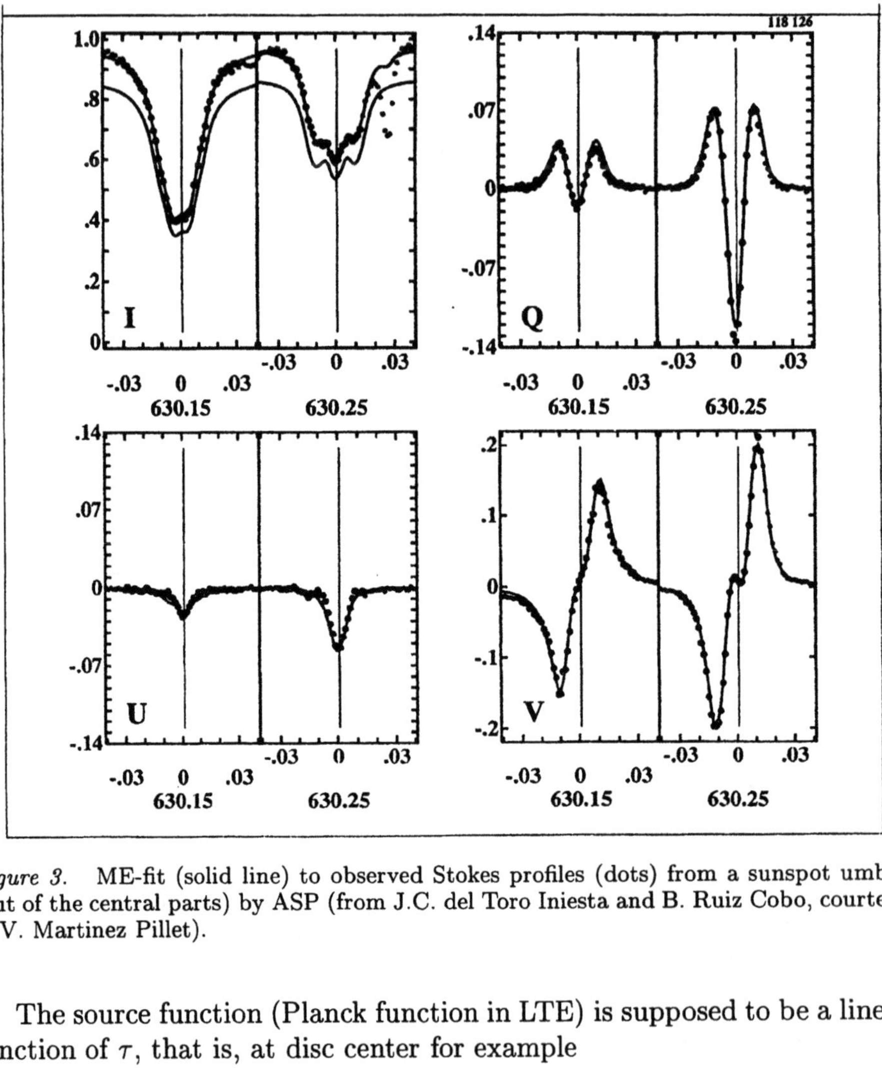

Figure 3. ME-fit (solid line) to observed Stokes profiles (dots) from a sunspot umbra (out of the central parts) by ASP (from J.C. del Toro Iniesta and B. Ruiz Cobo, courtesy of V. Martinez Pillet).

The source function (Planck function in LTE) is supposed to be a linear function of τ, that is, at disc center for example

$$B_T(\tau) = B_0 + B_1\tau.$$

The transfer equation can be written

$$\frac{d\vec{I}}{d\tau} = K(\vec{I} - \vec{B})$$

whith $\vec{I} = (I, Q, U, V)^\dagger$
$\vec{B} = B_T\vec{U} \qquad \vec{U} = (1, 0, 0, 0)^\dagger$

The Unno-Rachkovsky solution for emergent Stokes parameters is

$$\vec{I} = \vec{I_0} + \vec{I_1}\tau$$

with

$$\vec{I_1} = B_1\vec{U}$$

$$\vec{I_0} = B_0\vec{U} + B_1K^{-1}\vec{U}$$

Iterations are necessary to fit the observed profiles by least squares ([14],[15]). The figure (3) shows a Milne-Eddington-fit as used typically to analyse the data of the HAO Advanced Stokes Polarimeter.

5.1.2. *Multilayer models (NLTE)*
The radiative transfer equation is solved numerically ([16],[17],[18]). In NLTE case, iterations are necessary.

The inversion can be performed either by least square iterative procedure, or by grid of models and PCA method.

5.1.3. *Inversion using Response functions*
By using response functions, the least square inversion is fast. This method is used by the IAC group to determine gradients across the atmosphere ([19]).

5.1.4. *Flux tubes and canopies*
The Zurich group developped inversion techniques using mainly the I and V Stokes profiles. As an example we mention the analysis of magnetic structure in sunspot and canopy, deduced from observations in the infrared line FeI 15468Å([20]).

5.1.5. *Flows below Flux Tubes*
The paper [21] is an example of NLTE analysis, taking into account the expansion of flux tubes. In hydrostatic equilibrium, the magnetic pressure inside the tubes equals the difference between external and internal gas pressure

$$B^2(z)/8\pi = P_{ext}(z) - P_{int}(z)$$

Since P decreases with z, B decreases also, and the flux tubes expand. The analysis is performed along vertical rays, partly in the quiet atmosphere, partly in the magnetic tubes. To follow the rays into high levels, lines under study are formed up to the low chromosphere (MgI 5173,

NLTE). Stokes I (quiet + magnetic regions) and V (magnetic regions) profiles are fitted. A filling factor is taken into account. Velocity distributions are deduced from the observations.

5.1.6. *Multicomponent atmospheres (MISMA)*
The MIcro-Structured Magnetic Atmosphere is a very promising model of stochastic structures ([22]).

5.2. SOME EXAMPLES OF NLTE STRUCTURES

5.2.1. *Structures Photosphere-Chromosphere: grids + Fourier coefficients:*
To mimic gradients of temperature fluctuations across the atmosphere, it is possible to use a combination of different standard models in low and high levels (with continuity of pressure). Such models have been used to build a grid and fit Fourier expansions of profiles ([23]).

5.2.2. *Flare modelling*
A good example of model fitting can be found in the problem of chromospheric evaporations during flares.

Blue asymmetries are observed in Ca XIX (soft X-ray), while red asymmetries are visible in H_α ([24]).

6. Isolated structures, prominences and filaments

6.1. GEOMETRY

We call "isolated structures" the structures of cool material standing inside the hot corona: prominences (or filaments), chromospheric mottles, spicules,... They are generally supported or driven by the magnetic field. In the visible range, they are separated from the chromospheric background by optically thin media, so that the illuminating atmosphere plays an important role in their physical structure.

Theoretical models assume different geometries, which can be simple (1D horizontal or vertical slabs), or more sophisticated (2D horizontal or vertical prisms with rectangular sections, for example).

6.2. CLOUD MODELS

6.2.1. *Beckers Cloud Model (on the disc):*
The figure (4) shows a sketch of the so-called "cloud model" proposed by Beckers in 1964 ([25]). This model assumes that the Source function S, the Doppler width $\Delta\lambda_D$ and the Doppler shift defined by the bulk velocity v are constant. The cloud is characterized by 4 parameters:

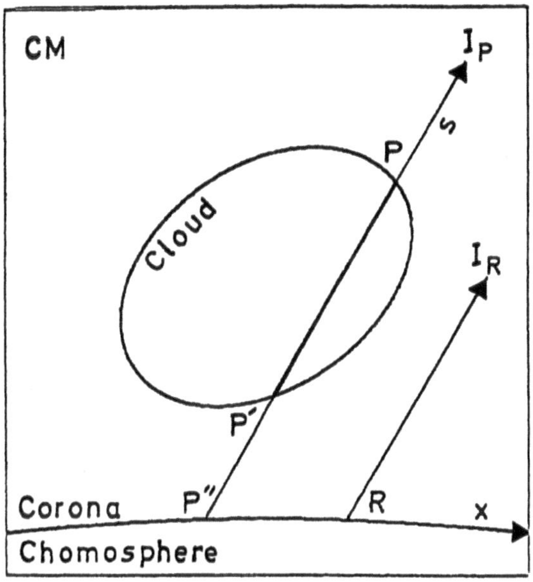

Figure 4. Sketch of the Beckers cloud model.

S, $\Delta\lambda_D$, v, τ_{max} where τ_{max} is the total opacity at line center along the L.O.S.

The observed intensity I_P is connected to the chromospheric background intensity I_R by

$$I_P(\lambda) = I_R e^{-\tau(\lambda)} + \int_P^{P'} S e^{-t(\lambda)} dt(\lambda).$$

6.2.2. *Contrast profile:*
It is very useful to introduce the "contrast profile"

$$C(\lambda) = \frac{I_P(\lambda) - I_R(\lambda)}{I_R(\lambda)}$$

$$C(\lambda) = (\frac{S}{I_R(\lambda)} - 1)(1 - e^{-\tau(\lambda)})$$

with

$$\tau(\lambda) = \tau_0 e^{-(\frac{\lambda - \lambda_0 - v\lambda_0/c}{\Delta\lambda_D})^2}$$

$$\Delta\lambda_D = \frac{\lambda_0}{c}\sqrt{\xi^2 + 2kT/M}.$$

If S/I_R and τ_0 are small, the contrast profile is proportional to the absorption coefficient

$$C(\lambda) \sim \tau_0 e^{-(\frac{\lambda - \lambda_0 - v\lambda_0/c}{\Delta\lambda_D})^2}$$

and estimations of τ_0, v, $\Delta\lambda_D$ become straightforward.

6.2.3. *Chromospheric mottles, AFS and fibrils (on the disc)*
Many works were devoted to chromospheric structures by using cloud models.

$I_R(\lambda)$ is assumed to be a spatial average of line profiles, for example in the interior of supergranular cells ([27]).

The figure (5) shows statistical results concerning Arch Filament Systems ([28]).

6.2.4. *Background effects: differential cloud model*
In many cases, and especially if the cloud is optically thin, $I(\lambda)$ is very sensitive to fluctuations of $I_R(\lambda)$ across the chromosphere. It is no more possible to assume that I_R is a mean chromospheric intensity.

In the "differential cloud model" ([26]), I_R is observed through the cloud (close to I_P) instead of outside. The equations are similar, but τ_0 is replaced by the opacity difference $\delta\tau_0$ of the cloud between the LOS corresponding to P and R.

6.2.5. *Cloud-model inversion with NLTE (theoretical) variable source function*
If the "cloud" is thick, the variation of S with depth cannot be neglected. The figure (6) shows theoretical results for H_α in 1D horizontal slabs.

The following inversion technique ([29],[30]) uses theoretical determinations of the source-function:

- assume approximate τ_0 and v
- compute theoretically $S(\tau)$ which depends on τ_0 and v
by taking into account the incident radiation from the chromosphere
- determine τ_0, v, $\Delta\lambda_D$ from the observed line profile
- iterate with new τ_0 and v values

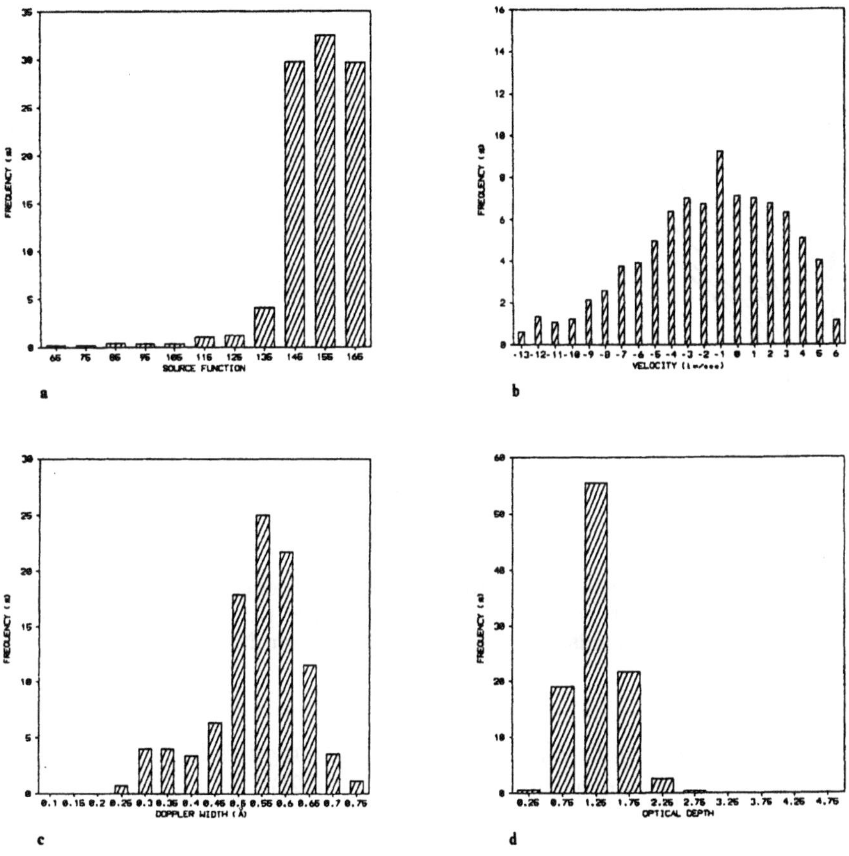

Figure 5. Histograms of source function (a), velocity (b), dopplerwidth (c) and optical depth (d) derived with the cloud model in Arch Filament Systems (From Alissandrakis et al., 1990).

6.2.6. *Non-LTE cloud models of mottles*

Contrast profiles and densities in mottles can be found in ([31]).

6.3. PROMINENCES AND FILAMENTS

Prominences are filaments seen at the solar limb. Both can be observed from ground and space in the visible and UV range.

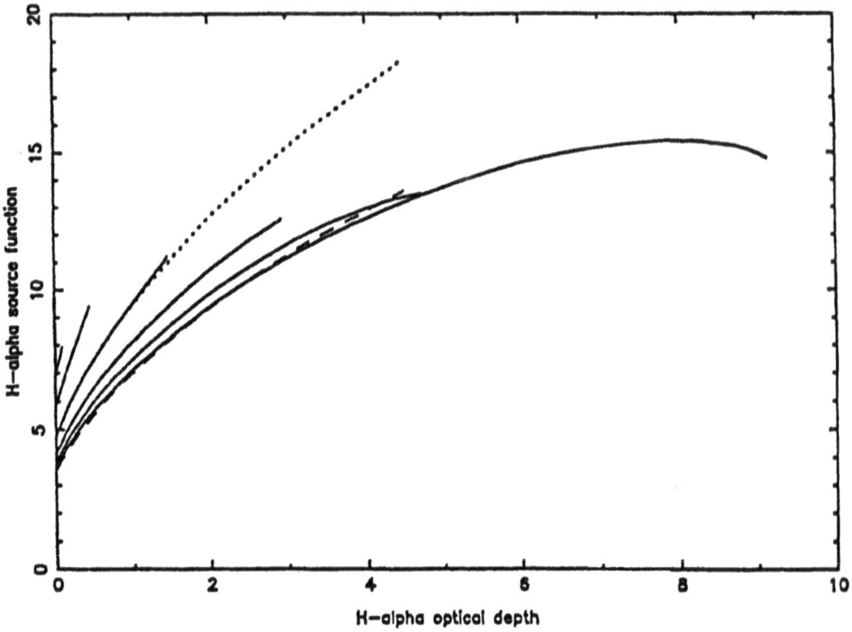

Figure 6. H_α source function in percents of the disc-center continuum versus line-center optical depth for 6 models with increasing gas pressures (from 0.02 to 0.30 dyn cm^{-2}). The dotted curve corresponds to a plage irradiation (from N. Mein et al., 1996, Heinzel et al., 1999).

6.3.1. *Diagnostics of T, N_e, v, B; orders of magnitude*
A review (with references) can be found in ([32]).

- **Temperature:**

The methods use generally line widths of two elements with different M:

$$\Delta\lambda_D = \frac{\lambda_0}{c}\sqrt{\xi^2 + 2kT/M}$$

It is also possible to use color or brightness temperatures. The results are generally between 5000 and 8000K.

- **Densities:**

The Stark effect in wings of Balmer lines, as well as line ratios, or depolarization by collisions (Balmer lines) lead to $10^{10} - 10^{11} rmcm^{-3}$.

- **Velocities:**

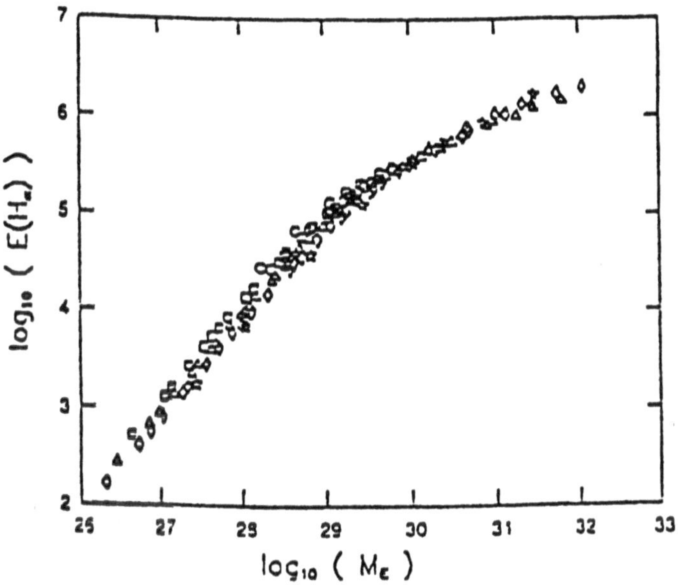

Figure 7. Correlation plot between H_α intensity and Emission Measure obtained from a set of 140 models of GHV (Vial, 1998, and Gouttebroze et al., 1993).

Dopplershifts show vertical velocities on the disc, and horizontal velocities at the limb. Proper motions lead to horizontal ones on the disc and vertical ones at the limb.

The values obtained for quiescent prominences are between 1 and 10 km/s, while the velocities of active region prominences can reach 50 km/s. Helical motions are very frequent in prominences.

- Magnetic field:

The magnetic field modifies the polarization of the light scattered by prominences (Hanle effect). Polarization measurements lead to magnetic field determinations: typically 10G for quiescent prominences (generally inverse configuration) and 50G for active region prominences (normal configuration).

6.3.2. *1D modelling*

A large set of models (1D slabs, constant temperature and pressure) was used to predict observed profiles of hydrogen lines. Correlations between observed quantities (example H_α vs. Emission measure $N_e^2 D$, see figure 7) were investigated ([34]).

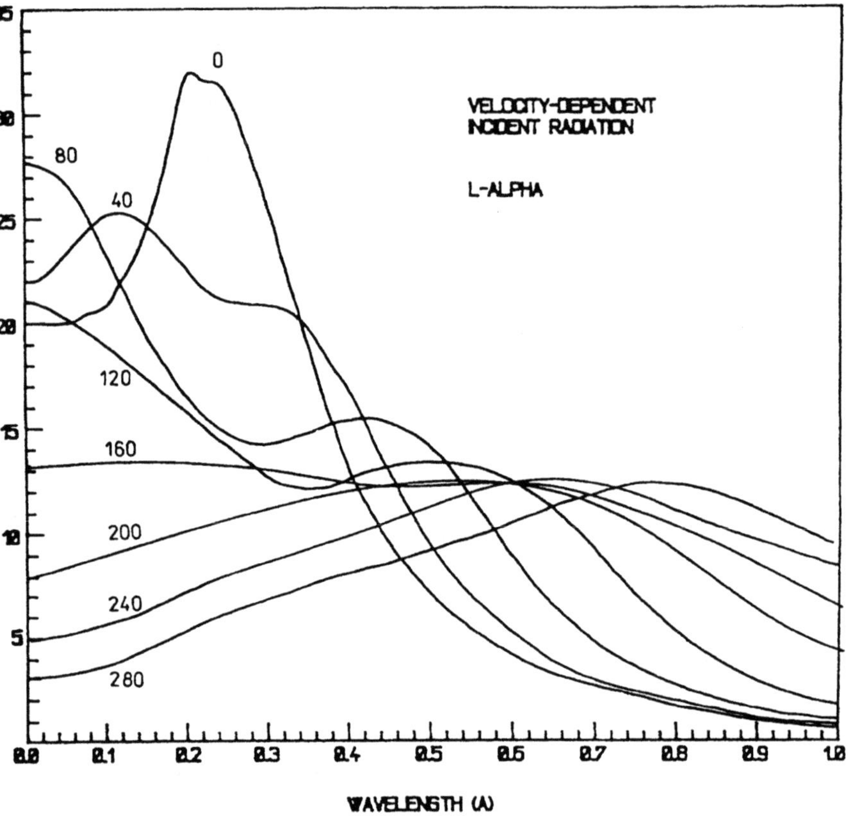

Figure 8. Mean intensities of the incident $L\alpha$ emission profiles as seen by a moving prominence. The profiles are displayed for the velocities ranging from 0 to 280 kms^{-1} (from Heinzel and Rompolt, 1987).

6.3.3. *Example of Lyman lines in filaments:*
SOHO observations of Lyman lines were used to estimate the temperature rise in the Prominence Corona Transition Region ([35]).

6.3.4. *Velocity effects: Doppler dimming and brightening*
The illumination of prominences by the atmosphere is sensitive to doppler effects (figure 8). As a consequence, dimming or brightening of prominence lines are observed, according to the velocity and the type of profile (emission/absorption) ([36],[37]).

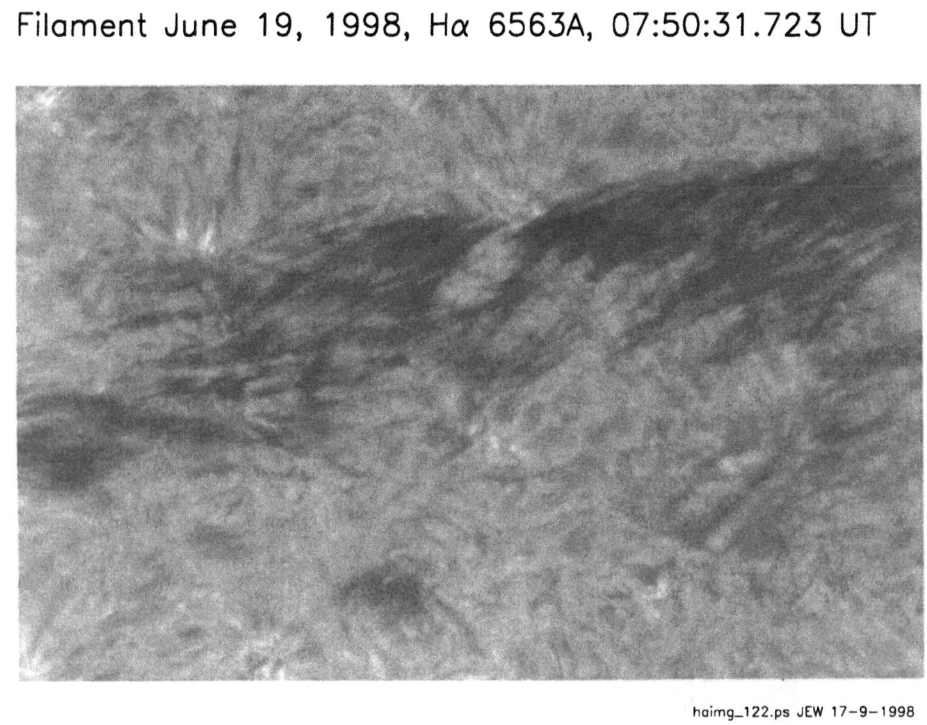

Figure 9. Filament fine structures observed with the SVST telescope (*Courtesy of O. Engvold*).

6.3.5. *2D Geometries*

Theoretical source functions and line profiles in prominences have been computed in 2D models ([38]).

6.3.6. *Threads and Multislabs; stochastic models*

High resolution images of filaments show many fine structures called "threads", as it can be seen on figure (9), obtained with the SVST telescope (La Palma, Canary Islands). They are not always resolved, especially in spectroscopy using large integration times (seeing effects). The number N of **unresolved threads** along the LOS can be inferred from statistical observations. Several cases were investigated:

1) Prominences:

By using average line intensity (CaII K) versus line shift [39] finds $N < 5$, with velocity dispersion around 5 km/s.

H_α contrasts lead to $N < 15$ or 20 with velocity dispersion smaller than 10 km/s [40].

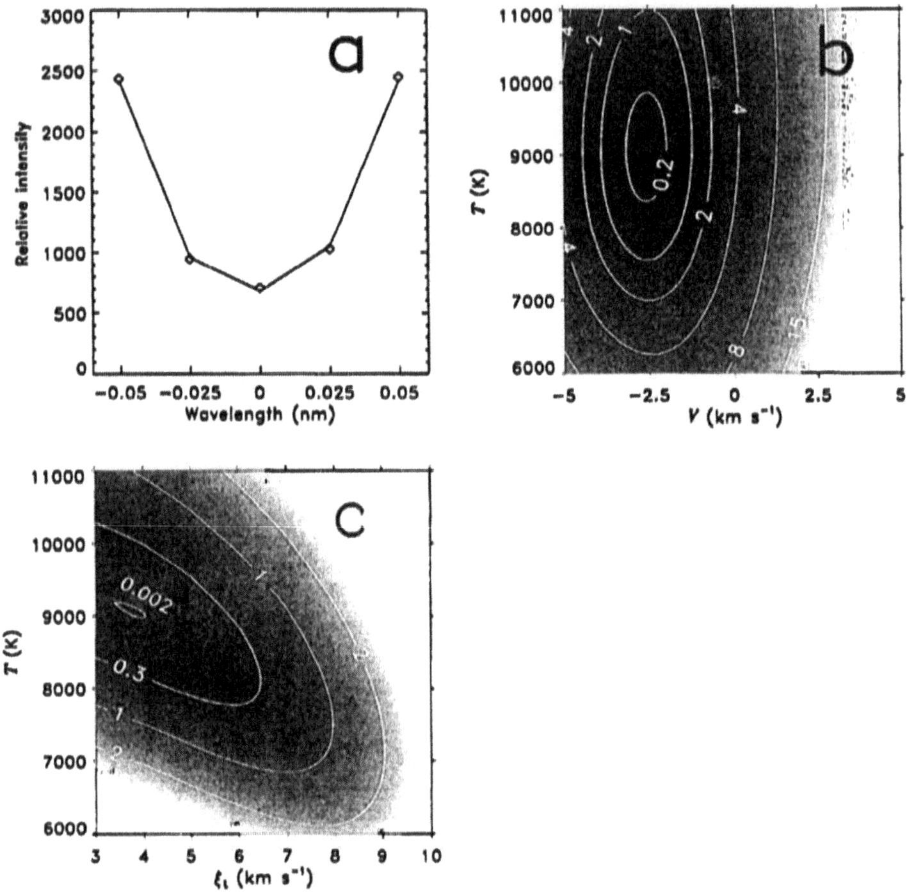

Figure 10. Inversion of MSDP data of a filament by a grid of models. a) Example of H_α profile (diamonds) and corresponding MALI grid profile (solid line). b) V-T cut through the distribution of the χ^2 merit function around the solution. c) $\xi_t - T$ cut (from Molowny et al., 1999).

Statistical results of H_α velocity, line width and width dispersions are analysed by [41] in terms of $N1$ lines of $N2$ threads unresolved around each line-of-sight (typically $N_1 \times N_2$ between 2 and 20). Different results are obtained in the body and the edges of the prominences.

In Ly_α and CIII, distributions of line intensities have been used by [42], leading to $N < 20$.

2) Filaments:

Profiles of the H_α line, especially intensities and dispersion of intensities, dopplershifts and dispersions of dopplershifts, lead to $N < 25$ ([43]). "Velocity threads" are suggested to be bundles of "opacity threads", which means that neighbouring opacity threads have similar velocities.

6.3.7. *Multi-slab modelling:*
By using energy balance between radiative losses and conduction, and taking into account ambipolar diffusion, [44] shows that the profiles of Ly_α, Ly_β and H_α are compatible with $N = 100$.

6.3.8. *Inversion using grids of models:*
Grids of NLTE models have been computed for filaments. By comparison with H_α observed profiles, temperature, microturbulence, LOS velocity and emission measure ($N_e^2 D$, with D=geometrical thickness of the filament) have been derived ([45]). For any given data noise, the grid method is able to specify the error ranges in the parameter-space. Figures (10, b and c) present iso-contours of the merit function χ^2 which characterizes the departures between observed and calculated profiles.

7. CONCLUSION

In this brief review, we mentioned many solar structures, in different geometries and physical conditions. Many inversion techniques are evolving because of the increasing speed of computers and sophisticated transfer techniques, and also because of the increasing volume of observed data cubes, especisally in the case of 2D-spectroscopy.

New data from large solar telescopes should allow to determine $\vec{B}$, $\vec{V}$, T, N_e,... with high resolution in space and time. New codes are needed to take simultaneously into account Stokes parameters, many line profiles, 2D or 3D geometries, NLTE conditions, unstable structures,...

A wide field of research!

References

1. Stenflo J.O. (1994) Solar magnetic fields. Polarized radiation diagnostics. *(Kluwer Academic Publications).*
2. Leroy, J-L. (1998) La polarisation de la lumière et l'observation astronomique, *(Gordon and Breach ed.).*
3. Landi Degl'Innocenti, E. (1992) Magnetic field measurements, *First Canary Islands Winter school* (F. Sanchez, M. Collados and M. Vazquez ed.), p 73
4. Mein, P. (1992) Dynamics of the solar atmosphere, *First Canary Islands Winter school* (F. Sanchez, M. Collados and M. Vazquez ed.), p 181
5. Dravins, D., Lindegren, L. and Nordlund, Å. (1981), *A&A* **96**, 345
6. Cavallini, F., Ceppatelli, G. and Righini, A. (1985), *Astron. Astrophys.* **143**, 116

7. Mein, P., Mein, N., Malherbe, J.-M., and Damé, L. (1987), *Astron. Astrophys.* **177**, 283
8. Mein, P. (1971), *Sol. Phys.* **20**, 3
9. Beckers, J.M. and Milkey, R.W. (1975), *Sol. Phys.* **43**, 289
10. Caccin, B., Gomez, M.T., Marmolino, C. and Severino, G. (1977), *Astron. Astrophys.* **54**, 227
11. Ruiz Cobo, B. and del Toro Iniesta, J.C. (1994), *Astron. Astrophys.* **283**, 129
12. Rees, D., Mulcahy, D. and Thatcher, J. (1998), A Database technique for fast spectral line inversion, *IASTED Intern. Conf., Signal processing and communications, Canary Islands*, p 408
13. del Toro Iniesta, J.C. and Ruiz Cobo, B. (1996), Forum Themis: Science with Themis (*Obs. de Paris, N. Mein and S. Sahal-Bréchot ed.*), p93
14. Skumanich, A. and Lites, B.W. (1987), *Astrophys. J.* **322**, 473
15. Lites, B.W., Skumanich, A., Rees, D.E. and Murphy, G.A. (1988), *Astrophys. J.* **330**, 493
16. Rees, D.E., Murphy, G.A. and Durrant, C.J. (1989), *Astrophys. J.* **339**, 1093
17. Semel, M. and Lopez Ariste, A. (1999), *Astron. Astrophys.* **342**, 201
18. Lopez Ariste, A. and Semel, M. (1999), *Astron. Astrophys.* **350**, 1089
19. Ruiz Cobo, B. and del Toro Iniesta, J.C. (1992), *Astrophys. J.* **398**, 375
20. Solanki, S.K., Rüedi, I. and Livingston, W. (1992), *Astron. Astrophys.* **263**, 339
21. Briand, C. and Solanki, S.K. (1998), *Astron. Astrophys.* **330**, 1160
22. Sanchez Almeida, J. and Landi Degl'Innocenti, E. (1996), *Sol. Phys.* **164**, 203
23. Briand, C., Mein, P. and Malherbe, .J-M. (1996), Forum Themis: Science with Themis (*Obs. de Paris, N. Mein and S. Sahal-Bréchot ed.*), p121
24. Canfield, R., Zarro, D.M., Metcalf, T.R. and Lemen, J.R. (1990), *Astron. Astrophys.* **348**, 333
25. Beckers, J.M. (1964), Thesis Utrecht, *AFCRL Env. Res. Papers* **49**.
26. Mein, P. and Mein, N. (1988), *Astron. Astrophys.* **203**, 162
27. Grossman-Doerth, U. and von Uexkull, M. (1973), *Sol. Phys.* **28**, 319
28. Alissandrakis, C.E., Tsiropoula, G. and Mein, P. (1990), *Astron. Astrophys.* **230**, 200
29. Mein, N., Mein, P., Heinzel, P., Vial, J.-C., Malherbe J.-M. and Staiger, J. (1996), *Astron. Astrophys.* **309**, 275
30. Heinzel, P., Mein, N. and Mein, P. (1999), *Astron. Astrophys.* **346**, 322
31. Heinzel, P. and Schmieder, B. (1994), *Astron. Astrophys.* **282**, 939
32. Vial, J.-C. (1998), *IAU colloq.* **167**, p175
33. Heinzel, P., Gouttebroze, P. and Vial, J-C. (1987), *Astron. Astrophys.* **183**, 351
34. Gouttebroze, P., Heinzel, P. and Vial, J.-C. (1993), *Astron. Astrophys. Suppl.* **99**, 513
35. Schmieder, B., Heinzel, P., Kucera, T. and Vial, J.-C. (1998), *Sol. Phys.* **181**, 309
36. Heinzel, P. and Rompolt, B. (1987), *Sol. Phys.* **110**, 171
37. Gontikakis, C., Vial, J.-C. and Gouttebroze, P. (1997), *Sol. Phys.* **172**, 189
38. Paletou, F. (1997), *Astron. Astrophys.* **317**, 244
39. Engvold, O., Jensen, E., Zhang, Y. and Brynildsen, N. (1989), *Hvar Obs. Bull.* **13**, 205
40. Zirker, J.B. and Koutchmy, S. (1990), *Sol. Phys.* **127**, 109
41. Mein, P. and Mein, N. (1991), *Sol. Phys.* **136**, 317
42. Pojoga, S., Nikoghossian, A.G. and Mouradian, Z. (1998), *Astron. Astrophys.* **332**, 325
43. Mein, N., Mein, P. and Wiik, J.E. (1994), *Sol. Phys.* **151**, 75
44. Fontenla, J.M., Rovira, M., Vial, J.-C. and Gouttebroze, P. (1996), *Astrophys. J.* **466**, 496
45. Molowny-Horas, R., Heinzel, P., Mein, P. and Mein, N. (1999), *Astron. Astrophys.* **345**, 618

SOUNDING THE SOLAR INTERIOR

W.A. DZIEMBOWSKI
Warsaw University Observatory
Al. Ujazdowskie 4, 00-478 Warszawa, Poland
and
Copernicus Astronomical Center, Polish Academy of Sciences
Bartycka 18, 00-716 Warszawa

1. Before helioseismology

Serious research on the solar internal structure began in the second half of nineteenth century. Attempts were made to asses physical parameters in the interior and, first of all, to explain the source of the energy radiated by the Sun. Already sir Arthur Eddington in his *The internal constitution of the stars* published in 1926 presents the hypothesis that the energy is liberated in building up helium nucleus from four protons, as most plausible explanation. By early 1940ies, mostly thanks to works of Hans Bethe, the basic chains of nuclear reactions leading to formation of helium nuclear reactions were known. At this time all basic physics needed to construct models of the Sun's interior became available. However, important improvements are being introduced to these days.

1.1. THE STANDARD SOLAR MODEL

Though numerical models of the Sun were constructed earlier, the concept of the standard solar model (SSM) was introduced in mid 1960ies. The goal was then to evaluate the neutrino flux for comparison with the forthcoming results from the chlorine experiments. The discrepancy between the prediction and measurements for many years stimulated interest in physics of the solar interior. Efforts were made both in improvements of the SSM and in construction of nonstandard solar models.

Basic assumptions underlying construction of the SSM are that the Sun began its life as a main sequence star $\tau_\odot$ years ago, where $\tau_\odot$ is set equal to the age of the oldest meteorites. Next, we assume that the Sun evolved, as

J.-P. Zahn and M. Stavinschi (eds.),
Advances in Solar Research at Eclipses from Ground and from Space, 245–264.

an equilibrium configuration, due to chemical changes caused by the nuclear reactions in the cores as well as the due to settling of helium and heavy elements. The latter effect is much less important and its inclusion is a recent innovation of the SSM. Models of the evolving Sun are built according to standard stellar theory with use of best available macroscopic physics data on equation of state, opacity, and nuclear reaction rate. These data are systematically improved. Most significant progress in the last decade is in the opacity calculations.

The input parameters for construction of the SSM are the age, $\tau_\odot$, total mass, M, surface luminosity, L, the surface radius, R, and the ratio between the heavy elements and hydrogen photospheric abundance, Z/X. The helium abundance, Y, and the parameter scaling the mixing length in the theory of convection, α_c, are treated as adjustable quantities.

1.2. DIFFERENTIAL ROTATION

Explaining differential rotation of the solar photosphere challenged theorists for a long time. Naturally this required modeling rotation in the whole convective envelope. Before we had data from helioseismic sounding there have been only hints about rotation beneath the photosphere from differences in the rate inferred from different tracers. How convection affects rotation is a formidable problem, which still cannot be regarded as solved. There is a whole monograph on the subject written by Rudiger (1989) before most important results from seismic sounding became available.

The law differential rotation is an essential ingredient of the dynamo theory. Before helioseismology provided data on solar rotation within the convective zone, the prevailing view was that the site of dynamo is the whole lower convective zone.

A considerable interest in rotation in deep interior was evoked by, as we now understand, an erroneous measurement of the solar oblateness by Dicke and Goldenberg (1967). This result indicated that there is a large contribution of the gravitational quadrupole moment to the oblateness. The explanation proposed was that the interior of the Sun rotates much faster than the atmosphere.

2. A brief history of solar oscillation observations

The history begins with the discovery of *the solar five minute oscillations* in the radial velocity data by Leighton, Noyes and Simon (1962). The oscillations were originally regarded as as local atmospheric phenomenon. The idea that the effect may be due to excitation of the global modes, first proposed by Ulrich (1970), was confirmed by observations made by Deubner (1975). He analyzed radial velocity measurements in the horizontal wave number *versus* frequency ($k_h - \omega$ diagram) and found well separated ridges, which he associated with different radial orders of the modes.

The modes responsible for the ridges were of rather high k_h, hence confined to the outer part of the solar envelope. Therefore they offered the probing potential only for this part of the solar interior. The discovery of radial and other modes of large horizontal scale (Claverie et al. 1979; Grec et al. 1980) opened prospects for seismic sounding of the whole interior.

The first significant result of the seismic sounding, namely the determination of the equatorial rate of rotation (Duvall et al. 1984), was based on measurements of Harvey and Duvall (1984).The subsequent five years were perhaps the most important in the history of helioseismology. Precise frequencies measurements for a large number of modes were obtained with several independently developed ground based instruments (Duvall et al. 1986; Brown & Morrow 1987; Libbrecht 1988, 1989). The frequencies were made available to a number of researchers who carried inversions revealing information about the Sun's interior.

A new direction in helioseismology was opened with the work of Duvall et al. (1993), who proved that the sound travel time between points in the solar photosphere may be measured. This discovery gave us a tool for sounding local flows and structures beneath photosphere.

Further development in the field were connected with the deployment of new instruments. The GONG network of automatic ground-based telescopes began operation in October 1995. Three helioseismic instruments: MDI, GOLF and VIRGO on board of the SOHO spacecraft began operation in April 1996.

3. Waves and modes in the solar model

3.1. AN APPROXIMATE TREATMENT

Oscillations observed in the Sun are acoustic waves propagating in the interior. The usual dispersion relation, $\omega^2 = c^2|k|^2$ applies approximately and proves useful even at quantitative level. In this relation ω is the angular wave frequency. Here I will use the cyclic frequency $\nu = \omega/2\pi$, as now preferred by observers. Solar oscillation modes used for seismic sounding have ν in the range (1-4.5) mHz

Since the Sun is nearly spherically-symmetric object, it is convenient to express the three-dimensional wave number in the following form

$$|k|^2 = k_r^2 + k_h^2 = k_r^2 + \frac{\ell(\ell+1)}{r^2}, \tag{1}$$

where ℓ is the spherical harmonic degree.

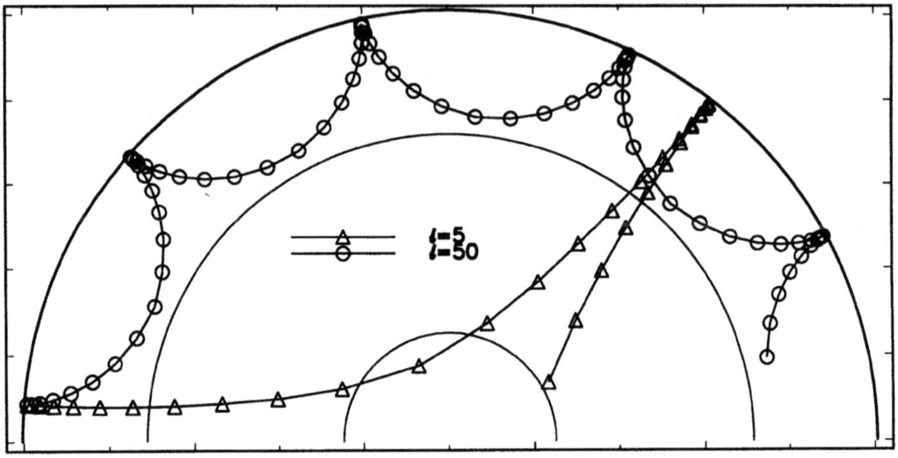

Figure 1. Propagation of acoustic rays at two indicated degrees, ℓ, and the same frequency $\nu = 5$mHz, in the Sun's interior. The time span between the consecutive symbols is 5 minutes.

Between the photosphere and the center the sound speed c increases by a factor ~ 600. Effects of stratification are most important for understanding qualitative properties of acoustic waves in the Sun.

Let us consider ω and ℓ as specified and use the dispersion relation to evaluate k_r. We get

$$k_r = \sqrt{\left(\frac{\omega}{c}\right)^2 - \frac{\ell(\ell+1)}{r^2}}, \tag{2}$$

and we see immediately that the wave cannot propagate below the *lower turning point*, $r = r_t$, where the square root vanishes. This simple dispersion is not valid near the surface because for its validity the model parameters should not vary too much over the wavelength. Quantitatively, the criterion may be expressed in the following form

$$\left|k_r \frac{dr}{d\ln p}\right| \sim \frac{\omega c}{g} \gg 1. \tag{3}$$

For the derivation, we could ignore the ℓ-dependent term because it is not important in the layers where c is small. If we replace $\gg 1$ with ~ 1 in the condition we get an approximate criterion for acoustic propagation. Sound cannot propagate through the layers where pressure and density change rapidly. Calculation shows that waves with

$$\nu < \nu_{ac} \approx 5.3 \text{ mHz}$$

are reflected somewhere below the temperature minimum.

I show in Figure 1 two examples of the path of acoustic signals propagation in a cross-section of the solar interior. Signals characterized by the central frequency ν and the central horizontal degree ℓ are called *acoustic rays*. If the dispersion relation (1) applies, the ray path, $\mathbf{r}(r, \theta, \phi)$, is determined by the group velocity, $\mathbf{v}_g$, through the following equation.

$$\frac{d\mathbf{r}}{dt} = \mathbf{v}_g = \frac{\partial \omega}{\partial \mathbf{k}}. \tag{4}$$

The initial propagation from the surface inwards is slow and nearly vertical. The increase of the sound-speed causes that the distance traveled in a unit time increases and the ray path bends (the role of the ℓ–dependent term increases). At $r = r_t$ it turns back toward the surface, where it is reflected again. The time and distance separating points of consecutive emergence of the ray at the surface called *one-skip time* and *one-skip distance*, respectively, are the observables for the time-distance helioseismology.

The position of lower turning point depends on the value of parameter

$$d = \frac{\ell(\ell+1)}{\nu^2}.$$

Lower d implies lower r_t. We see in Figure 1 that at $\nu = 3$ mHz, which is mean frequency of the observed oscillation, the $\ell = 50$ ray does not enter the radiative interior and that the $\ell = 5$ just barely enters the energy producing core. This emphasizes the role of low degree oscillations in sounding the deep interior of the Sun.

In contrast to terrestrial seismology, the main helioseismic observables are the eigenmode frequencies $\nu_{\ell,n}$, which corresponds to constructive interference of the waves. The number n designates the *radial degree*. The interference condition leads to the following expression for the radial phase:

$$\Phi = \int_{r_t}^{R} k_r dr = \int_{r_t}^{R} \sqrt{\left(\frac{\omega}{c}\right)^2 - \frac{\ell(\ell+1)}{r^2}} dr = (n+\alpha)\pi. \tag{5}$$

The phase shift α cannot be determined because the adopted approximation is invalid near the surface. The value of α may be regarded as independent of ℓ, but its dependence on ω cannot be ignored. The uncertainty of the exact location of the upper turning point is contained in α. We should note that in this approximation, the oscillation frequencies are fully determined by the two functions $c(r)$ and $\alpha(\nu)$.

3.2. EXACT THEORY OF ADIABATIC OSCILLATIONS

The equation governing the oscillations are obtained by linearization of the general equations of fluid dynamics, about hydrostatic equilibrium. In our application, the static equilibrium is the standard solar model. Perturbation of all parameters are expressed in terms of the the displacement of a fluid element $\vec{\xi} = \vec{x} - \vec{x}_0$. All terms beyond linear in $\vec{\xi}$ are ignored. In particular, for the acceleration we have

$$\frac{d\vec{v}}{dt} = \begin{cases} \frac{\partial^2 \vec{\xi}}{\partial t^2} & \text{if rotation neglected} \\ \frac{\partial^2 \vec{\xi}}{\partial t^2} + 2\Omega \frac{\partial^2 \vec{\xi}}{\partial t \partial \phi} & \text{if effects of rotation included up to } \Omega^1 \end{cases} \tag{6}$$

The angular dependence of the perturbations is represented in terms of *spherical harmonics* Y_ℓ^m, where m is the *azimuthal order* and ℓ, as before, is mode degree. The time dependence is assumed in the form $\exp(-i\omega t)$. Then, we may write,

$$\vec{\xi} = r[y(r)\vec{e}_r + z(r)\vec{\nabla}_H + w(r)\vec{e}_r \times]Y_\ell^m(\theta, \phi) \exp(-i\omega t),$$

where the three scalar functions, $y(r)$, $z(r)$, $w(r)$, together with ω are determined by a numerical integration. The last component may be ignored in the case of acoustic oscillations in slowly rotating stars.

Effects of rotation on solar oscillations effects are indeed very small (we have $\Omega/\omega \sim 10^{-5}$) and a good approximation for ω is obtained neglecting effects of rotation altogether. Then, the equation for oscillation for oscillations may be written in the following compact form.

$$\omega^2 \rho \vec{\xi} = \vec{L}(\vec{\xi}), \tag{7}$$

where ρ is density and $\vec{L}$ is the perturbed volume force. Mathematically, $\vec{L}$ is a Hermitian operator. Thus, its eigenvalues, ω^2, are real. In the adopted approximation, $\vec{L}$ does not depend on m: we have the $2\ell+1$ fold degeneracy of frequencies relative to m.

Solutions of Eq.(7), $\vec{\xi}_{n,\ell}$, $\omega_{n,\ell}$, describes oscillations of spherical stars. Acoustic modes discussed in the previous section are traditionally called *p-modes* and denoted p_n. Equation (1) gives a useful approximation if $n+\ell$ is sufficiently large, which is the case for all p-modes detected in the Sun. If $n \gg \ell$ then Eq.(5) implies

$$\omega_{n,\ell} \propto (n + \ell/2 + \text{const}). \tag{8}$$

At $\ell > 80$, in addition to p-modes, *f-modes* are also detected in the Sun; these are a close analogs of the surface gravity waves in deep water and – to a good approximation – obey the same dispersion relation, which is

$$\omega = \sqrt{gk_h} = \sqrt{\frac{\sqrt{\ell(\ell+1)}GM}{R^3}}. \tag{9}$$

The modes not yet detected in the Sun are *g-modes.* The large interest in searching for them follows from the fact that their frequencies are much more sensitive than p-modes to the core structure. So far the searches has not been successful. None of several claims of the discovery was confirmed.

Eq. (7) defines a *variational principle*, which yields the following expression for small differences in frequency $\delta\omega$ in terms of corresponding small changes in $\delta\rho$ and $\delta\vec{L}$:

$$\delta\omega = \frac{\int \vec{\xi}^* \cdot [\delta\vec{L}(\vec{\xi}) - \omega^2 \delta\rho\vec{\xi}\, d^3\vec{x}}{2\omega I}, \tag{10}$$

where

$$I = \int |\vec{\xi}|^2 \rho\vec{\xi}] d^3\vec{x},$$

is a moment of inertia; $\vec{\xi}$ and ω are solutions of Eq.(7).

This integral expression for small corrections to ω is a very important tool in helioseismology. The small differences $\delta\vec{L}$ may represent forces not included in $\vec{L}$, as well as, effects arising from various inadequacies of the reference solar model. An important application of Eq.(10) is to evaluate the frequency shifts caused by rotation. From Eqs.(6) and (10) we get

$$(\delta\omega)_{\rm rot} = \frac{\int \Omega[m|\vec{\xi}|^2 - i\vec{\xi}^* \cdot (\vec{e}_z \times \vec{\xi})]\rho d^3\vec{x}}{I}. \tag{11}$$

The degeneracy in ω is now lifted. The difference $(\delta\omega)_{\rm rot}(m)$ is an antisymmetric function of m. Higher order effects in Ω and magnetic field give rise to frequency shifts which are symmetric functions of m.

4. Measuring solar oscillations

Since their discovery, the prevailing way of detecting and measuring oscillations in the Sun is an analysis of the radial velocity data. Of the ongoing experiments only VIRGO uses intensity data.

The whole disc measurements, such as ones with GOLF, yield averaged values, $\overline{v_{\rm rad}(t)}$, in which only the low degree modes contribute to the signal. Modes manifest themselves as peaks in the amplitude spectra. The peaks are well separated, so that identification of the ℓ and n values is unambiguous. On the other hand, the m components of the ℓ, n multiplets are usually not resolved into separate peaks.

Spatially resolved observations, such as ones done with MDI and GONG instruments, yield temporal two-dimensional maps (*Doppler images*) of the radial velocity field, $v_{\rm rad}(\theta, \phi,)$ for subsequent analysis. After eliminating

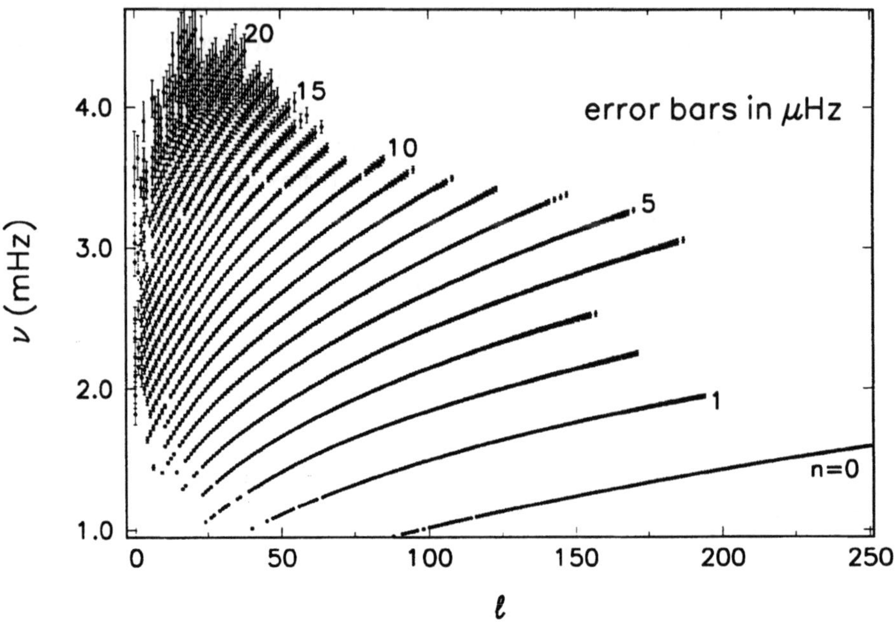

Figure 2. Frequencies, $\bar{\nu}_{\ell,\nu}$ with errors measured with the SOHO/MDI instrument are plotted against mode degree, ℓ. Numbers at selected branches indicate mode degree n. The f-mode branch is denoted with $n = 0$. The error bars correspond to standard deviations multiplied by 1000.

the rotational and granular contributions, the remaining signal is represented in terms of spherical harmonics and then the peaks in amplitude spectra for each ℓ and m determine $\nu_{n,\ell,m}$.

The acoustic travel times τ between two points in solar disc, $\mathbf{x}_1, \mathbf{x}_2$, are determined by analyzing cross-covariance of the oscillation signal from narrow ranges of the parameter d, which determines the ray path under the surface. The values $\tau(d, \mathbf{x}_1, \mathbf{x}_2)$ are determined as maxima in the cross-correlation function $\Psi(\tau)$. With data on the mean solar structure, we may translate d to the depth. In this way, measurements of sound travel time enable a three-dimensional probing of subphotospheric flows and structures changing in a relatively short time. This method of sounding the invisible interior, like in geophysics, is called *seismic tomography.* The method is applicable only to the outer layers of the Sun.

For deep sounding of the Sun we have to rely on the frequency data. Following suggestion of Ritzwoller and Lavely (1991), the data from spatially-resolved observations are now represented in the form

$$\nu_{\ell,\nu}(m) = \bar{\nu}_{\ell,\nu} + \sum_{k=1} a_{k,\ell,\nu} \mathcal{P}(m)_k^{\ell}, \tag{12}$$

where $\mathcal{P}(m)_k^\ell$ are orthogonal polynomials of degree k, with $k \leq 2\ell$. With this representation the mean multiplet frequencies, $\bar{\nu}_{\ell,\nu}$, probe the spherical structure. Individual a_{2j} coefficients reflect distortion from the spherical shape proportional to Legendre polynomial $P_{2j}(\cos\theta)$. The part of the distortion, which is not symmetric about the equator is not measurable by this method. Individual a_{2j-1} coefficients correspond to individual coefficients in the expansion of rotation rate $\Omega(r,\theta)$ in terms of the first derivatives of the Legendre polynomials of the odd orders.

The mean frequencies for 2050 modes determined by Rhodes et al. (1998) from measurements done with MDI instrument are shown in Figure 2. The 1σ error bars would not be visible if not multiplied by factor 1000. This illustrates the accuracy of the data we have at hand. We see that errors increase with frequency. The data for $\nu < 4$mHz are not really useful for seismic sounding.

Corresponding values of the a_1 coefficients are given in Figure 3. This is largest of all a coefficients. If there were no θ-dependence in the rotation rate all the remaining odd-order coefficients would vanish. In general, the a_1 coefficients are measure of the angular averages of Ω weighted with $\sin^2\theta$. The fact that a_1 coefficients are dominant and that they do not show any visible mode-dependence indicates that the solar rotation is close to uniform. The values show that the mean rate of rotation is close to that in the photosphere.

The a-coefficients of even orders exhibit large variations with the solar cycle. These variations are undoubtedly connected to the variability of the magnetic field. However, we do not have yet a theory that would allow us to use these data for a quantitative probing.

5. Probing the radial structure

5.1. METHODS

Duvall (1982) demonstrated that the solar oscillation frequencies obey, to a good approximation, the asymptotic relation given in Eq.(5) upon a suitable adjustment of a constant α. Gough (1984) showed that the formula may be analytical inverted to yield $c(r)$ in the interior. First application (Christensen-Dalsgaard et al. 1985) revealed that for $r > 0.4R$ there is a good agreement in $c(r)$ between the Sun and its model. At greater depth, the method was inadequate. Improved methods employing the asymptotic approximation (Brodsky & Vorontsov 1988; Christensen-Dalsgaard et al. 1988) extended downward the reliability of the sounding. Nevertheless none of them were well suited for probing the energy producing core.

Perhaps the most important result of the inversion based on asymptotic approximation was measuring the depth of the convective envelope

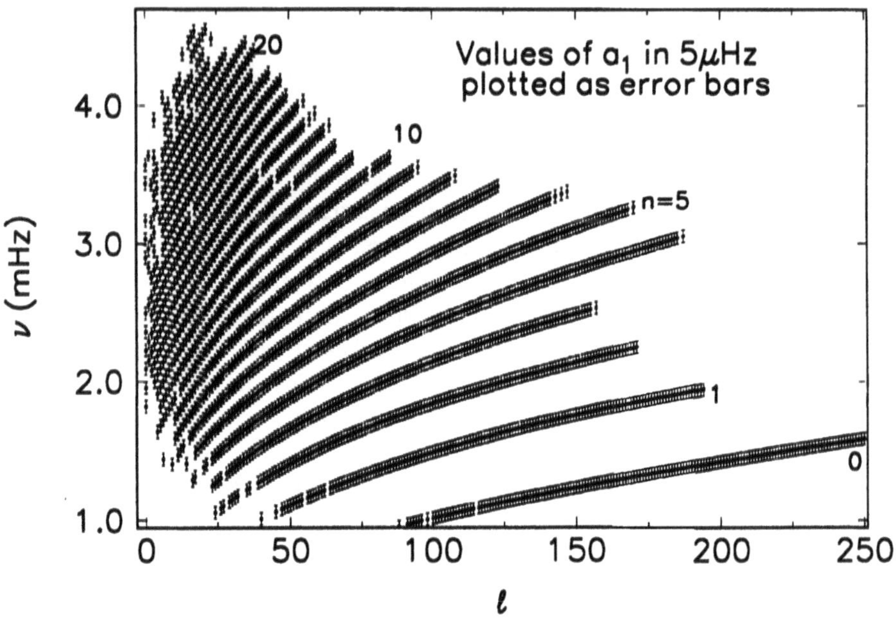

Figure 3. The same as Fig. 2, but instead of magnified standard deviation, the error bars show values of the a_1 coefficients magnified by factor 200

(Christensen-Dalsgaard et al. 1991). How this may be applied to understand the mechanism of magnetic activity will be discussed it next section. The results are also important in the context of the photospheric lithium abundance problem.

A method which did not employ the asymptotic approximation was developed by Dziembowski et al. (1990). The basic underlying assumption was that the differences in density, $\delta\rho/\rho$, pressure, $\delta p/p$, and the adiabatic exponent, $\delta\Gamma_1/\Gamma_1$, between the Sun and its model are small in the whole solar interior. The hydrostatic equilibrium condition links $\delta\rho$, δp. The two quantities may be conveniently expressed in terms of $\delta u/u$, where $u = p/\rho$. The differences in $\delta\Gamma_1$ may be expressed in terms of δu and the difference in helium abundance, δY. Since significant values of $\delta\Gamma_1$ are expected only in chemically homogeneous outer layers, the value of δY may be regarded constant and referring to the convective zone and the atmosphere.

In this way we may express the $\delta\vec{L}$ in terms of the unknown function $\delta u(r)$ and the parameter δY. The variational principle (Eq. 9) may be used to link the difference in mode frequencies to differences to δu and δY. However, the resulting integral expression cannot be directly applied to solar data because there are additional differences between the sun and the model which are not included in $\delta\vec{L}$. These differences are localized close to the surface and are due to inadequate treatment of the convection and nonadiabatic effects in the oscillations, which were not taken into account in Eq.(7). The localization implies that we may express these additional differences in the form $F_{\text{surf}}(\nu)/I_i$. Thus, finally, we get

$$\left(\frac{\delta\nu}{\nu}\right)_i = \int_0^1 \mathcal{K}_{u,i}\frac{\delta u}{u}dx + \mathcal{J}_i\delta Y + \frac{F_{\text{surf}}(\nu)}{I_i}, \qquad (13)$$

where $i \equiv (\ell, n)$ is and $x = r/R$. The kernels $\mathcal{K}_{u,i}$ and coefficients $\mathcal{J}_i$ are evaluated with the $y_i(x)$, $z_i(r)$ eigenfunctions calculated numerically for the reference model.

There are two distinct methods for determining $\delta u(x)$. One of them – known as the the least-square regularized method – consists in expressing $\delta u/u$ in terms of known functions. The coefficients together with the coefficients in the parametrization of $F_{\text{surf}}(\nu)$ and δY are determined by a minimization of

$$\sum_i \left(\frac{\delta\nu_i}{\sigma_i}\right)^2 + \lambda\int_0^1 \left[\frac{d}{dx}\left(\frac{\delta u}{u}\right)\right]^2 dx,$$

where σ_i are errors and the second term with an adjustable parameter λ is introduced to regularize the solution. Some regularization is needed because the problem is ill-posed and without regularization or if λ is too small, one obtains solutions which exhibit artificial oscillations.

The alternative is the optimal averaging method. In this case we abandon the idea of finding the function; instead we try to determine mean values of $\delta u/u$ weighted with possibly narrow kernels centered at selected x_0-values. We consider linear superpositions of individual kernels for the modes present in the data set:

$$K_u(x_0, x) = \sum_i c_i(x_0)\mathcal{K}_{u,i}(x) \qquad (14)$$

and try to determine coefficients c_i at selected points. In the *SOLA* version of this method, which was developed by Pijpers and Thompson (1992), one tries to construct kernels $K_u(x_0, x)$ which are possibly close to Gaussians $G(\frac{x-x_0}{w})$ characterized by their half-width at half-maximum, w. Subject to least-square minimization is the quantity

$$\int [K_u - G(\frac{x-x_0}{w})]^2 dx + \beta\sum_{i=1}^{J} c_i^2\sigma_i^2.$$

The second term, with a trade-off parameter β, is added to avoid large error magnification. For a specified w, larger β leads to smaller errors in

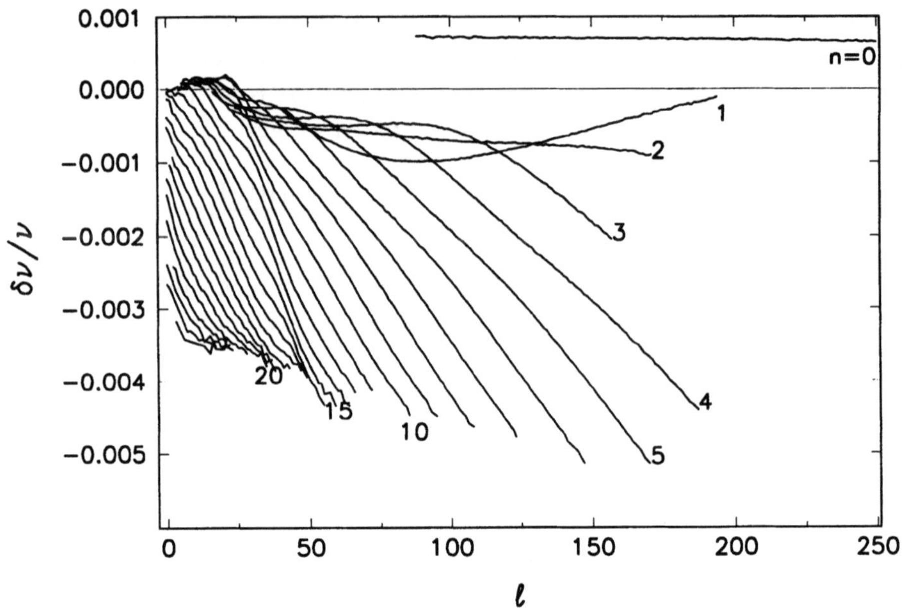

Figure 4. The relative differences in frequencies between the Sun and its model.

the localized mean but the kernels may differ significantly from Gaussians and therefore w cannot be regarded as the measure of localization.

5.2. RESULTS

Modern models reproduce solar frequencies with a relative accuracy better than 5×10^{-3}. This is illustrated in the Fig. 4, where I show, as continuous lines, the differences between solar frequencies and model frequencies. Solar frequencies are the same as shown in Figure 2. The calculated frequencies are for the standard solar model of Christensen-Dalsgaard et al.(1996). Similar agreement is found for data from other sources and for other solar models.

The nearly constant relative frequency differences in the case of f-mode ($n = 0$) lead Shou et al. (1997) to conclusion that there is a small difference between the solar radius and the radius of the model, from direct photometric measurements. With the use of equation like Eq.(9) here, they found that

$$\frac{R_{\mathrm{model}} - R_{\odot}}{R_{\odot}} = 4.5 \times 10^{-4}.$$

For the p-modes the dominant part of the frequency scales with $1/I_i$, which implies that most of it is due to the last term in Eq. (11). It is only the small remaining part which reflects the difference in the interior structure

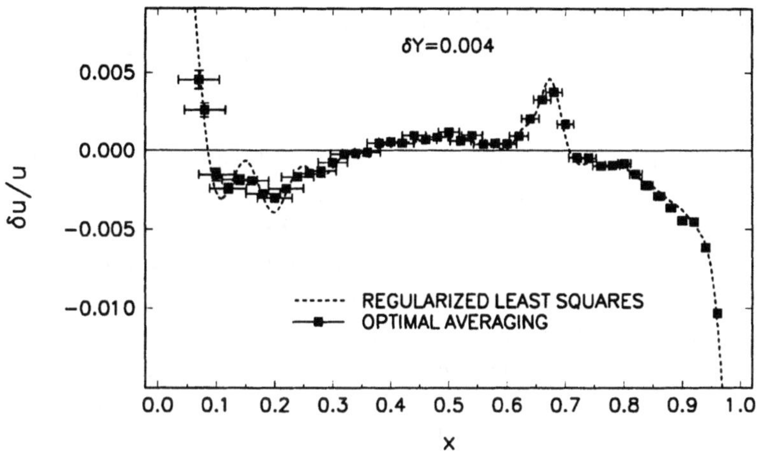

Figure 5. The values of the relative corrections to $u = p/rho$ in the model of solar interior and the correction to the helium abundance, Y, in the convective zone inferred from the p-mode frequency differences shown in Figure 4. The vertical error bars correspond to the width of the Gaussian kernels. The horizontal error bars reflect the errors in the frequency data.

and in the helium abundance between the Sun and the model. Figure 5 shows the result of the inversion of the frequency differences plotted in Figure 4. The two methods of the inversion were used, and we see that the results agree. We see also that the seismic correction everywhere , except the outermost and innermost layers, are less than 0.01. For $r < 0.05R$ the inversion is unreliable because p-modes have very weak sensitivity there. Only data for g-modes could change the situation.

With the use of a linearized mechanical equilibrium condition, the small corrections to u may be translated into small corrections to ρ and p. For the temperature, assuming the perfect gas law, we have

$$\frac{\delta T}{T} = \frac{\delta u}{u} + \frac{\delta Y}{1.58 - Y}. \tag{15}$$

Thus, only for the outer part of the Sun, which may be regarded as homogeneous, we have a direct seismic measurements of the temperature.

It seems justified to say that helioseismic sounding supports the standard solar evolutionary model of the Sun. The remaining differences in the radiative interior may be explained by suitable changes in the opacity coefficient which are within the uncertainty of its calculation (Tripathy et al. 1998). The differences in the convective zone may be eliminated by introducing acceptable modification in modeling the energy transport in this zone. Although seismology does not provide a direct measure of the temperature in the core, which is most important for evaluation of the neutrino fluxes, the good agreement between the standard model and the Sun demonstrated in Figure 5, effectively rules out the possibility that the neutrino deficit problem may be solved by certain modifications of the solar model.

The important addition to the standard solar model construction prompted by the results from seismic sounding is taking into account the downward drift of the elements heavier than hydrogen. This slow movement, if not inhibited, reduces the photospheric helium abundance by some 10 percent. When the first helioseismic measurements of the helium abundance were made (Vorontsov et al. 1991; Dziembowski et al. 1991) the element segregation has not been included in the reference solar models and the determined corrections δY was about 0.03. Shortly after Christensen-Dalsgaard et al. (1993) showed that inclusion of the effect of the element segregation in the model significantly reduces also differences in the sound speed.

6. Measuring the rotation rate in the Sun's interior

6.1. METHODS

If we substitute in Eq.(11) $\Omega(r,\theta)$ in the form of the following series

$$\Omega(r,\mu) = \sum_{k=0} \tilde{\Omega}_{2k+1}(r)\frac{dP_{2k+1}(\mu)}{d\mu}, \tag{16}$$

where $\mu = \cos\theta$, then a one-to-one correspondence between the a_{2k+1} coefficients and the $\tilde{\Omega}_{2k+1}(r)$ coefficients follows (Ritzwoller & Lavely 1992). The even-order term in the expansion of Ω, if present, average out. If there is equatorial symmetry, which we believe is approximately true, then the determination of $\Omega(r,\mu)$ may be reduced to an independent inverse problem of the following form

$$\int_0^1 \mathcal{K}_{\mathrm{rot},i}(x)\tilde{\Omega}_{2k+1}(x)dx = 2\pi a_{2k+1,i}. \tag{17}$$

An analytical expression for the kernels $\mathcal{K}_{\mathrm{rot},i}$ in terms of the eigenfunctions may be found in Schou et al. (1994) or Pijpers (1997). Any of the two methods outlined in the previous section may be employed to determine $\tilde{\Omega}_{2k+1}$.

Alternatively, the independent one-dimensional problems may be cast to a two-dimensional inverse problem

$$\int_0^1 dx \int_{-1}^1 d\mu \mathcal{M}_{k,i}(x,\mu)\Omega_(x,\mu)dx = 2\pi a_{2k+1,i}, \tag{18}$$

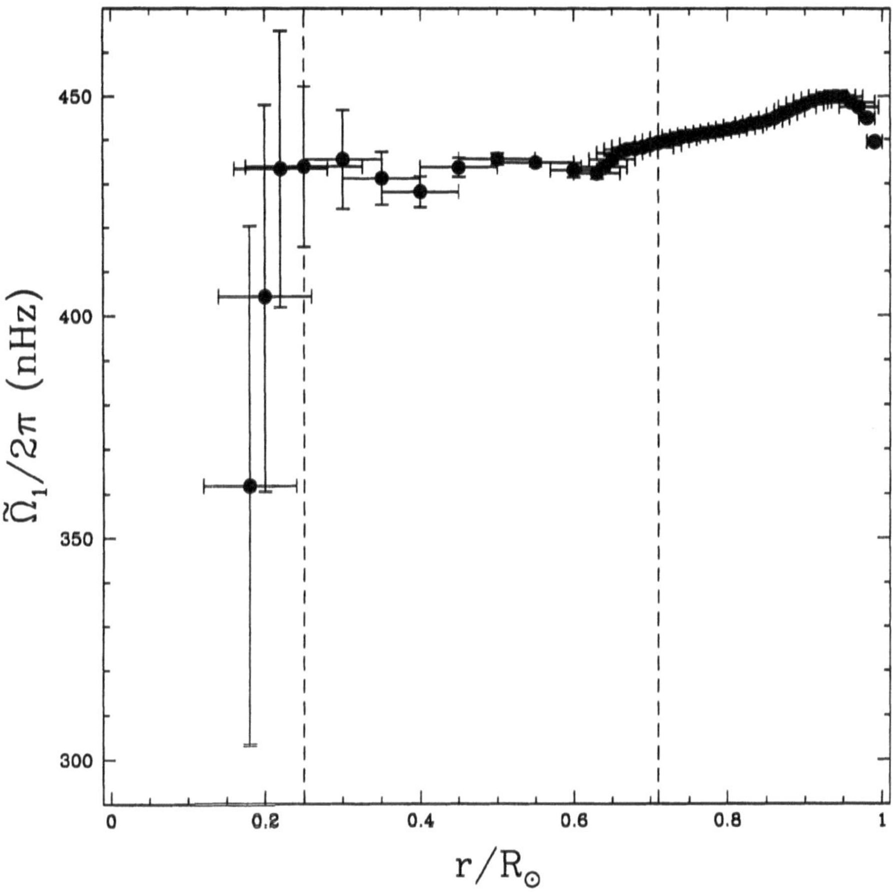

Figure 6. $\tilde{\Omega}_1$ determined by DiMauro et al. (1998) from the SOHO/MDI a_1 data. The vertical lines mark positions of the core edge and the bottom of the convective envelope.

where

$$\mathcal{M}_{k,i}(r,\mu) = \mathcal{K}_{k,i}(x)\frac{dP_{2k+1}}{d\mu}\frac{1-\mu^2}{(2k+1)(2k+2)}.$$

The basic methodology of solution is similar to that in the one-dimensional case. Again the choices are the regularized least-square and the optimal averaging methods. A detailed discussion is given by Schou et al. (1994, 1998).

6.2. RESULTS

Showing that there is no significant depth-dependence in the equatorial rotation rate by Duvall et al. (1984) was the first significant result from helioseismic sounding. Though still crude, the data on the rotation in the interior were sufficient to rule out large contribution of the gravitational quadrupole moment to the solar oblateness postulated by Dicke and Goldenberg (1967).

Not long after we learned from Duvall and Harvey (1986) that in the whole convective envelope the latitude dependence of the rotation rate is similar to that determined for the photosphere. During the next few years new measurements of the splitting coefficients (Brown and Morrow 1987; Libbrecht 1989) gave us more detailed information about the behavior of solar rotation in the convective zone and in the outer part of the radiative interior. It was shown that there is a narrow layer, called after Spiegel and Zahn (1992) *the solar tachocline*, where the transition from the latitude dependent to uniform rotation takes place.

The tachocline is regarded as the most likely site of the solar dynamo (e.g. Gilman et al. 1989). Kosovichev (1996a) showed that it is located predominantly in the radiative interior. This result is very important for understanding magnetic field generation of magnetic field in the Sun and other stars. However, we still do not have a fully satisfactory theory.

Two dimensional inversions of data from the MDI instrument (Schou, 1998) resulted in a fairly detailed map of the rotation rate for the outer ($r \geq 0.5R$) part of the Sun. One result worth emphasizing is much slower rotation in the polar region than implied by the traditional three-term expansion in $\mu^2 = \cos^2\theta$ or derivatives of P_{2k+1}, as in Eq.(16).

The plot in Fig. 6 shows the behavior of

$$\widetilde{\Omega}_1 = \frac{3}{2}\int_0^{\pi} \Omega \sin^2\theta d\theta, \tag{19}$$

as found by means of the optimal averaging method of inversion applied to the a_1 coefficients shown in Figure 3. Note that $\frac{2}{3}\widetilde{\Omega}_1 r^2 \rho dr$ gives the local contribution to the angular momentum. The plot shows that the mean rotation rate increases with depth, as suggested long time ago by the rotation rate of photospheric tracers. We also note that the region of nonuniform rotation extends into the radiative interior by about $0.06R$. The sharp change in the tachocline is better seen in Figure 7, which displays $\widetilde{\Omega}_3$ in the outer part of the Sun.

Below $r = 0.4R$ the errors of determination increase rapidly. It seems fair to say that we do not know how the solar core rotates. There is a hint that the angular velocity in the core may be lower than in the envelope but this is a controversial result found only from inversions of certain data sets. Unfortunately, p-modes have only very weak sensitivity to the core rotation. I am afraid that the result will remain controversial until we obtain rotational splitting data for g-modes.

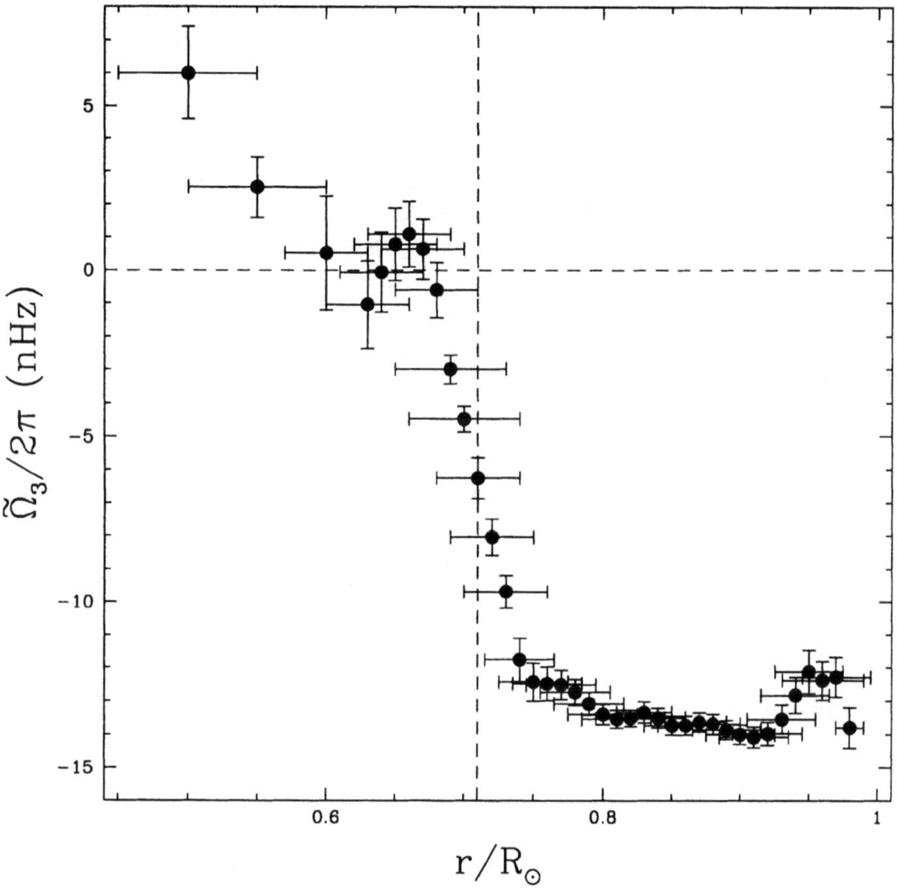

Figure 7. $\tilde{\Omega}_3$ determined by DiMauro et al. (1998) from the SOHO/MDI a_3 data.

7. Time-distance helioseismology

One advantage of time-distance helioseismology is its ability to probe structures and flows which are not symmetric about the equator and whose signatures are averaged out in mode frequencies. For extracting the information contained in the $\tau(d, \mathbf{x}_1, \mathbf{x}_2)$ data we rely on the *Fermat principle*, which allows us to use ray path in the standard solar model, for evaluating changes induced in τ by flows, magnetic fields and local perturbations of the sound speed. A signature of the flows is $\tau(d, \mathbf{x}_1, \mathbf{x}_2) \neq \tau(d, \mathbf{x}_2, \mathbf{x}_1)$. Disentangling two other effects requires modeling.

Kosovichev (1996b) developed a three-dimensional method of inversion of the $\tau(d, \mathbf{x}_1, \mathbf{x}_2)$ data and applied it to infer properties of convection in various regions of the Sun. In his work he relied on the ray theory approximation, which is not applicable in the outermost layers because condition (3) is not satisfied. Bogdan (1997) discussed the consequences of using a more accurate description.

The time-distance analysis made by Duvall et al. (1996) on the oscillation data of Duvall et al. (1993) lead to detection of downflows around the sunspots with velocities reaching up to 1.5 km/s. Inversion of the time-distance data from the same set revealed the existence of the sound-speed perturbations beneath the active region down to 32 Mm from the photosphere. In another application of the time-distance analysis Duvall et al. (1997) determined the vertical extent of the supergranular velocity field for 8Mm. This points to a superficial origin of the rather than a connection with the HeII ionization zone, as it has been suggested. The data used in this analysis were from the MDI instrument

Long searched meridional circulation has been discovered by Giles et al. (1997), who also carried out a time-distance analysis on the MDI data. The poleward flows with velocities ~ 20 km/s were measured down to the depth of 26 Mm. This discovery was confirmed by Basu et al. (1998), who used the ring diagram analysis – an alternative method of local seismology – on data from the GONG network.

8. Final remarks

The chances for a significant progress in the accuracy of testing models of the Sun's interior depend on detecting g-modes and measuring their frequencies. The same applies to prospects for measuring rotation in the inner core. However, I see nothing that could suggest that the simple picture of Sun's interior we have today could be challenged by such new data.

Whether or not we such data, helioseismology will remain an important tool of solar physics because the Sun is a variable object. The outstanding unsolved problem is the mechanism of solar magnetic activity varying in the eleven-year cycle. I believe that helioseismology will play a crucial role in searching the solution. There is a role for local seismology in providing three-dimensional maps of constantly changing outer layers of the Sun. The long term monitoring of changes in frequencies, on the other hand, is the best tool for searching the deep seated causes of the spectacular changes in the outer layers.

For further reading on helioseismology, I suggest a comprehensive introduction to this field written by Christensen-Dalsgaard et al. (1999) on the basis of several lectures given by the four coauthors at NATO ASI in Ceşme (Turkey) in 1998.

References

1. Basu, S., Antia, H.M. and Tripathy, S.C. (1998) Ring diagram analysis of the velocity fields within the solar convective zone, in *Structure and dynamics of the Sun and Sun-like stars, Proc. SOHO 6/GONG 98 Workshop*, eds. Korzennik, S.G. and Wilson, A., (ESA: Nordwijk), 705–710
2. Bogdan, T.J. (1997) A comment on the relationship between the modal and time-distance formulations of the local helioseismology, *Astrophys. J.* **477**, 475–484
3. Brodsky, M.Y. and Vorontsov, S.V. (1988) On the technique of the inversion of helioseismological data, in *IAU Symposium No 123, Advances in helio- and asteroseismology* eds. Christensen-Dalsgaard, J. and Frandsen, S., (Reidel: Dordrecht), 137–140
4. Brown, T.M. and Morrow, C.A. (1987) The depth and latitude dependence of solar rotation, *Astrophys. J.* **314**, L21–L26
5. Christensen-Dalsgaard, J., Däppen, W., Dziembowski, W.A. and Guzik, J.A. (1999), Introduction to helioseismology, in *Variable Stars as Essential Astrophysical Tool*, ed. Ibanglu, (Kluver), 00–00
6. Christensen-Dalsgaard, J., Duvall, T.L., Gough, D.O., Harvey, J.W. and Rhodes, E.J. (1985) Speed of sound in the solar interior, *Nature* **315**, 378–382
7. Christensen-Dalsgaard, J., Gough, D.O. and Thompson, M.J. (1989) Differential asymptotic sound-speed inversions, *Mon. Not. Roy. Astr. Soc.* **238**, 481–502
8. Christensen-Dalsgaard, J., Gough, D.O. and Thompson, M.J. (1991) The depth of the solar convective zone, *Astrophys. J.* **378**, 413–437
9. Christensen-Dalsgaard, J., Proffitt and Thompson, M.J. (1993) Effects of diffusion on solar models and their oscillation frequencies, *Astrophys. J.* **403**, L75–L78
10. Christensen-Dalsgaard, J. et al. (1996), The current state of solar modeling, *Science* **272**, 1286–1292
11. Claverie, A., Isaak, G.R., McLeod, C.P., van der Raay, H.B,, and Roca Cortes, T. (1979) Solar structure from global studies of the 5-minute oscillations , *Nature* **282**, 591–594
12. Deubner, F.-L. (1975) Observations of low wavenumber nonradial eigenmodes of the Sun, *Astron. Astrophys.* **72**, 371–375
13. Dicke, R.H. and Goldenberg, H.M. (1967) Solar oblateness and general relativity, *Phys. Rev. Lett.* **18**, 313–316
14. DiMauro, M.P., Dziembowski, W.A. and Paternó, (1998) Rotation in the solar interior: new results by helioseismic data inversion, in *Structure and dynamics of the Sun and Sun-like stars, Proc. SOHO 6/GONG 98 Workshop*, eds. Korzennik, S.G. and Wilson, A., (ESA: Nordwijk), 759-762
15. Duvall, T.L. (1982) A dispersion law for solar oscillations, *Nature* **300**, 242–243
16. Duvall, T.L., Dziembowski, W.A., Goode, P.R., Gough, D.O., Harvey, J.W. and Leibacher, J. (1984) The internal rotation of the Sun, *Nature* **310**, 22–25
17. Duvall, T.L. and Harvey (1984) Rotational frequency splitting of solar oscillations, *Nature* **310**, 19–22
18. Duvall, T.L., Harvey, J.W. and Pomerantz, M.A. (1986) Latitude and depth variation of solar rotation of the Sun, *Nature* **321**, 500–501
19. Duvall, T.L. et al. (1997), Time-distance helioseismology with the MDI instrument: initial results, *Solar Phys.* **170**, 63–73
20. Dziembowski, W.A., Pamyatnykh, A.A. and Sienkiewicz, R. (1990) Solar model from helioseismology and the neutrino flux problem, *Mon. Not. Roy. Astr. Soc.* **244**, 542–550
21. Dziembowski, W.A., Pamyatnykh, A.A. and Sienkiewicz, R. (1991)Helium content in the solar convective zone from helioseismology , *Mon. Not. Roy. Astr. Soc.* **249**, 602–605
22. Gilman, P.A., Morrow, C.A. and DeLuca, E.E. (1989) Angular momentum transport and dynamo action in the Sun: implications from recent oscillation measurements, *Astrophys. J.* **338**, 528–537

23. Giles, P.M., Duvall, T.L., Scherrer, P.H. and Bogart, R.S. (1997) A subphotospheric flow from the Sun's equator to its pole, *Nature* **390**, 52–54
24. Gough, D.O. (1984) Toward solar model, *Mem. Soc. Astron. Ital.* **55**, 13–35
25. Grec, G., Fossat, E. and Pomerantz (1980) Solar oscillations: full disk observations from the geographic South Pole , *Nature* **288**, 541–544
26. Kosovichev, A.G. (1996a) Helioseismic constraints on the gradient of the angular velocity at the base of the solar convective zone, *Astrophys. J.* **469**, L61–L64
27. Kosovichev, A.G. (1996b) Tomographic imaging of the Sun's interior, *Astrophys. J.* **471**, L55–L57
28. Leighton, R.B., Noyes, R.W. and Simon, G.W. (1962) Velocity fields in the solar photosphere I. Preliminary report, *Astrophys. J.* **135**, 474–499
29. Libbrecht, K.G. (1988) Solar p-mode phenomenology, *Astrophys. J.* **334**, 510–516
30. Libbrecht, K.G. (1989) Solar p-mode frequency splitting, *Astrophys. J.* **336**, 1092–1097
31. Pijpers, F. P. (1997) Solar rotation inversions and relationships between a-coefficients and mode splittings , *Astron. Astrophys.* **326**, 1235–1240
32. Pijpers, F. P. and Thompson (1992) Faster formulation of the optimally localized averages method for helioseismic inversions, *Astron. Astrophys.* **262**, L33–L36
33. Rhodes, E.J., Kosovichev, A.G., Schou, J., Sherrer, P.H. and Reiter, J. (1997) Measurements of frequencies of solar oscillations from the MDI medium-ℓ program, *Solar Phys.* **175**, 287–310
34. Ritzwoller, M.H. and Lavely, E.M. (1991) A unified approach to helioseismic forward problem, *Astrophys. J.*, **369**, 557–566
35. Rudiger, G. (1989) *Differential Rotation and Stellar Convection.* Academie-Verlag, Berlin.
36. Schou, J., Christensen-Dalsgaard, J. and Thompson, M.J. (1994) On comparing helioseismic inversion methods, *Astrophys. J.* **433**, 389-416
37. Schou, J. et al. (1998) helioseismic studies of differential rotation in the solar envelope by the Solar Oscillation investigation using Michelson Doppler Imager, *Astrophys. J.* **505**, 390-417
38. Schou, J., Kosovichev, A.G., Goode, P.R. and Dziembowski, W. A. (1997) Determination of the Sun's seismic radius from the SOHO Michelson Doppler Imager, *Astrophys. J.* **489**, L197-L200
39. Spiegel, E. A. and Zahn, J.-P. (1992) The solar tachocline, *Astrophys. J.* **489**, L197-L200
40. Tripathy, S.C. Basu and Christensen-Dalsgaard, J.(1998), Helioseismic determination of opacity corrections, in *IAU Symposium, Sounding Solar and Stellar Insteriors* (poster volume), eds. Provost, J. and Schmider, F.X., (Université de Nice), 129–130.
41. Ulrich, R.K. (1970) The five-minute oscillations on the solar surface, *Astrophys. J.* **162**, 993–1001
42. Vorontsov, S.V., Baturin, V.A. and Pamyatnykh, A. (1991) Seismological measurement of the solar helium abundance, *Nature* **349**, 49–51

WHAT WE KNOW ABOUT THE SOLAR INTERIOR

J.-P. ZAHN
Observatoire de Paris
F–92195 Meudon, France

Abstract. Because the interior of the Sun is opaque to electromagnetic waves, the radiation we receive from it on Earth is emitted in the outermost layers: the photosphere. Therefore our knowledge of the solar interior is based solely on theoretical models, which are built by making some plausible assumptions about the physical conditions and processes that are likely to prevail there. Fortunately, a powerful technique – helioseismology – has been developed in the last twenty years, which permits to probe directly the solar interior by means of acoustic waves, and this had a tremendous impact on solar physics because it provides tight observational constraints on our models. We shall illustrate this here by a few examples of recent advances in modeling the Sun.

1. Main properties

Until recently, all what we knew about the Sun came from observing it in visible light. Its distance and radius $R_\odot$ have been determined long ago by triangulation of the solar system, although the precision was increased significantly by using the modern technique of radar echoes; the result is $R_\odot = 6.9626\,10^{10}$ cm. In the early 19th century, spectroscopy of the solar light revealed the gaseous nature of the emitting layers and the presence of many chemical elements known on Earth, with the notable exception of helium, which was detected on the Sun before it was discovered on Earth. Later, with the advent of atomic physics, it became possible to decrypt the spectral lines, and to assess the temperature and composition of the surface layers: hydrogen turned out to be most abundant species, followed by helium (about 25% in mass), the other elements representing only 2% altogether (see Table 1 for a recent determination of these element abundances). The mass of the Sun, $M_\odot = 1.9891\,10^{33}$g, is deduced from the

J.-P. Zahn and M. Stavinschi (eds.),
Advances in Solar Research at Eclipses from Ground and from Space, 265–288.

orbital motions of the planets and from precise laboratory measurements of the gravitational attraction between two bodies. Finally, the heat flux received on Earth yields the total amount of thermal energy radiated by the Sun into space, i.e. its luminosity: $L_\odot = 3.845\,10^{33}$ ergs/s.

TABLE 1. Element abundances in the solar photosphere and in meteorites, expressed in number of atoms in a logarithmic scale (Grevesse & Sauval 1998)

Element	Photosphere	Meteorites	Element	Photosphere	Meteorites
H	12.00	–	Cr	5.67	5.69
He	[10.93]	–	Mn	5.39	5.33
Li	1.10	3.31	Fe	7.50	7.50
Be	1.40	1.42	Co	4.92	4.91
B	2.55	2.79	Ni	6.25	6.25
C	8.52	–	Cu	4.21	4.29
N	7.92	–	Zn	4.60	4.67
O	8.83	–	Ga	2.88	3.13
F	[4.56]	4.48	Ge	3.41	3.63
Ne	[8.08]	–	As	–	2.37
Na	6.33	6.32	Se	–	3.41
Mg	7.58	7.58	Br	–	2.63
Al	6.47	6.49	Kr	–	3.31
Si	7.55	7.56	Rb	2.60	2.41
P	5.45	5.56	Sr	2.97	2.92
S	7.33	7.20	Y	2.24	2.23
Cl	[5.5]	5.28	Zr	2.60	2.61
Ar	6.40	–	Mo	1.92	1.97
K	[6.40]	–	Sn	2.0	2.14
Ca	6.36	6.65	Te	–	2.24
Sc	3.17	3.10	Xe	–	2.17
Ti	5.02	4.94	Ba	2.13	2.22
V	4.00	4.02	Pb	1.95	2.06

With a few 'reasonable' assumptions, these global parameters permit to readily derive some important properties of the solar interior – on the back of an envelope, so to speak.

1.1. CENTRAL TEMPERATURE

The first of these assumptions is that of *hydrostatic equilibrium*, which states that everywhere in the solar interior there is an exact balance between gravity and the pressure forces. This allows to evaluate the central pressure, which is the weight of a column of matter of unit area; it is given by

$P_c = \overline{\rho}\,\overline{g} R_\odot$ where $\overline{\rho}$ and $\overline{g}$ are suitable averages of the density and the gravity along the radius. As crude estimates, one may take the median value of all involved quantities:

$$\overline{\rho} \approx \frac{\rho_c}{2} \qquad \overline{g} \approx \mathcal{G}\frac{M_\odot/2}{(R_\odot/2)^2}$$

with ρ_c being the central density and $\mathcal{G}$ ($= 6.673\,10^{-8}$ in cgs units) the gravitational constant, which leads to

$$P_c \approx \rho_c \frac{\mathcal{G} M_\odot}{R_\odot}.$$

One may further assume that the solar material consists mostly of ionized hydrogen, as suggested by its surface composition, the molecular mass thus being $\mu = 0.5$ g, and that it obeys the perfect gas law $P = \mathcal{R}\rho T/\mu$, with $\mathcal{R} = 8.315\,10^7$ (in cgs) being the gas constant. This yields an estimate of the central temperature which depends only on the mass and the radius:

$$T_{\text{central}} \approx \frac{\mu}{\mathcal{R}} \frac{\mathcal{G} M_\odot}{R_\odot} \approx 10^7\,\text{K}, \tag{1}$$

a crude figure which is not that far from the value predicted by the best models of the present Sun.

1.2. HEAT TRANSPORT

This central temperature is rather high compared to that of the surface layers, whose effective temperature T_{eff}, defined as that of the black body which radiates the same luminosity as the Sun: $L_\odot = 4\pi\sigma R_\odot^2\, T_{\text{eff}}^4$, is approximately 5,800 K (σ being the Stefan constant). It tells us that the thermal energy radiated into space is produced in the deep interior. One way to transport this energy to the surface is by radiation: the photons travel a short distance, are either scattered in another direction, or absorbed and re-emitted, and so on, until they finally reach the outermost layers from which they escape. Their mean-free path is small enough to guarantee that the solar interior is locally in thermal equilibrium and thus that radiation is close to that of the ideal black body, meaning that the energy density only depends on temperature, and that it scales as T^4. This allow to estimate the *optical thickness* τ of the Sun, along its radius, which scales also as T^4:

$$\tau \approx \left(\frac{T_{\text{central}}}{T_{\text{eff}}}\right)^4 \approx 10^{13}.$$

This huge figure indicates that the Sun is extremely opaque: on the average the mean free path of a photon is of order $R_\odot/\tau \approx 10^{-2}$ cm.

If it did not interact so strongly with matter, a photon emitted in the central region would cross the Sun in 2.3 seconds, but this very opaque medium forces it into a random walk: instead of τ interactions with matter, it will undergo τ^2 of them and therefore it will take τ times longer to escape, which amounts to 10^6 years.

But closer inspection of the solar surface shows that it is the seat of turbulent motions which will also contribute to the heat transport. Indeed, near the surface, where hydrogen and helium are partly ionized but still excited enough, the opacity is so high that the thermal stratification becomes unstable: a bubble of solar matter which is lifted there from its equilibrium position becomes lighter than its environment, and hence it continues to rise. This *convective instability* occurs whenever the temperature gradient exceeds a critical value, which is the adiabatic gradient (i.e. the gradient that would be experienced by the bubble if it did not exchange heat with the surrounding medium). The instability condition is usually expressed in terms of logarithmic temperature gradients with respect to pressure:

$$\frac{d \log T}{d \log P} > \left(\frac{\partial \log T}{\partial \log P} \right)_{\text{adiabatic}}$$

Detailed models indicate that this criterion is satisfied in the outer third of the Sun (in radius), the convection zone. This turbulent region is the cause of the various phenomena which characterize the solar activity. The inner two thirds of the Sun are in stable equilibrium, and there the thermal energy is transported by radiation.

1.3. THE ENERGY SOURCE

For a long time, the source of solar energy remained a puzzle. All energy sources which were considered in the 19th century predicted a life expectancy for the Sun which was much smaller than the estimated age of some rocks on Earth. The problem was solved only in the 1920's, as one became aware that nuclear reactions could provide the necessary energy. When hydrogen is transformed into helium, 0.7% of the mass is changed into energy, according to Einstein's formula $E = mc^2$. One can easily estimate the Sun's life-expectancy t_{exp} by equating the rate at which energy is emitted with the total amount that is available:

$$t_{\text{exp}} = 0.1 \times 0.007 M_{\odot} c^2 / L_{\odot} \approx 10^{10} \text{ years}$$

where we have assumed that only 1/10 of the solar material will be changed into energy, namely that which is located in the innermost region, where the temperature is high enough to sustain nuclear reactions. This figure is compatible with the age of the solar system, which is now determined by precise radiochemistry of its oldest components, namely the meteorites, and which is found to be 4.6 billion years.

The specific nuclear reactions which are responsible for the energy release in the Sun have been identified later on. The main contribution comes from the p-p chain, which starts with the fusion of two protons into a nucleus of deuterium: $p + p \rightarrow {}^2H + e^+ + \nu_e$. This reaction involves the weak interaction and therefore emits a neutrino; since it is very slow, it regulates the whole energy production. Other reactions follow, until ^{4}He is produced.

The next step of nuclear burning will transform helium into carbon, but with ten times lesser efficiency per unit mass processed. It explains why most stars are observed in the hydrogen burning phase, on the so-called *main sequence.*

The production rate of nuclear energy in the solar interior is very sensitive to temperature, because the fusion of two nuclei can only occur when their electrostatic repulsion is overcome by their kinetic energy: the higher the temperature, the higher the release of nuclear energy. This has an important consequence: assume that for some reason the production of nuclear energy slightly increases: additional heat is then deposited in the outer layers, which for this reason expand; but through hydrostatic balance that increase in radius leads to a decrease of the central temperature (see eq. 1), which has the effect of reducing the nuclear reaction rate. This self-regulation ensures that main-sequence stars, such as the Sun, are in stable thermal equilibrium.

2. Modeling the solar interior

A stellar model describes the internal constitution of a star, namely its pressure and temperature stratification, its mass distribution, and also its chemical composition, as a function of depth, at a given age. Most models are based on the assumption of hydrostatic equilibrium, although small departures from it are easily accounted for by including for instance the turbulent pressure in the convection zone. They also assume spherical symmetry, a simplification which is well justified for the solar interior, where the centrifugal force and the magnetic stresses may indeed be neglected, compared to gravity.

2.1. THE GOVERNING EQUATIONS

The first of these equations expresses the hydrostatic equilibrium:

$$\frac{dP}{dr} = -\rho \frac{d\Phi}{dr}.$$

the gravitational potential Φ being derived from Poisson equation

$$\nabla^2\Phi + 4\pi\mathcal{G}\rho = 0 \quad \Longrightarrow \quad \frac{d\Phi}{dr} = \frac{\mathcal{G}m(r)}{r^2}.$$

The two combine into

$$\frac{dP}{dr} = -\rho\frac{\mathcal{G}m(r)}{r^2} \tag{2}$$

with

$$\frac{dm}{dr} = 4\pi r^2\rho. \tag{3}$$

If the pressure P is a function of density ρ only, in which case the star is a *barotrope*, those two equations (2) and (3) suffice to determine its structure. But in general P is also a function of temperature T, and in that case we need to draw some information from the conservation of energy, which we may express as

$$\rho T\frac{dS}{dt} = -\frac{1}{r^2}\frac{d}{dr}F(r) + \rho\varepsilon \tag{4}$$

(S is the specific entropy). In the Sun the energy flux has mainly two components: the radiation flux and the convective flux:

$$F = F_{\mathrm{rad}}(r) + F_{\mathrm{conv}}(r).$$

At sufficient optical depth, we may use Eddington's approximation to relate the radiation flux with the temperature gradient:

$$F_{\mathrm{rad}} = -\chi\frac{dT}{dr} \qquad \text{with} \quad \chi = \frac{16\sigma T^3}{3\rho\kappa}, \tag{5}$$

χ being the radiative conductivity and κ the opacity. We shall see below how the convective flux is expressed likewise in terms of the entropy gradient.

These equations are completed with an equation of state which gives the pressure (gas pressure + radiation pressure) as a function of density, temperature and composition, $P = P(\rho, T, X_i)$, the nuclear energy production rate $\varepsilon = \varepsilon(\rho, T, X_i)$, and the opacity $\kappa = \kappa(\rho, T, X_i)$, with X_i being the mass fraction of the chemical species which are taken in account to build the model.

We thus have a 4-th order differential system in r, the dependent variables being P, T, m and L.[1] To solve it, 4 boundary conditions are required, which are the obvious $m = 0$ for the mass and $L = 0$ for the luminosity

[1] Actually, one prefers to take the mass m as the independent variable.

at center $r = 0$; at the surface, defined by some optical depth around unity, the value of the pressure P and that of the temperature T are usually borrowed from an atmosphere model.

This would be sufficient to build a model if the chemical composition were known at each depth. But this composition varies with time, due to turbulent mixing, microscopic diffusion and nuclear reactions, and it too must be calculated in a consistent way. For that purpose one has to add a set of n differential equations, of first order in time and second order in space, which follow the evolution of each chemical species under the effect of microscopic diffusion and nuclear reactions:

$$\frac{d}{dt}\rho X_i(r) = \frac{1}{r^2}\frac{d}{dr}\left[r^2 \rho D_t \frac{dX_i}{dr}\right] - \frac{1}{r^2}\frac{d}{dr}[r^2 \rho X_i U_i] + \sum_j \rho s_{ij}(\rho, T, X_i, X_j), \tag{6}$$

where D_t is the turbulent diffusivity (or more generally the macroscopic diffusivity), U_i the microscopic diffusion velocity, which may be drawn from Boltzmann's equation, and the s_{ij} represent the sources and sinks due to nuclear reactions. The associated boundary conditions simply state that the flux of each species vanishes at center and surface.

In the first stellar models built in the 1950's, one was able only to track the transformation of hydrogen into helium, but nowadays n is no longer limited by the computer performances, and one deals currently with a dozen or more species. The time integration of this differential system is started at an early stage, when the star is still entirely convective, with specified uniform composition.

2.1.1. *The convective flux*

As we have seen in §1.2, when the temperature becomes too steep, the medium becomes convectively unstable, and this occurs whenever

$$\frac{d\log T}{d\log P} > \left(\frac{\partial \log T}{\partial \log P}\right)_{\text{adiabatic}} \quad \text{or in short} \quad \nabla > \nabla_{\text{ad}}. \tag{7}$$

This instability condition is known as the Schwarzschild criterion. It predicts that solar-type stars possess an outer convection zone, because near their surface the opacity strongly increases due to the recombination of the two major constituents, hydrogen and helium.

At sufficient optical depth, the transport of energy through the convective motions is usually treated as a diffusion process; the convective flux is taken proportional to the entropy gradient:

$$F_{\text{conv}} = -D_t \rho T \frac{dS}{dr}.$$

The turbulent diffusivity D_t may be expressed in terms of the velocity v and mean-free path ℓ of the turbulent eddies: $D_t = k v \ell$, with k being a coefficient of order unity (it would be 1/3 for isotropic turbulence). An equivalent, more convenient form for the convective flux is

$$F_{\rm conv} = D_t \frac{C_P \rho T}{H_P} \left[\nabla - \nabla_{\rm ad}\right], \tag{8}$$

where C_P is the heat capacity at constant pressure, $H_P = P/\rho g$ the pressure scale height, $\nabla = d\ln T/d\ln P$ and $\nabla_{\rm ad} = (\partial \ln T/\partial \ln P)_{\rm ad}$.

The turbulent velocity is estimated by letting an eddy accelerate through buoyancy in the ambient medium, over its mean free path ℓ:

$$\frac{1}{2}v^2 = -\int_0^\ell g\frac{\delta\rho}{\rho}\,dr \approx -g\int_0^\ell \left[\left(\frac{d\ln\rho}{dr}\right)_{\rm eddy} - \left(\frac{d\ln\rho}{dr}\right)_{\rm ambient}\right] \ell' d\ell'.$$

The usual prescription is to take this mean free path, usually called *mixing-length*, proportional to the pressure scale height: $\ell = \alpha H_P$, with α being a parameter of order unity (Vitense 1953).

Undoubtly, this treatment of convection must appear extremely crude to any physicist, and indeed this remains one of the weakest points of modeling the solar interior.

2.1.2. *The need to calibrate*

Mathematically, the problem is well defined: the differential system is of order $4 + 2n$ in space with an equal number of imposed boundary conditions. Its numerical integration presents no fundamental difficulty. Thus it would seem that it suffices to run a stellar structure code with specified mass $M_\odot$ to the present age of the Sun in order to obtain an acceptable solar model. One could then validate such a model by comparing the predicted radius and luminosity with the observed ones. This would indeed be the case if the initial composition were known with high precision and if the physical processes coming into play were all translated into perfect numerical prescriptions. Unfortunately, these two conditions are not satisfied yet.

The abundance of helium determined by spectroscopy is rather inaccurate: most of its spectral lines are hard to interpret because they are produced in the chromosphere and in the corona, where there are large departures from local thermodynamic equilibrium. Furthermore, what one needs to start the time integration is the initial abundance Y_0, which may differ significantly from the present one because helium, being heavier than hydrogen, may have settled below the observed layers. For this reason Y_0 is taken as an arbitrary parameter (chosen in some reasonable range), to be adjusted by comparing the model with the actual Sun.

Another major uncertainty arises from the poor description of the convective heat transport. As we have seen, the recipe which is still commonly used, for lack of something better, involves also an arbitrary parameter, which is the ratio α between the mixing length and the pressure scale height.

For these reasons, the time integration of the constitutive equations is started with trial values for the two parameters Y_0 and α, and the resulting radius and luminosity, at the age of the Sun, are compared with the actual values $R_\odot$ and $L_\odot$. This procedure is iterated until the calibration is judged satisfactory, considering the other uncertainties.

2.1.3. *The standard solar model*

The standard model may be defined as the model which uses the best physical data that are available (equation of state, opacities, nuclear reaction rates) and which agrees best with the observations, while including the minimum of physical processes to achieve this result. As the observational constraints become more severe, the standard model is bound to include physical processes which have been neglected before. At the time of this writing, the last process implemented in the standard model was the microscopic diffusion of helium and the heavy elements in the radiation zone, and this addition was required by the helioseismic profiles which clearly indicated that helium has settled out of the convection zone (see below).

2.1.4. *How good are standard models?*

Standard models have produced very decent results, when compared to the observations. They account for the observed main-sequence, in the Hertzsprung-Russel (HR) diagram[2]: most stars are gathered there because they are in their hydrogen burning stage, which lasts longer than any other phase because the transformation of hydrogen into helium produces by far the largest amount of energy, compared to the other nuclear reactions which occur later on as the star evolves.

Standard models also predict the evolution from the main-sequence to the giant branch. The theoretical isochrones fit fairly well the observed HR diagram of star clusters (such stars have all the same age). When perturbed and submitted to linear stability analysis, they reproduce the main properties of observed variable stars.

There are also some weak points, however. Careful comparison of cluster diagrams with theoretical isochrones show minor, but real discrepancies, in particular for massive stars, which seems to indicate that in these stars some mixing occurs beyond the convective core according to the Schwarzschild

[2]In this diagram, a star is represented by its luminosity L versus its effective temperature $T_{\rm eff}$

criterion (7). Also, certain chemical elements produced in the core of stars are observed at their surface, which implies that they have been transported through the radiation zone, and that is not allowed in standard models.

Moreover, in the later stages of stellar evolution, when the star expands to a red giant, it loses a significant amount of matter, and this mass loss is not accounted for in standard models.

For all these reasons, it is important to check the validity of our models, and the Sun provides the best opportunity for that, thanks to its proximity.

3. Probing the solar interior

Since the observed values of the solar luminosity and radius are needed to calibrate the initial helium abundance Y_0 and the mixing length parameter α, one has to use other means to check the accuracy of solar models. Indirect evidence can be gleaned from observing the properties of other stars, for which models have been built with the same prescriptions, but this would not be sufficient to constrain our models. Fortunately we now dispose of other means, which allow us to probe directly the solar interior.

3.1. SOLAR NEUTRINOS

Unlike photons, neutrinos interact extremely little with matter and therefore most of those emitted by the fusion of hydrogen in the central region of the Sun do escape and a very small fraction of these can be detected on Earth. No less than six different experiments have measured the flux of solar neutrinos, and they brought the proof that the solar energy does arise from nuclear fusion. Originally, the hope was that these experiments would provide precious constraints on the central conditions of the Sun, particularly on its temperature. However the measured flux turned out to be systematically smaller than that predicted by the standard model, by a margin which cannot be reconciled by adjusting the sensitive parameters.

The reason of this discrepancy has been the matter of a long standing debate. Is it due to serious imperfections of the standard model? We shall see below that non-standard models do not fare much better, although they include additional physical processes, if they are constrained by helioseismology. Or is there something wrong with the experiments and their interpretation, as suggested by the incompatibility of their results? Or else is the observed deficit due to some unknown property of the neutrino, beyond the electroweak theory? Recent results obtained at SuperKamiokande, in Japan, seem to favor the latter explanation; their straightforward explanation is that μ-neutrinos have a non-zero mass. If that is true, the e-neutrinos would probably also have a mass, and those emitted in the solar core would then oscillate between two states, of which only one is detected by the

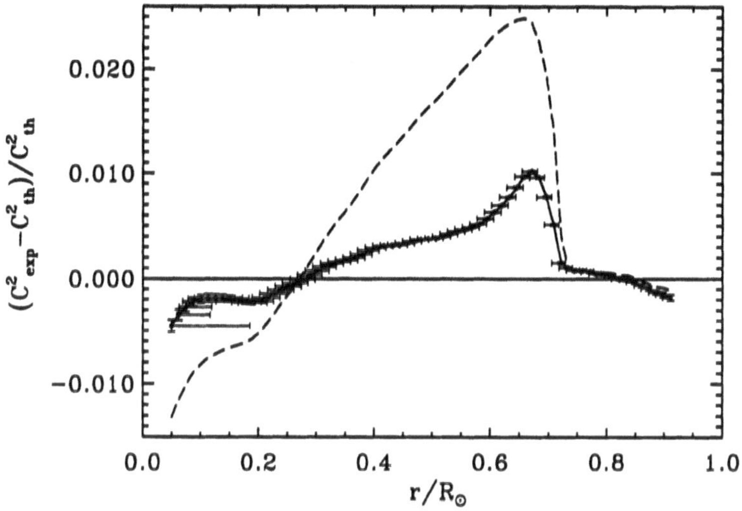

Figure 1. Sound speed difference between solar models and the profile obtained through helioseismology (Brun, Turck-Chièze & Morel 1998) without (dashed line) and with microscopic diffusion (full line with inversion error bars) (courtesy *Astrophys. J.*).

present experiments. Thus it is quite possible that in a not so distant future, once this problem has been settled by particle physicists, the observed neutrino flux can be used as a diagnostic of the physical conditions in the solar core.

3.2. HELIOSEISMOLOGY

Fortunately another technique has been developed in the meanwhile for the same purpose of sounding the solar interior, and it turned out to be extremely powerful: called helioseismology, it makes use of the elastic properties of the Sun which allow for the propagation of waves. The sound waves observed at the surface of the Sun penetrate to a depth which depends on their frequency and on their wavelength; through an appropriate treatment inspired by seismology on Earth, they reveal some properties of the layers which they traverse (see the article in this volume by W. Dziembowski, which is devoted to that subject).

The vertical profile of the sound velocity determined by this technique has now reached a precision of order 10^{-4}, thus calling for models built with at least the same numerical accuracy, and also for a refined evaluation of the solar age. Helioseismology has permitted to locate the base of the convection zone at 0.713 $R_\odot$, with a relative precision of 10^{-3}, and it proves unquestionably that helium has settled below it. Fig. 1 compares the result of helioseismic inversion with two solar models: one built by ignoring microscopic diffusion, and the other including it (Brun, Turck-Chièze & Morel 1998). Note the significant improvement when accounting for the settling of helium and other heavy elements.

Moreover, the characteristics of high degree modes allow to evaluate the helium content of the convection zone. Last but not least, the rotational splitting of the mode frequencies yields the internal rotation velocity of the Sun as a function of depth and latitude (Brown et al. 1990).

To conclude, the present standard model, including microscopic diffusion, fits the helioseismic data rather well. Undoubtly this excellent agreement will further improve with better opacities, better nuclear reaction rates, some refinements of the equation of state, and more accurate determinations of the chemical composition. However there are some observations which call for the implementation, in the model, of additional physical processes which have been neglected until now. And we have stressed already that the treatment of convection is rather rudimentary, which is felt especially near the surface, where the superadiabatic gradient predicted by the mixing-length theory has not the steepness required by the observed mode frequencies.

We shall now report on some recent advances to remedy these shortcomings.

4. The treatment of convection

In the last decade, significant progress has been made in modeling stellar convection. Present day computers allow numerical simulations in three dimensions which describe the convective motions near the surface of the Sun quite in a realistic way, matching the observed granulation pattern and predicting spectral line profiles which fit extremely well the observed ones (Stein & Nordlund 1998). Although they are limited to the uppermost layers of a star, these simulations give some clues on what is occurring deeper down. For instance, they suggest that stellar convection is highly intermittent, with strong, long-lived downflows; these are present in all simulations of stratified convection, as confirmed again by the most recent ones (Brummel et al. 1996; Freytag et al. 1996). These downdrafts are similar to the turbulent plumes observed in the Earth atmosphere, above heat sources such as chimneys, volcanos, etc. In the Sun they originate at the surface, in the 'cold' intergranular lanes. One of their characteristics is to entrain surrounding matter, thereby increasing their width as they proceed downwards; for a more complete account of their properties, see Rieutord and Zahn (1995).

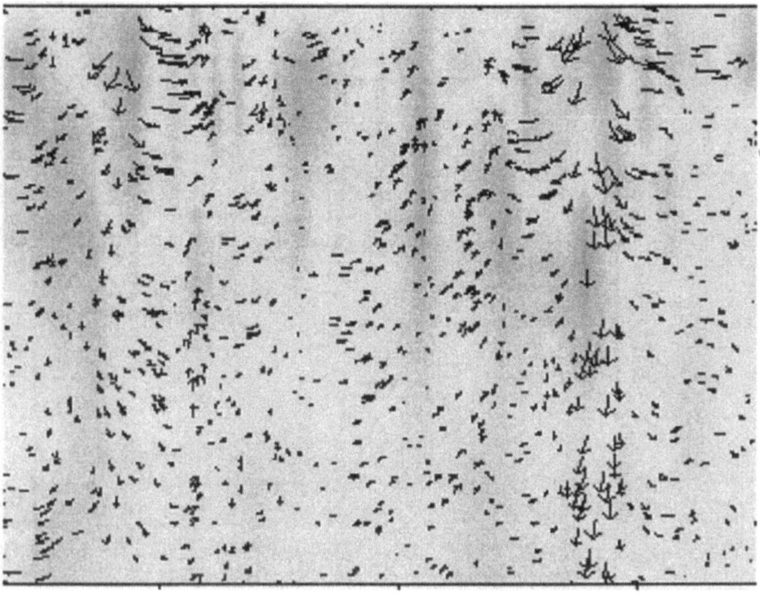

Figure 2. Time-distance tomography of subsurface layers of the Sun, with the MDI instrument on board of SOHO. The field displayed is stretched in the vertical direction: it measures 8 Mm in depth and 150 Mm in width. (http://sohowww.nascom.nasa.gov/gallery/MDI/mdi009.gif)

Such plumes have now been detected in the Sun, through time-distance tomography, a clever technique which uses the same data as helioseismology to probe the near surface layers (D'Silva et al. 1996). The result of such a tomography is displayed in Fig. 2. Plumes have certainly a strong impact on the dynamics of stellar convection zones, which cannot be reduced to a turbulent diffusion, as postulated in the mixing-length treatment (cf. eq. 8). Moreover they play probably a key role in the dynamo mechanism, since they are able to 'pump down' the magnetic field (Brandenburg et al. 1996; Tobias et al. 1998; Dorch 1998).

Plumes are likely to traverse the whole convection zone, and to penetrate somewhat into the stable radiation zone. Such convective penetration is observed in numerical simulations, in the laboratory and in the Earth atmosphere, but its description in the Sun is still not satisfactory. All models so far have predicted an extent of penetration which is too large to agree with the results of helioseismic inversion. The reason may be that the role of rotation has been largely neglected up to now, or that our simulations are still not realistic to capture the turbulent nature of plumes, and the almost adiabatic behavior of convective penetration.

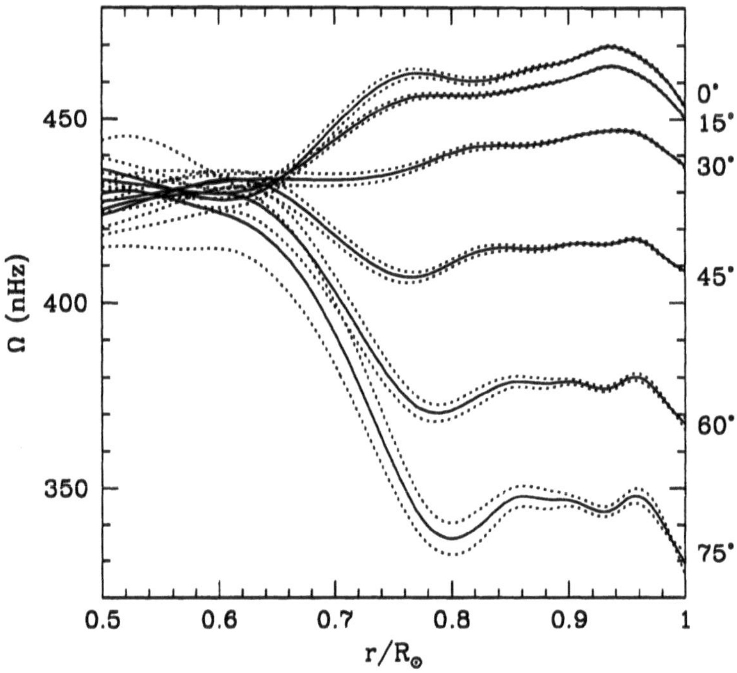

Figure 3. The solar rotation rate, as a function of radius and latitude, displayed with the error bars of helioseismic inversion. In the convection zone, the rotation varies with latitude, but little with radius, whereas in the radiation zone below, the rotation is almost uniform. The thickness of the shear layer at the base of the convection zone – called tachocline – is actually much thinner than this inversion suggests. (Based on data obtained through the GONG network; courtesy H.M. Antia.)

Numerical simulations also helped greatly to identify the excitation mechanism of acoustic waves in the Sun (Nordlund & Stein 1998; Georgobiani et al. 2000). Likewise they have been used to estimate the flux of internal gravity waves produced at the base of a convection zone (Kiraga et al. 1999), which have been invoked to extract angular momentum from the radiative interior (see below).

Another goal of numerical simulations is to explain the differential rotation of the Sun, namely why in the convection zone the rotation rate depends on latitude, but very little on depth (below a superficial shear layer). Early calculation always predicted that the angular velocity should be constant on cylinders centered on the rotation axis, which is contrary to what is observed through helioseismology (see Fig. 3). The coupling between convection and rotation has been investigated in the so-called f-plane approximation by Brummel, Hurlburt and Toomre (1998): convection is modeled locally, at a given latitude, in a cartesian slab. In these simulations, plumes depart from the vertical and tend to align with the rotation axis. The same team has engaged in an ambitious project which aims at modeling the whole solar convection zone, in spherical geometry (Elliott, Miesch & Toomre 1999); preliminary results reveal for the first time a significant departure from cylindrical rotation, as observed in the Sun. Here also work is in progress to increase the spatial resolution, in order to account for the turbulent nature of plumes.

But in spite of all these very significant advances, it has not been possible so far to derive a simple expression for the convective flux, to be implemented in stellar structure models. Contrary to claims made a few years ago, all mixing-length prescriptions contain at least one adjustable parameter; and when this parameter is calibrated on the Sun, there is no guarantee that its value can be applied to other stars. It actually depends on the location of the star in the HR diagram, as was shown by Ludwig, Freytag and Steffen (1999), who determined through hydrodynamical simulations the relevant parameter of the two most popular prescriptions, namely the classical mixing-length recipe and the so-called 'full turbulent spectrum' version of Canuto, Goldman and Mazzitelli (1996). The latter prescription is favored by some because the convective transport is less efficient in the subphotospheric layers than in the classical mixing-length treatment; therefore it produces a temperature stratification which is in closer agreement with the helioseismic data (Antia & Basu 1997). For this reason it also permits to fit with the same parameter the interior model of the Sun and the intensities of the Balmer lines (Bernkopf 1998). But when applied to the pre-main sequence evolution of solar-type stars, this FST predicts a very strong Li destruction which is incompatible with the observations (Ventura et al. 1997).

5. Mixing and angular momentum transport in the solar radiation zone

There are several signs that the radiation zone does not behave as assumed in the standard model, where it is deemed to be a stable medium, in which the chemical composition evolves just through nuclear reactions and microscopic diffusion. One is the fact that the largest difference in the profile of the sound velocity, between the standard model and the Sun, occurs right below the convection zone (cf. Fig. 1). Some mixing is clearly needed there to soften the profile of molecular weight below the convection zone; microscopic diffusion and gravitational settling alone would establish a gradient whose steepness is incompatible with the helioseismic data.

Another sign that some mixing is occurring below the convection zone is given by the lithium abundance, which is 160 times less at the surface of the Sun than it is in the meteorites (see Table 1). Such depletion is observed in all solar-type stars: the lithium abundance is 100 times higher at the surface of young stars of solar mass (members of the Pleiades cluster, for example) than it is on the Sun. The reason is certainly that this fragile element is destroyed inside these stars by the nuclear reaction $^7\mathrm{Li} + \mathrm{p} \rightarrow {}^4\mathrm{He} + {}^4\mathrm{He}$, but the temperature at the base of the convection zone is not high enough to explain such a large depletion. Therefore some mixing must occur within the radiation zone to carry lithium deeper down.

There are reasons to suspect that such mixing is related to the rotation. The region below the solar convection zone where such mixing is suspected is also the place where the rotation rate changes from differential, i.e. latitude dependent, in the convection zone to almost uniform in the deep interior (see Fig. 3). Also, the lithium depletion in solar-type stars seems to be correlated with their rotation, or rather with the loss of angular momentum: in the Hyades, tidally locked binaries, which have lost less angular momentum than single stars, have also kept more lithium than those. Therefore our models should also be able to describe the internal rotation of the Sun: in particular, they have to explain why the radiative interior appears to rotate so slowly and so uniformly. Before this property was established by helioseismology, the solar core was expected to rotate faster than the convective envelope, which is spun down by the solar wind. As we shall now see, steps have been taken to respond to these needs.

5.1. ROTATIONALLY DRIVEN MERIDIONAL CIRCULATION

One mechanism to transport matter and angular momentum in rotating stars is the thermally driven meridional circulation; it was first described by Eddington (1925) and Vogt (1925), and later rigorously assessed by Sweet (1950). In a rotating star, radiative equilibrium is no longer achieved in its radiative zone (in general), i.e.

$$-\nabla \cdot \mathbf{F}_{\mathrm{rad}} + \rho\varepsilon \neq 0;$$

thus motions will arise which contribute to the transport of heat, such as to satisfy the conservation of thermal energy as stated in (4). After an adjustment phase, a stationary state may be reached where heat is transported by a large scale circulation whose velocity $\mathbf{V}$, in the meridional plane, is given by

$$\rho T \mathbf{V} \cdot \nabla S = -\nabla \cdot \mathbf{F}_{\mathrm{rad}} + \rho \varepsilon. \tag{9}$$

This circulation is very slow; its turnover time, called the Eddington-Sweet time, is of order

$$t_{\mathrm{ES}} = t_{\mathrm{KH}} \left(\frac{\Omega^2 R^3}{\mathcal{G} M} \right)^{-1}, \qquad \text{where } t_{\mathrm{KH}} = \frac{\mathcal{G} M^2}{RL} \tag{10}$$

is the Kelvin-Helmholtz time characterizing the thermal adjustment of the star, and Ω the angular velocity. This time is exceedingly long in the present Sun (some 10^{12} years!), but it could have been less than 10^9 years when the Sun was young, and was presumably rotating much faster. Therefore, in the early phases of solar evolution, this meridional circulation was able to extract angular momentum from the radiative interior, to be carried away by the wind, which is responsible for the solar spin-down.

5.1.1. *Transport of angular momentum*

By transporting angular momentum, the circulation modifies the rotation profile, thus producing differential rotation, and such differential rotation is liable to shear instabilities, particularly in the horizontal direction, where the motions are not constrained by the buoyancy force. It is thus natural to assume that this shear in latitude produces produces a very anisotropic turbulence, indexanisotropic turbulence with much larger diffusive transport in the horizontal than in the vertical direction, by which it smoothes out the differential rotation which is the cause of the instability.

This assumption greatly simplifies the problem by reducing it to one dimension in space, with the angular velocity depending only on the vertical coordinate: $\Omega = \Omega(r)$. The rotation profile evolves then according to (Zahn 1992)

$$\frac{\partial \Omega}{\partial t} \simeq -K \left(\frac{2\Omega}{N} \right)^2 r^2 \frac{\partial^4 \Omega}{\partial r^4}, \tag{11}$$

where $K = \chi / \rho C_P$ is the thermal diffusivity, and N the buoyancy frequency:

$$N^2 = \frac{g}{H_P} \left[\left(\frac{\partial \ln T}{\partial \ln P} \right)_{\mathrm{ad}} - \frac{d \ln T}{d \ln P} \right]. \tag{12}$$

It shows that the characteristic time of the meridional circulation may be much shorter than indicated by the Eddington-Sweet time (10), in regions where there are steep gradients of Ω. When applied to the Sun, this equation predicts that the innermost core is rotating much faster than the surface (Matias & Zahn 1997), which is contrary to what is observed through helioseismology. It proves that meridional circulation alone is insufficient to extract angular momentum form the solar interior, at least in the later phases of evolution, and that there must be some other mechanism responsible for it.

Two physical processes are under consideration, namely magnetic torquing due to a fossil field (Gough & McIntyre 1998), and transport of angular momentum through gravity waves generated at the base of the convection zone (Press 1981; Kumar & Quataert 1997; Zahn, Talon & Matias 1997; Kumar, Talon & Zahn 1999). Work is in progress to assess the efficiency of these mechanisms, and to check whether they are compatible with other observational constraints. What they have in common is that apparently neither of these is able to produce mixing within the radiation zone.

5.1.2. *Mixing through meridional circulation*

On the other hand, meridional circulation will transport chemical elements, thus producing horizontal composition gradients which in turn may inhibit the circulation; this feed back has been invoked by Mestel (1953) to explain why rotating stars are not thoroughly mixed in a Eddington-Sweet time, but evolve towards the giant branch. On closer examination, the effect of composition inhomogeneities may not be as radical as suggested by Mestel (Zahn 1992; Talon & Zahn 1997; Maeder & Zahn 1998), but the transport of chemicals by the circulation will be hindered mainly because advection is partly eroded by the horizontal diffusion due to the anisotropic turbulence, as was shown by Chaboyer and Zahn (1991). The result is that the mixing behaves as a vertical diffusion, whose coefficient scales as

$$D_{\text{eff}} \propto \frac{(rU)^2}{D_h} \tag{13}$$

with D_h being the horizontal diffusivity. Putting this coefficient D_{eff} in the transport equation (6), one can predict how the radiative interior of the Sun is mixed by this rotationally induced circulation.

Unfortunately, there is no reliable prescription available yet giving the magnitude of the horizontal diffusion coefficient D_h, and one has to turn to observational constraints to fix its magnitude. One of such constraints is that lithium has been depleted by a factor of order 100 since the formation of the solar system, which can easily be achieved by this meridional circulation, but but one has to take in account that beryllium appears to have been spared from such depletion (Balachandran & Bell 1997). Since these elements are destroyed respectively at 2.5 and $3.5\,10^6$K, it indicates that the mixing occurs rather close to the base of the convection zone.

The conclusion is that meridional circulation, which digs very deep into the solar interior, may be partly responsible for the depletion of lithium, but cannot account for all of it – otherwise it would also destroy beryllium.

5.2. MIXING IN THE TACHOCLINE

We mentioned already that helioseismology has detected a layer of strong shear between the differentially rotating convection zone of the Sun and its uniformly rotating interior, which is now called *tachocline* (see Fig. 3). Its thinness – less than 5% of the solar radius (Antia, Basu & Chitre 1998; Corbard et al. 1999) – was quite unexpected. Here is why.

The latitude-dependent rotation $\widehat{\Omega}(\theta)$ imposed by the convection zone on the top of the radiative interior causes there a horizontal temperature gradient $\nabla T(\theta)$ through the thermal wind balance, and this temperature gradient tends to spread inwards on a thermal time-scale, due to radiative diffusion. If there were no stratification, that differential rotation would diffuse as

$$\frac{\partial \widehat{\Omega}}{\partial t} \simeq K \frac{\partial^2 \widehat{\Omega}}{\partial r^2}. \tag{14}$$

However the stable stratification hinders the spread, because angular momentum can only be transported by meridional advection in the absence of any turbulence (the microscopic viscosity plays a negligible role), and the vertical component of this flow works against buoyancy. The progression of the tachocline is slowed down and it turns into a 'hyper-diffusion', such as encountered earlier in the solar spin-down problem (Howard, Moore & Spiegel 1967):

$$\frac{\partial \widehat{\Omega}}{\partial t} \simeq -K \left(\frac{2\Omega}{N}\right)^2 r^2 \frac{\partial^4 \widehat{\Omega}}{\partial r^4}. \tag{15}$$

This equation similar to (11), but it is applied here to the differential rotation $\widehat{\Omega}$. In other words the thickness of the layer increases now as

$$\frac{h}{r} \simeq \left(\frac{t}{t_{\rm ES}}\right)^{1/4}, \tag{16}$$

with a time-scale which is the local Eddington-Sweet time

$$t_{\rm ES} = \frac{r^2}{K} \left(\frac{\Omega}{N}\right)^2.$$

According to this prediction, the tachocline would have spread as far as $r \simeq 0.3 R_\odot$ in the present Sun.

5.2.1. *What stops the spread of the tachocline?*

Two mechanisms have been invoked to stop the radiative spread of the tachocline. One is the effect of magnetic stresses due to a fossil field, as advocated by Gough and McIntyre (1998). Their model is promising, but it has not been worked out in detail yet: it remains to be seen how the poloidal field threads into the convection zone, and avoids imposing differential rotation throughout the radiative interior.

Another is *anisotropic turbulence*, as suggested by Spiegel and Zahn (1992), which has already been mentioned above when dealing with the meridional circulation. If isotropic, turbulence would contribute also to the deepening of the layer; it would operate on a viscous time-scale r^2/ν, with ν being the turbulent diffusivity. But the radiative spread can be stopped through anisotropic turbulence, provided that the contrast between horizontal and vertical transport is greater than the square of the aspect ratio: $\nu_h/\nu_v \gg (r/h)^2$, h being the thickness of the layer. Then a stationary regime is achieved, in which the advection of angular momentum is balanced by horizontal diffusion, such that

$$K\left(\frac{2\Omega}{N}\right)^2 r^2 \frac{\partial^4 \widehat{\Omega}}{\partial r^4} \simeq \frac{\nu_h}{r^2}\frac{1}{\sin^3\theta}\frac{\partial}{\partial\theta}\left(\sin^3\theta\frac{\partial\widehat{\Omega}}{\partial\theta}\right). \tag{17}$$

The thickness of the tachocline is then

$$\frac{h}{r} \simeq \left(\frac{\Omega}{N}\right)^{1/2}\left(\frac{K}{\nu_h}\right)^{1/4} \quad \text{or} \quad h \simeq 20,000\left(\frac{K}{\nu_h}\right)^{1/4} \text{km}; \tag{18}$$

thus a horizontal viscosity of about 10^6 cm^2 s^{-1} is sufficient to comply with the observed upper limit of $h \leq 35,000$ km.

More realistic numerical simulations have been carried out by Elliott (1997); they confirm the main properties derived from boundary layer theory, and they yield roughly the same estimate for the thickness of the tachocline.

5.2.2. *Element mixing within the tachocline*

In the tachocline model of Spiegel & Zahn, it would seem at first sight that the mixing is achieved by the large-scale circulation, which is divided in two cells in each hemisphere, and that the tachocline would be thoroughly mixed after a time h/U, where U is the amplitude of the vertical velocity. But we have seen above that anisotropic diffusion erodes the advective transport by a large scale circulation, resulting in a vertical diffusity whose expression is given in (13), which may be implemented in (6) to predict the evolution of the chemical composition within the tachocline.

This has been done by (Brun, Turck-Chièze and Zahn 1999), who have also taken into account the variation in time of both the rotation rate Ω, and the amplitude of differential rotation $\delta\Omega$. since the effective diffusivity scales with these quantities as

$$D_{\text{eff}} \propto D_h^{1/2}\Omega\left(\frac{\delta\Omega}{\Omega}\right)^2 \propto \delta\Omega^{5/2}\,\Omega^{-1},$$

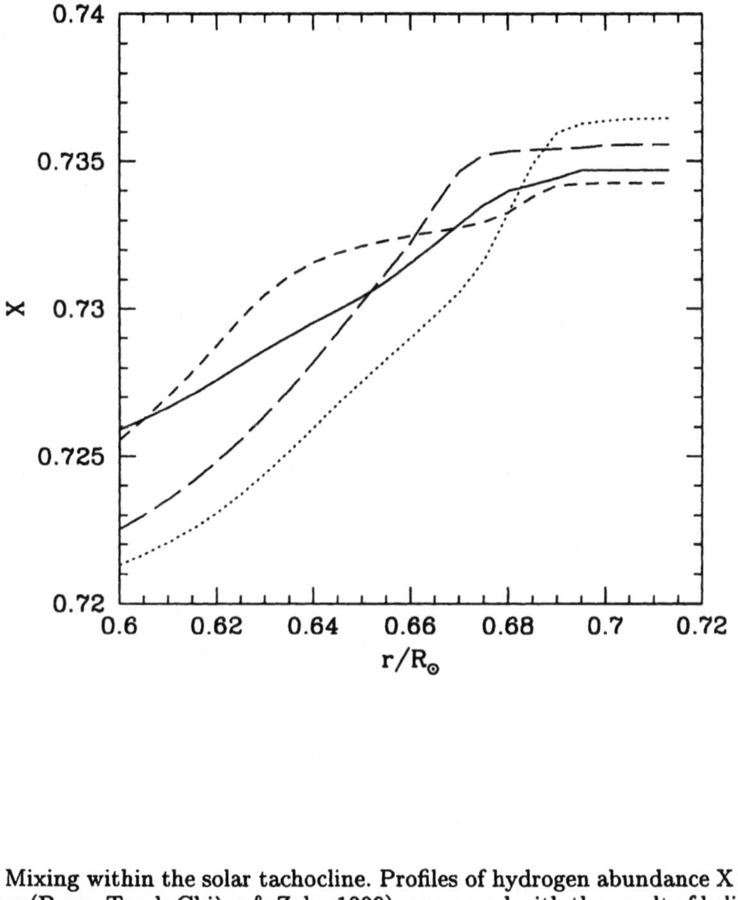

Figure 4. Mixing within the solar tachocline. Profiles of hydrogen abundance X for various solutions (Brun, Turck-Chièze & Zahn 1999), compared with the result of helioseismic inversion, in continuous line (Antia & Chitre 1998).

assuming that $D_h = \nu_h \propto \delta\Omega$. To link that differential rotation with the rotation rate, they used the relation drawn from the observations by Donahue et al. (1996): $\delta\Omega \propto \Omega^{0.7\pm0.1}$, and the solar spin-down was described by Skumanich's law (1972): $\Omega \propto t^{-1/2}$.

The result is represented in Fig. 4, where several solutions for different values of the relevant parameters (presumed thickness of the tachocline, buoyancy frequency at the top of the radiation zone) are compared with the profile of hydrogen abundance obtained by helioseismic inversion (Antia & Chitre 1998). Work is in progress to refine these theoretical models, to better fit the observations.

6. Conclusion

These were just a few examples of coordinated efforts to improve the physical description of the solar interior, beyond the standard model. Other non-standard models have been built to produce a lesser flux of neutrinos, in an effort to match the observed values; in general they introduce some extra mixing in the solar core, which modifies the chemical composition and therefore the network of nuclear reactions and the emission of neutrinos (cf. Morel & Schatzman 1996). But so far such models have failed to comply with the helioseismic constraints. These leave very little room for speculation, in contrast with the situation we knew some twenty years ago. Solar physics has entered an era where models have to fit the observations with a precision of order 10^{-4} – which is quite unusual in astrophysics!

Appendix – Until when will total eclipses occur?

Quite naturally, this question came up as we were watching the eclipse in Bucharest, on the 11th of August, 1999.

There are two reasons why such eclipses will not go on for ever. One is the increase of the Earth-Moon distance, by roughly 2 cm par year, due to tidal dissipation. Thus the angular size of the Moon decreases with the rate $r_{\rm moon} = 2/3.8\,10^{10}$, expressed in radian/yr. The second reason is that, as the Sun evolves, its radius increases by about 2 cm per year, according to our models, meaning that its angular size increases with the rate $r_{\rm sun} = 2/7\,10^{10}$. Let us assume that these rates will stay constant in the future – a condition which is not necessarily fulfilled for $r_{\rm moon}$.

Presently, the longest eclipses last about 7.5 mn, and the transit time of the leading (or trailing) edge of Moon over the solar disk is about 90 mn; therefore the Moon is larger than the Sun by about 1/12, in the most favorable case. We may thus conclude that total eclipses will occur for another $t_{\rm eclipses}$, where

$$[r_{\rm moon} + r_{\rm sun}]\, t_{\rm eclipses} = \frac{1}{12} \quad \Rightarrow t_{\rm eclipses} \approx 10^9 \text{ years.}$$

If there are still intelligent creatures on Earth, by then, they will certainly envy us for having envoyed such spectacular events!

References

1. Antia, H.M. and Basu, S. (1997) Effect of turbulent pressure on solar oscillation frequencies, *Proc. SCORe'96; ed. F.P. Pijpers, J. Christensen-Dalsgaard and C.S. Rosenthal, Kluwer Acad. Publ., Astrophys. Space Sci. Library* **225**, 51
2. Antia, H.M., Basu, S. and Chitre, S.M. (1998) Solar internal rotation rate and the latitudinal variation of the tachocline, *Mon. Not. Roy. Astron. Soc.* **298**, 534
3. Antia, H.M. and Chitre, S.M. (1998) Determination of temperature and chemical composition profiles in the solar interior from seismic models, *Astron. Astrophys.* **339**, 239
4. Balachandran, S. and Bell, R.A. (1997) The "Missing UV opacity" and the solar beryllium abundance, *American Astron. Soc. Meeting* **191** #74.08
5. Bernkopf, J. (1998) Unified stellar models and convection in cool stars, *Astron. Astrophys.* **332**, 127
6. Brandenburg, A., Jennings, R.L., Nordlund, Å, Rieutord, M., Stein, R. F. and Tuominen, I. (1996) Magnetic structures in a dynamo simulation, *J. Fluid Mech.* **306**, 325
7. Brown, T.M, Christensen-Dalsgaard, J., Dziembowski, W.A., Goode, P., Gough, D.O. and Morrow, C.A. (1989). Inferring the Sun's internal angular velocity from observed p-mode frequency splittings, *Astrophys. J.* **343**, 526
8. Brummell, N.H., Hurlburt, N.E. and Toomre, J. (1996) Turbulent compressible convection with rotation. I. Flow structure and evolution, *Astrophys. J.* **473**, 494
9. Brummell, N.H., Hurlburt, N.E. and Toomre, J. (1998) Turbulent compressible convection with rotation. II. Mean flows and differential rotation, *Astrophys. J.* **493**, 955
10. Brun, A.S., Turck-Chièze, S. and Morel, P. (1998) Standard solar models in the light of new helioseismic constraints. I. The solar core, *Astrophys. J.* **506**, 913
11. Brun, A.S., Turck-Chièze, S. and Zahn, J.-P. (1999) Standard solar models in the light of new helioseismic constraints. II. Mixing below the convection zone, *Astrophys. J.* **525**, 1032
12. Canuto, V.M., Goldman, I. and Mazzitelli, I. (1996) Stellar turbulent convection: a self-consistent model, *Astrophys. J.* **473**, 550
13. Chaboyer, B. and Zahn, J.-P. (1992) Effect of horizontal turbulent diffusion on the transport by meridional circulation, *Astron. Astrophys.* **253**, 173
14. Corbard, T., Blanc-Féraud, L., Berthomieu, G. and Provost, J. (1999) Non linear regularization for helioseismic inversions. Application for the study of the solar tachocline, *Astron. Astrophys.* **344**, 696
15. Donahue, R.A., Saar, S.H. and Baliunas, S.L. (1996) A relationship between mean rotation period in lower main-sequence stars and its observed range, *Astrophys. J.* **466**, 384
16. Dorch, B. (1998) Thesis, http://www.astro.ku.dk/~dorch
17. D'Silva, S., Duvall, T.L. Jr., Jefferies, S.M. and Harvey, J.W. (1996) Helioseismic tomography, *Astrophys. J.* **471**, 1030
18. Eddington, A.S. (1925) Circulating currents in rotating stars, *Observatory* **48**, 73
19. Elliott, J.R. (1997) Aspects of the solar tachocline, *Astron. Astrophys.* **327**, 1222
20. Elliott, J.R., Miesch, M. and Toomre, J. (1999) Large-eddy simulations of turbulent solar convection and its coupling with rotation, *Proc. SOHO-9 workshop*, 54
21. Freytag, B., Ludwig, H.-G. and Steffen, M. (1996) Hydrodynamical models of stellar convection. The role of overshoot in DA white dwarfs, A-type stars, and the Sun, *Astron. Astrophys.* **313**, 497
22. Georgobiani, D., Kosovichev, A. G., Nigam, R., Nordlund, Å. and Stein, R. F. (2000) Numerical simulations of oscillation modes of the solar convection zone, *Astrophys. J.* **530**, 139
23. Gough, D.O. and McIntyre, M.E. (1998) Inevitability of a magnetic field in the Sun's radiative interior, *Nature* **394**, 755
24. Grevesse, N. and Sauval, A.J. (1998) Standard solar composition, *Space Sci. Rev.* **85**, 161

25. Howard, L.N., Moore, D.W. and Spiegel, E.A. (1967) Solar spin-down problem, *Nature* **214**, 1297
26. Kiraga, M., Zahn, J.-P., Stepien , K., Jahn, K., Różyczka, M. and Muthsam, H.J. (1999) Hydrodynamical simulations of penetrative convection and generation of internal gravity waves, *ASP Conf. Ser.* **173**, 269
27. Kumar, P. and Quataert, E. (1997) Angular momentum transport by gravity waves and its effect on the rotation of the solar interior, *Astrophys. J.* **493**, 412
28. Kumar, P., Talon, S. and Zahn, J.-P. (1999) Angular momentum redistribution by waves in the Sun, *Astrophys. J.* **520**, 859
29. Ludwig, H.-G., Freytag, B. and Steffen, M. (1999) A calibration of the mixing-length for solar-type stars based on hydrodynamical simulations. I. Methodical aspects and results for solar metallicity, *Astron. Astrophys.* **346**, 111
30. Maeder, A. and Zahn, J.-P. (1998) Stellar evolution with rotation. III. Meridional circulation with μ-gradients and non-stationarity, *Astron. Astrophys.* **334**, 1000
31. Matias, J. and Zahn, J.-P. (1997) The internal rotation of the Sun, *IAU Symposium 181, poster volume (ed. G. Berthomieu and F.-X. Schmieder)*
32. Mestel, L. (1953) Rotation and stellar evolution, *Mon. Not. Roy. Astron. Soc.* **113**, 716
33. Morel, P. and Schatzman, E. (1996) Diffusion near the solar core, *Astron. Astrophys.* **310**, 982
34. Nordlund, Å. and Stein, R.F. (1998) The excitation and damping of p-modes, *Proc. IAU Symp. 185: New Eyes to See Inside the Sun and Stars; ed. F.-L. Deubner, J. Christensen-Dalsgaard and D. Kurt*, 199
35. Press, W. H. (1981) Radiative and other effects from internal waves in solar and stellar interiors, *Astrophys. J.* **245**, 286
36. Rieutord, M. and Zahn, J.-P. (1995) Turbulent plumes in stellar convective envelopes, *Astron. Astrophys.* **296**, 127
37. Skumanich, A. (1972) Time scales for CaII emission decay, rotational braking, and Lithium depletion, *Astrophys. J.* **171**, 565
38. Spiegel, E.A. and Zahn, J.-P. (1992) The solar tachocline, *Astron. Astrophys.* **279**, 431
39. Stein, R.F. and Nordlund, Å. (1998) Simulations of solar granulation. I. General properties, *Astrophys. J.* **499**, 914
40. Sweet, P.A. (1950) The importance of rotation in stellar evolution, *Mon. Not. Roy. Astron. Soc.* **110**, 548
41. Talon, S. and Zahn, J.-P. (1997) Anisotropic diffusion and shear instabilities, *Astron. Astrophys.* **317**, 749
42. Tobias, S.M., Brummell, N.H., Clune, T.L. and Toomre, J. (1998) Pumping of magnetic fields by turbulent penetrative convection, *Astrophys. J.* **502**, 177
43. Ventura, P., Zeppieri, A., Mazzitelli, I. and D'Antona, F. (1998) Pre-main sequence lithium burning: the quest for a new structural parameter, *Astron. Astrophys.* **331**, 1011
44. Vitense, E. (1953) Die Wasserstoffkonvektionzone der Sonne, *Zeitschrift f. Astrophys.* **32**, 135
45. Vogt, H. (1925) Zum Strahlungsgleichgewicht der Sterne, *Astron. Nachr.* **223**, 229
46. Watson, M. (1981) Shear instability of differential rotation in stars, *Geophys. Astrophys. Fluid Dynam.* **16**, 285
47. Zahn, J.-P. (1992) Circulation and turbulence in rotating stars, *Astron. Astrophys.* **265**, 115
48. Zahn, J.-P., Talon, S. and Matias, J. (1997) Angular momentum transport by internal waves in the solar interior, *Astron. Astrophys.* **322**, 320

THEORY OF SOLAR LUMINOSITY VARIATIONS

H.C. SPRUIT
Max-Planck-Institut für Astrophysik
D-85740 Garching, Germany

Abstract. The theory of stellar structure can be used to identify the most plausible mechanisms for the irradiance variations associated with the solar cycle. Changes in surface emissivity, i.e. the reduced cooling in spots and enhanced emission by small scale magnetic fields, are the most effective mechanisms and account for most of the observed variation. Helioseismology will soon be able to test the consequences of changes in surface emissivity, and distinguish them from other scenarios for irradiance variability.

1. Introduction

When a sunspot group appears on the surface of the Sun, it reduces the brightness locally. Does this lead to an actual reduction of the Sun's brightness as measured on Earth (the so-called irradiance)? Or is the 'missing flux' re-emitted elsewhere on the solar surface? Faculae are bright areas, visible especially near the limb of the Sun. Do they contribute to the irradiance? Does the irradiance vary only because of spots and faculae, or are there other effects as well, and can these be large enough to influence climate on Earth?

Some of these questions have been answered by actual measurements with space-based radiometers over the past 20 years (Figure 1). The data show a clear variation, obviously correlated with the solar cycle: the solar magnetic field does indeed cause the Sun's brightness to vary a bit, by about 0.1% between minimum and maximum of the cycle. The record shows prominent sharp dips; these are associated with the passage of individual sunspot groups across the disk. When a large group appears, the Sun actually does become dimmer by a tiny bit. The smoothed data (thick line in the Figure) shows, however, that the *average* trend is the opposite: at maximum activity, the Sun is brighter.

J.-P. Zahn and M. Stavinschi (eds.),
Advances in Solar Research at Eclipses from Ground and from Space, 289–305.

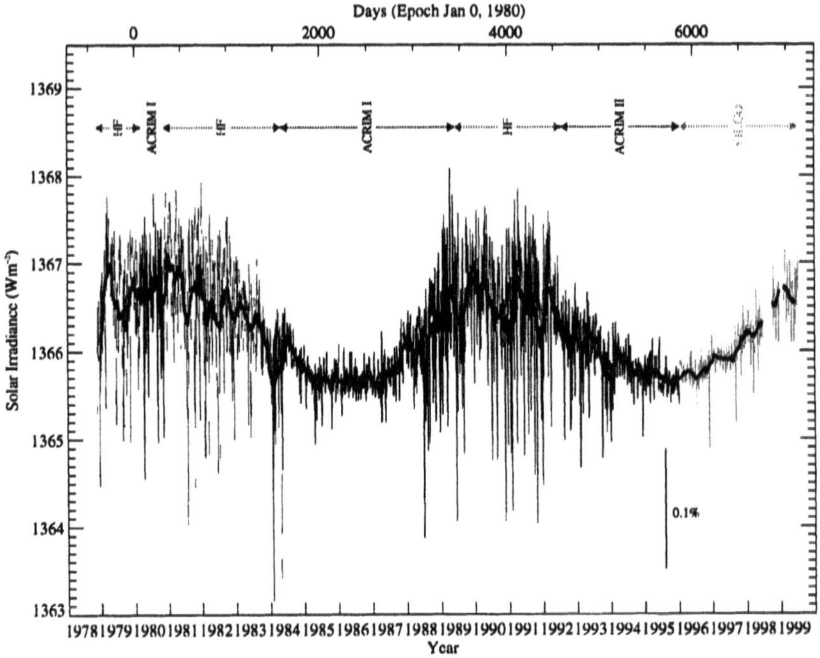

Figure 1. Solar irradiance record, showing the variation due to the solar cycle. From Fröhlich and Lean, 1998 (see also `http://www.pmodwrc.ch/solar_irr/sola_irr.html`)

The level of variation seen in Figure 1 is not large, and by general opinion in the climate modeling community, insufficient to cause variations in the Earth's climate, by a factor of 10 or so (see the contributions in Nesme-Ribes et al. 1994; Friis-Christensen et al. 2000). Even if the Sun were to settle into a state of activity as seen during the minima in Figure 1, the average reduction of irradiance by 0.05% would have negligible effects, and probably would not explain events like the so-called 'little ace age'[1].

For climate modeling, the record of Figure 1 is too short. One would like to know what the irradiance has been in the past, during the last 10000 years, for example, so comparisons can be made with climate records such as have been obtained from ice cores in Greenland and the Antarctic (see the contributions in Friis-Christensen et al. 2000). Is it conceivable, for example, that Figure 1 shows only a fraction of what the solar magnetic field is capable of doing, and that variations of, say, 10 times larger amplitude could have been caused by the solar magnetic field, or something else that has not yet been discovered on the Sun? These are questions that can excite

[1] A period during the 17th century when the climate in Europe was colder by a degree C or so. This cold spell appears to have been limited mostly to Western Europe, however.

the imagination, including that of funding agencies.

In the absence of reliable independent indicators of the past brightness of the Sun, it is of some importance to understand how the variations seen in Figure 1 come about. If theoretical understanding is sufficient, and accepted by the community, it may be possible to assess with more certainty whether irradiance variations larger than those seen so far are likely to have occurred in the past, and perhaps influence the climate.

Until the advent of helioseismology (see Dziembowski, in this volume), our knowledge of the internal structure of the Sun was based on the theory of stellar structure and evolution. Since this theory is reasonably simple, confidence in the models was high even before helioseismological tests became possible. The main uncertainty had to do with the exact values of the opacity of stellar plasma. Increasingly accurate values for this were computed from first principles (atomic physics) by extensive numerical work. The basic correctness of the theory was then confirmed by helioseismology, to such an extent even that one is now confident that the 'solar neutrino problem' has its cause in neutrino physics, not inaccuracies in the solar models (e.g. Bahcall 1996; Richard et al. 1996; Brun et al. 1998).

Though these solar models are spherical and only slowly evolving in time, the theory behind them also applies to time dependent and non-spherical disturbances. In this form, the theory makes definite predictions about the thermal behavior of the Sun, that is, the evolution in time and position in the star of perturbations that might be related to the observed irradiance variations. In the following I present my view how this theory can be used to locate the most plausible source of irradiance variations. After this, I discuss how the new helioseismological measurements currently being made can be used to test the predictions of proposed mechanisms of irradiance variation.

Conceptually, one can separate the thermal effects of magnetic fields into three types:

1. 'Sources and sinks'. The generation of a magnetic field involves the conversion of energy of motion into magnetic energy. Since the motions in the solar envelope are thermally driven, this ultimately means conversion of thermal into magnetic energy: building up a magnetic field produces a thermal sink somewhere. The opposite happens when the field decays: magnetic energy is converted into heat. These thermal effects exist only during *changes* in the magnetic energy content of the envelope.

2. 'Shadows', or changes in the heat transport coefficient. Magnetic fields interfere with convection, causing a reduction in the efficiency of heat transport in the envelope. In contrast to (1), these changes last as long as the magnetic field itself is present.

Both these sources and sinks, and changes in the transport coefficient cause thermal perturbations, varying with magnetic activity, which propagate through the envelope and cause variations in surface energy flux. Related to the second class of perturbations are:

3. The effects of magnetic fields at the surface of the star. Sunspots, being dark, radiate less than the surrounding photosphere, while the small elements that make up plages and the network have an excess emission. In addition, it is conceivable that the magnetic elements have an indirect effect surface by modifying the convective flow in their surroundings slightly (for which there is some observational evidence, see section 3.6).

2. Time scales

The response of the solar envelope to thermal perturbations is not governed by a single time scale, but by a wide range of time scales. The longest of these is the thermal time scale of the Sun as a whole, called the Kelvin-Helmholtz time scale. It is about 10^7 years, namely the thermal energy content of the Sun, of the order 10^{41} Joule, divided by its luminosity of $4\,10^{26}$ W. Hence, on time scales of human interest, the Sun has an extremely large thermal inertia. If the central heat source of the Sun were switched off, the internal structure and the luminosity would start to change only on this long time scale.

The Kelvin-Helmholtz time scale is a global time scale for the Sun as a whole; more generally, we can define the *thermal time scale* $\tau_{\rm t}$ as a function of depth:

$$\tau_{\rm t}(z) \equiv U(z)/L(z) \approx \frac{1}{L}\int_{R-z}^{R} 4\pi r^2 u dr, \qquad (1)$$

where L is the luminosity at depth z, U the thermal energy of the envelope down to a depth z, and u the thermal energy per unit volume, approximately (for an ideal gas of constant γ) given by $u = P/(\gamma - 1)$. This is the time scale on which the structure of the envelope, and the observed luminosity, would start changing when the heat flux in the star were interrupted, by some magical means, at depth z. Some rough values for this quantity are $\tau_{\rm t} \sim 10^5$yr at $z = 2\,10^5$ km (depth of the convection zone), 10 yr at $20,000$ km (the size of a supergranule), 10 hrs at 2000 km (size of a granule). This shows that the thermal time scale depends rather strongly on depth in the Sun. As a result, the thermal response of the Sun also depends critically on the location of the disturbance. Near-surface disturbances have much larger effects than ones originating near the base of the convection zone.

A second kind of time scale involved in thermal readjustments is the *diffusive* time scale. This is the time scale on which differences in entropy between different parts of the convection zone are ironed out by the convective flows. This is easily computed in a 'mean field' or diffusion model for convection, in which the local convective heat flux is assumed to be a function of only the local conditions (thermodynamic state of the gas and the temperature gradient). The validity of this model is questionable, since it is now known that stellar envelope convection is in fact extremely non-local, with almost all flows generated by cooling at the stellar surface (e.g. Nordlund 1986; Stein and Nordlund 1989, for reviews see Nordlund and Stein 1996; Spruit 1997). The diffusion model has the advantage, however, that it is simple to apply, and is (unfortunately) still almost universally accepted. For these historical and practical reasons, I still use it for the estimates in this section. Estimates in the context of the new nonlocal view of envelope convection have not yet been developed to the same level of detail. I return to this in section 4, where the examples of heat flux blocking by sunspots and 'thermal shadows' are considered in the nonlocal convection picture. I argue there that the new nonlocal view only strengthens the conclusions based on the diffusion model.

For small perturbations, the heat flux in the local model is proportional to the entropy gradient, and time dependent perturbations are governed by a diffusion equation. The 'turbulent' diffusion coefficient in this equation is $\kappa_{\rm t} \approx \frac{1}{3} l_{\rm c} v_{\rm c}$ where $l_{\rm c}$ and $v_{\rm c}$ are the convective length scale and velocity. In mixing length models of the solar convective envelope, this quantity varies only weakly, at a value of the order 10^{13} cm^2/s. In this model thermal inhomogeneities (more precisely: entropy inhomogeneities) of length scale d are smoothed by turbulent diffusion on a time scale

$$\tau_{\rm d} = d^2/\kappa_{\rm t}. \tag{2}$$

For $d = 2\,10^5$ km this is about 1 yr, for $d = 2000$ km about 1 hr. Comparing $\tau_{\rm t}$ and $\tau_{\rm d}$, we see that they are of similar magnitude close to the surface (to be precise: in the surface boundary layer where convection is not efficient enough to keep the stratification close to adiabatic). In deeper layers, the thermal time scale is much *longer* than the diffusive time scale, by a factor of up to 10^5.

The two time scales measure different types of thermal adjustment process. These same processes appear in the thermal behavior of, say, a chunk of metal, aluminum for example, heated from the inside and suspended in space. The thermal time scale is the time scale on which its temperature adjusts to a change in the heat input, such that the heat radiated from the surface into space balances the heat input again. It is determined by the heat capacity [U in eq. (1)] and the power level (L). The time scale on which different parts of the chunk equilibrate to the same temperature is governed by a different process, namely thermal conduction (the equivalent of the turbulent diffusion in the Sun). The diffusion (conduction) time scale is much shorter than the thermal time scale, because of the large heat conductivity of Al. In the Sun, it is the very large turbulent diffusivity in the bulk of the convection zone that causes the very short diffusive time scale compared with the thermal time scale.

How do these different time scales come into play when the convection zone is thermally perturbed by, say, the storage of energy in a growing magnetic field? Such perturbations can be computed in detail, either by numerical methods (Endal et al. 1985) or more analytically. We can, for example, consider the initial value problem in which a perturbation is allowed to evolve in time by heat transport in the convection zone. In general this evolution has components on all the time scales of the problem, including the very long thermal time scale. Detailed calculations of this problem are discussed briefly in the next subsection (for a more complete analysis, see Spruit 1982ab, 1991; Arendt 1992).

2.1. MODES OF THERMAL RELAXATION

In the mixing length approximation for convective energy transport the heat flux is given by:

$$\mathbf{F} = -\kappa_t \rho T \nabla S, \tag{3}$$

where $\mathbf{F}$ is the heat flux, κ_t the turbulent diffusivity, ρ the density, T the temperature and S the entropy. The energy equation, in the absence of sources, is given in terms of the entropy by

$$\rho T \frac{\mathrm{d}S}{\mathrm{d}t} = -\mathrm{div}\mathbf{F}. \tag{4}$$

As an example, consider the one dimensional problem in which all quantities depend on depth z only. In this case, combining (3) and (4), and neglecting flows we get

$$\frac{\partial S}{\partial t} = \frac{F}{\rho T H} + \kappa \frac{\partial S}{\partial z^2}, \tag{5}$$

where $H = \partial \ln \rho T / \partial z^{-1}$ is the pressure scale height. For perturbations in which the left hand side is balanced by the first term on the right, the time scale is of the order $F/(c_p \rho T H)$, which is just a thermal time scale (cf. 1). For modes in which the left hand side is balanced by the second term, we evidently have an ordinary diffusion problem, hence these modes evolve on diffusive time scales. The first term has appeared as a consequence of the stratification of pressure, i.e. the inhomogeneity of the convection zone. In a homogeneous medium (as in a chunk of Al), the thermal time scale comes in only through the radiating surface boundary condition (which one could also consider as a form of inhomogeneity). If T_s is the surface temperature, this condition is

$$\sigma T_s^4 = F. \tag{6}$$

Since the perturbations in entropy outside spots and the magnetic elements making up the small scale field are small, the response of the convection zone can be computed by linearizing eqs. (3),(4),(6). An arbitrary initial perturbation may then be decomposed in the standard way in terms of the eigenmodes of the problem. These are found by setting $\delta S = e^{\eta t} f(\mathbf{r})$, and solving for $f(\mathbf{r})$ with the decay rate η as eigenvalue (the problem thus defined has only decaying solutions). Specializing to the one dimensional case ($f(z)$) one finds a series of modes ordered by the number of nodes n of f. The fundamental ($n = 0$) has a decay rate of the order $\eta_0 \sim \tau_t^{-1}$ that is, this mode decays on the thermal time scale. The higher modes decay with rates of the order $\eta_n \sim n^2\tau_d^{-1}$, they decay on diffusive time scales.

The extreme separation of time scales, 10^5yr for the thermal time scale and 1yr for the longest diffusive time scale (with $n = 1$) leads to a simple picture. The high heat conductivity tends to keep the convection zone entropy uniform, so that sources and sinks of energy are shared by a large part of the convection zone. This large part of the convection zone has a large heat capacity, however, so that sources and sinks tend to spend their efforts in slow heatings and coolings of large amounts of mass, rather than propagating to the surface.

How do these different time scales come into play when the convection zone is thermally perturbed by, say, the storage of energy in a growing magnetic field? The evolution of such perturbations can be computed in detail, either by numerical methods (Endal et al. 1985; Gilliland 1988) or more analytically. We can, for example, consider the initial value problem in which a perturbation is allowed to evolve in time by thermal transport in the convection zone. In general this evolution has components on all the time scales of the problem, including the very long thermal time scale. Formal aspects of this problem have been discussed elsewhere (Spruit 1982ab, 1991; Arendt 1992). In the following, the basic conclusions of these analyses are summarized.

3. Expected level of luminosity variations

For quantitative estimates, the strength of the field and its filling factor in the convection zone have to be specified. Assume that we have a layer of field with strength of the order of 10,000G (equipartition with the convective flows as estimated by a mixing length model), one scale height deep, near the base of the convection zone (where most of the magnetic flux is probably located). I summarize here some results, discussed in greater detail elsewhere (Spruit 1991).

3.1. SOURCES-AND-SINKS

If the energy needed to build up the solar magnetic field during one half of the solar cycle is taken out of the thermal energy near the base of the convection zone, the calculations show that a surface luminosity variation of only $\delta L/L \sim 10^{-7}$ results. This is due to the very large heat capacity of the lower convection zone. The effect is stronger if the source of the magnetic field is assumed to be closer to the surface, but is still much smaller than the observed effect. Recent models for the emergence of magnetic flux from the base of the convection zone (D'Silva & Choudhuri 1993; D'Silva & Howard 1993; Caligari et al. 1998) indicate that the actual field strength at the base of the convection zone is probably about 10 times higher than the equipartition estimate, of the order 10^5 G. The magnetic energy per unit of magnetic flux is then also 10 times higher, but this still does not lead to a significant luminosity effect.

3.2. SHADOWS

If magnetic fields interfere with convection, a thermal perturbation develops as well. If at some depth below the surface there is an area where magnetic fields reduce the convective flows, the surroundings will be affected thermally, and one might hope to see a 'thermal shadow' at the surface. This can again be computed in the diffusion assumption for convection (Spruit 1977). The amplitude of the effect depends crucially on the 'covering factor'. If a reduction of convective efficiency is assumed that uniformly covers a horizontal surface at some depth z, magnetic fields comparable to equipartition with convective flows can have stronger effects than the source-and-sink perturbations (Gilliland 1988). A measurable effect, however, is predicted only if the field is located close below the surface. At face value, this would make changes in convective efficiency a reasonable candidate for irradiance variations.

In practice, the effect is very strongly reduced, however, if the change in convective efficiency does not extend over an entire horizontal surface. If there are 'holes' in this cover, the effect is much smaller. This is because the turbulent heat conductivity in the convection zone is so high that the heat flux is easily 'shunted' past blocking objects below the surface (Spruit, 1977). Compare this with efforts to thermally insulate homes. To be effective, it is not enough that *most* of the building's outer surface is very well insulated. A more modest insulation but without any leaks is more effective. For this reason, thermal shadows of subsurface fields are probably not very important (see also section 4.3).

3.3. SURFACE EFFECTS

By far the most effective way in which a magnetic field influences the irradiance is by its effect on the *net surface emissivity.* The reduced emission from a sunspot area shows up directly in the irradiance records as a dip tracking the passage of the area across the disk. The only complication is that one might expect, depending on one's prejudices, that part of this reduction could be compensated by a brightening elsewhere. For example in the form of a 'bright ring' surrounding the spot. Evidence of such bright rings is absent for most spots that have been studied for this effect, and where observation of a ring is reported, it compensates for only a small fraction of the spot blocking (e.g. Rast et al. 1999). Brightening of the photosphere around a spot is hard to measure because it is usually masked by facular emission, which is a different effect. Such faculae are part of all active regions. Though their excess emission is roughly of the same order as the spot deficit, this is a coincidence that does not hold any more on closer inspection. For example, they are present in the same amount in active regions with or without sunspots, and their contribution to the irradiance variation does not correlate very well with spot areas (Foukal & Lean 1986; Lean et al. 1998).

The absence of bright rings is understood in terms of a turbulent diffusion model for the heat flux in the convection zone (Spruit 1977, 1982b; Foukal et al. 1983; Chiang et al. 1984). The 'blocked heat flux', for the most part, does not reappear elsewhere on the surface, but stays inside the convection zone, being stored/released on the very long thermal time scale of the convective envelope. This conclusion holds, in the diffusion model, as long as the blocking effect of the spot extends to a depth of at least 1000 km, a mild requirement given that the observed Wilson depression of the umbra of a spot is already of the order of 500 km.

3.4. SPOT BLOCKING: STOVE TOP ANALOGY

The absence of bright rings in the diffusion model can be understood by a kitchen analogy (for an improved, but more qualitative, model based on the new nonlocal view of convection, see section 4.1). Imagine a stove top with an electric heater plate. Assume this plate to be made of a massive block of a metal with a high thermal conductivity, Aluminum or Copper, say. Such a block, like the solar convection zone, has two time scales, the thermal time scale, given by the heat content divided by the heating rate, and the thermal conduction time scale given by the thermal diffusion time across the block. These are different physical quantities and their values can be quite different. If the thermal conductivity is large, the diffusion time scale can be much shorter than the thermal time scale. In the solar convection zone this is the same except that the ratio of time scales ($\sim 10^5$) is much more extreme.

Suppose the heater plate has been on and has reached a steady temperature, so that heat input balances the conduction/convection losses at the top. Suppose also that the ratio of time scales is sufficiently large. Then put a piece of thermally insulating material on top, covering part of the surface of the plate (as a model for the effect of a spot). Where does the blocked heat flux go? The answer is of course that it does not go anywhere, the blocked heat stays in the block, heating it up slowly, instead of being radiated at some other part of the surface. On account of the large thermal conductivity, the temperature in the block stays uniform, and almost no 'bright ring' appears around the spot. When the insulating spot is removed after a while, the heat flux returns to its old value instantaneously. The temperature has increased only slightly during the blocking phase, and this slight increase disappears again on a thermal time scale.

3.5. EXCESS EMISSION BY SMALL SCALE MAGNETIC FIELDS

Apart from the highly visible sunspots, the Sun has a small scale magnetic field consisting of a very large number of tiny magnetic elements. They are called faculae if their density is high, and 'network' at lower density, and collectively contain much more magnetic flux than sunspots. They are visible especially near the solar limb, showing that they radiate more isotropically than the limb-darkened normal solar atmosphere.

This excess emission has two components, one originating in somewhat higher atmospheric layers, in the form of line- and UV-continuum emission. This component probably is due to dissipation of mechanical (wave) energy carried by the magnetic field. In addition, there is a component seen in white light, originating in the surface layers of the magnetic structures. This second component is understood theoretically as a radiative transfer effect (Spruit 1977): the radiating surface in a magnetic element is a little 'dimple' in the surface. The large number of small magnetic elements effectively corrugates the surface, and makes it radiate both more effectively and more isotropically, just like a rough surface has a higher emissivity than a polished surface. This effect increases the radiative flux from any part of the surface where small scale magnetic fields are present. It is a net effect (predicted before observation, Spruit 1977) that is not balanced by a deficit somewhere else on the surface. Since the magnetic elements are so small, they are hardly resolved in the observations. Recent high-resolution observations (Topka et al. 1997), however, verify the picture predicted by flux tube models (Spruit 1976 figure 11; Steiner et al. 1997) for the appearance of small magnetic elements.

Since the effect of small scale fields is an increased *emissivity*, it increases the *cooling rate* at the surface.

3.6. CHANGES IN GRANULATION IN ACTIVE REGIONS

The changes of surface emissivity by magnetic fields discussed above are due to the different radiative properties of small magnetic elements ('flux tubes'). More indirect effects may play a role as well. The shape of granules appears to be different in magnetic regions (Macris and Roesch 1983, Muller and Roudier 1984, Muller 1986, Title et al. 1992). They are smaller, more irregular, and the measured flow speeds are lower. These effects are present *in addition* to the magnetic elements themselves, which also change the appearance of granulation by filling in the intergranular lanes, especially when seen at low spatial resolution. The changes may be the result of the geometric constraints the magnetic tubes put on the convective flow outside them.

Since the flow appears to be different, it would seem possible that the heat flux it carries is also different. This might contribute to the observed solar cycle variations of irradiance (Muller 1986, Kuhn et al. 1988). Limits on this contribution can be put by observations of the *colors* of the solar cycle variation signal. Solanki and Unruh (1998) find that the wavelength dependence of the irradiance signal argues against a dominant contribution from small changes of surface temperature over a larger areas, but do not quote quantitative limits. In order to contribute in the right sense to the solar cycle variation, the constraints imposed on the flow by the magnetic flux elements would have to lead to an *increase* in the heat flux carried by granulation. Direct (spatially resolved) measurements of a heat flux change in granulation in magnetic regions are probably difficult, since the effect would be small and hard to separate from the enhanced emission from the magnetic elements.

Changes in surface heat flux by modified granulation, if they exist, will affect the convective envelope in the same way as the excess emission from the small scale magnetic field. They can also be represented by an effective change of the emissivity of the solar surface. In particular, one does not expect these changes to be 'compensated' by opposite changes elsewhere on the surface (except, as before, on the 10^5yr thermal time scale of the envelope).

4. Improving on the turbulent diffusion picture

In the results quoted a diffusion model for convection was used. It assumes that convection can be modeled by a turbulent viscosity for momentum and a turbulent diffusivity for the transport of heat. While this was a simple and somewhat justifiable model of convection in the absence of detailed knowledge of the convective flows in a stellar envelope, we now know that it does not represent stellar convective envelopes well. Numerical simulations (Nordlund 1982, 1985ab, 1986, 1990; Nordlund & Dravins 1990; Nordlund & Stein 1990, 1991, 1996, Stein and Nordlund 1989, 1991, 1998, Steffen et al. 1989; Steffen 1993; Ludwig et al. 1999) show an extremely *nonlocal* picture. The convective flows are driven almost exclusively by cooling at the surface, with narrow fast moving downdrafts between slow almost isentropic upflows (for a discussion see Spruit 1997). In the present context, the most important property is that the convective flow at all depths is driven by cooling at the surface rather than by a local overturning process. The material in the cool downdrafts survives to large depths below the surface with little mixing into the upflows.

Given this extremely nonlocal picture, it is appropriate to ask how the thermal behavior of perturbations in the convection zone can be estimated without appealing to a turbulent diffusion model. I do this below for three of the effects discussed, the spot blocking, the facular emission, and the 'thermal shadow' problems. In all three cases, the conclusion is that thermal perturbations of the normal solar surface are even smaller than in the diffusion picture. This has to with the fact that the flows in the new non-local picture are caused by cooling at the surface, and that most of the changes are one-way (downward, away from the visible surface). The diffusion picture on the other hand communicates perturbations equally well in all directions.

4.1. SPOT BLOCKING

Below the spot (modeled as a region of reduced heat transport efficiency extending to some depth below the surface) the upflows have exactly the same temperature as upflows in the unspotted surroundings at the same level, namely that given by the entropy at the base of the upflows. In this sense, there is no 'pile up of heat below the spot'. Because of the reduced heat loss at the surface, however, the downflows below the spot will be less vigorous. The unspotted surface notices nothing of the spot's presence (except for an extremely narrow ring where lateral radiative exchange takes place, and except for the presence of a moat flow, see below). It continues to cool upwellings into downdrafts as before, since the entropy in the upflows has not changed. Thus, we expect again that bright rings will be absent around spots, but the reason is even simpler than in the diffusion model. The spot is a region at the surface where less heat is radiated away, and this is now independent of the depth of the spot below the surface (in contrast to the diffusion model, where the spot has to extend to a minimum depth of 1000 km for the explanation to work).

Still, one may wonder what happens to the amount of heat generated in the solar interior that now fails to be emitted at the surface. This part of the problem is the same as in the diffusion model: the imbalance causes a secular increase of the entropy in the entire convection zone until a new thermal equilibrium is reached. Because of the very long thermal time scale of the convection zone (10^5 yr), the effect is negligible on observable time scales. In a steady state, when the average number of spots does not change, the convection zone does not heat up, because its mean temperature is higher than it would be without spots (Spruit & Weiss 1986). Episodes of larger than average spot coverage cause heating, those of less than average spot coverage cause cooling on this time scale.

4.2. FACULAR EMISSION

The lower internal pressure in a small magnetic element ('tube') in the photosphere causes it to be more transparent: a surface of unit optical depth as seen from Earth has a little dimple at the location of the small magnetic element. Its side walls radiate an additional heat flux (Spruit 1976, 1977). This heat flux is supplied to the element by convection, i.e. by the granulation surrounding it, and conversely the side wall, by the extra radiation it emits, cools the surroundings. In a simple minded diffusion model for convection, this causes a dark ring around the element, which compensates for a part of the extra emission from the element. The cooling, however, also makes the photosphere next to the tube heavier, so that a circulation is set up: down along the tube walls, and towards the element at the surface. This effect has been studied with numerical simulations by Knölker et al. (1991, see also Steiner et al. 1998), who find a very fast downflow along the tube wall (km/s) due to the excess cooling. In a diffusion model, the information of the low temperature in the downdrafts diffuses back to the surface. In reality, the cool downdraft drops below the visible surface because it is heavy, and only a negligible fraction of it finds its way back to the surface. The excess emission is therefore expected to be stronger than in the diffusion model, and not compensated by a dark ring.

4.3. THERMAL SHADOWS

The thermal shadow expected from a blocking object below the surface (Spruit 1977), which is a small effect already in the turbulent diffusion model, is likely to be even smaller in the new picture of stellar envelope convection. Since all upward flowing gas has nearly the same entropy, a blocking object below the surface has no effect on what happens above it, as long as there is enough room above it for upflows to comfortably reach the solar surface. It may interfere somewhat with downflowing gas, but this has no consequence for the surface temperature above the blocking object, since only a small fraction of the downward moving gas is carried back up to the surface. As before, this is a consequence of the non-local nature of the convective flow.

5. Helioseismological tests

In the above I have concentrated on surface effects, and argued that temperature enhancements with origins deep inside the convection zone are less likely to contribute. Such deep origins have been advocated by Kuhn et al. (1988), and Kuhn and Stein (1996) who propose that a temperature enhancement with origin at the base of the convection zone causes a temperature enhancement of 0.5K in the active latitudes.

Potentially exhausting theoretical debates for or against these models can probably be circumvented since the models make different predictions for the associated perturbations below the surface that can be tested by observation. The excess radiation from small scale magnetic fields comes about through an increased surface emissivity, associated with the increased 'surface roughness'. This increases the average cooling rate at the surface and creates somewhat stronger and/or cooler downdrafts. It therefore causes slightly *lower* horizontally averaged temperatures below the surface. The two models thus predict opposite thermal effects below the surface.

The prospects for a test of these predictions are good, since helioseismological data are now quite sensitive, with the data obtained by SOHO/MDI and the GONG network (Kosovichev 1996, Kosovichev and Schou 1997, Duvall et al. 1998, Duvall and Kosovichev 1999). Changes in the p-mode frequencies related to the solar cycle were reported by e.g. by Libbrecht and Woodard (1990), Dziembowski et al. (1998), and Dziembowski (this volume). The dependence of the signal on wavenumbers (l, m) shows that most of the effect is due to a change in wave propagation in a layer very close to the solar surface, at the latitudes of magnetic activity. The sense of the effect, an increase of the frequencies, is the *opposite* (Goldreich et al. 1991) of what would be expected if the temperature in these latitudes were higher. Though a higher temperature increases the propagation speed (increasing the frequencies), it also expands the solar envelope at these latitudes, by vertical hydrostatic balance. This lengthens the path traveled and decreases the mode frequencies. The net effect turns out to be dominated by the the second effect, the path length increase, because it is linear in T, while the propagation speed increases only as $T^{1/2}$.

In addition to the thermal effect, the presence of the magnetic field also has an effect, since it increases the 'stiffness' of the gas (in terms of propagation, the relevant speed is the fast mode speed, which is larger than the sound speed). Since the contribution of this effect scales with the ratio of magnetic to gas pressure, it is important only in a thin layer near the surface. The observed changes in mode frequency can thus be due either to a somewhat lower temperature, or the effect of the magnetic field in the atmosphere. The observation that most of the effect is localized at the surface argues for a major contribution from the magnetic stiffness effect (Goldreich et al. 1991; Woodard et al. 1991). In any case, the observed sign of the effect is incompatible with enhanced temperatures being the main change in envelope structure in the active latitudes.

A second way of measuring variations in propagation conditions below the solar surface is through time-distance seismology (Duvall et al. 1996; Kosovichev 1996; Braun et al. 1997), in which one does not measure the mode frequencies but the wave travel times between points on the surface (much like seismology is used to measure irregularities in the Earth's interior). These measurements show shorter travel times in the active regions (Duvall et al. 1998; Duvall & Kosovichev 1998; Giles et al. 1997; Bogdan et al. 1998; Goode & Dziembowski 1999), and downflows below the surface. The shorter travel times are compatible with a contribution of the magnetic field to the propagation speed.

The lower temperatures expected by enhanced surface cooling in the small scale magnetic field has further observational consequences. In the absence of rotation, the increased downward buoyancy would just cause a downdraft below active regions. Associated with it a horizontal flow converging on the active region is expected. Effects like these are beginning to be explored by time-distance seismology.

References

1. Arendt, S. (1992) *Astrophys. J* **389**, 421
2. Bahcall, J. (1996) *Astrophys. J.* **467**, 475
3. Bogdan, T.J., Braun, D.C., Lites, B.W. and Thomas, J.H. (1998) *Astrophys. J.* **492**, 379
4. Braun, D.C. (1997) *Astrophys. J.* **487**, 447
5. Brun, A.S., Turck-Chièze, S. and Morel, P. (1998) *Astrophys. J.* **506**, 913
6. Caligari, P., Schüssler, M. and Moreno-Insertis, F. (1998) *Astrophys. J.* **502**, 481
7. Chiang, W.H. and Foukal, P.V (1984) *Solar Phys.* **97**, 9
8. D'Silva, S. and Choudhuri, A.R. (1993) *Astron. Astrophys.* **272**, 621
9. D'Silva, S. and Howard, R.A. (1993) *Solar Phys.* **148**, 1
10. Duvall, T.L., Jr., Kosovichev, A.G., Scherrer, P.H.; Bogart, R.S., Bush, R.I., De Forest, C., Hoeksema, J.T., Schou, J., Saba, J.L.R., Tarbell, T.D., Title, A.M., Wolfson, C.J. and Milford, P.N. (1997) *Sol. Phys.* **170**, 63
11. Duvall, T.L., Jr., Kosovichev, A.G. and Scherrer, P.H. (1998) in *Sounding solar and stellar interiors*, eds. J. Provost & F.-X. Schmider (IAU symposium 181, Nice, France, September 30 - October 3), Dordrecht: Kluwer Academic Publishers

12. Duvall, T.L., Jr. and Kosovichev, A.G. (1999) in SOHO-9 Workshop "Helioseismic Diagnostics of Solar Convection and Activity", Stanford, California, July 12-15
13. Dziembowski, W.A., Goode, P.R., Di Mauro, M.P., Kosovichev, A.G. and Schou, J. (1998) *Astrophys. J.* **509**, 456
14. Endal, A.S., Sofia, S. and Twigg, L.W. (1985) Astrophys. J. **290**, 748
15. Foukal, P.V., Fowler, P. and Livshits, M. (1983) *Astrophys. J.* **267**, 863
16. Foukal, P.V. and Lean, J. (1986) *Astrophys. J.* **302**, 826
17. Friis-Christensen, E., Fröhlich, C., Haigh, J., Schüssler, M. and von Steiger, R. (eds.) (2000) Proceedings ISSI workshop *Solar Variability and Climate, Space Sci. Rev.*, Kluwer, in press.
18. Fröhlich, C. and Lean, J. (1998) *Geophys. Res. Let.* 25, 4377–4380
19. P.M. Giles, T.L. Duvall Jr., P.H. Scherrer and R.S. Bogart, (1997) *Nature* **390**, 52–54.
20. Gilliland, R.L. (1988) in *Solar radiative output variation*, P. Foukal, ed., Cambridge Research & Instrumentation Inc., Cambridge, MA 02139, 239
21. Goldreich, P., Murray, N., Willette, G. and Kumar, P. (1991) *Astrophys. J.*, **370**, 752
22. Goode, P.R. and Kuhn, J.R. (1990) *Astrophys. J.* **356**, 310
23. Goode, P.R. and Dziembowski, W.A. (1999) in SOHO-9 Workshop "Helioseismic Diagnostics of Solar Convection and Activity", Stanford, California, July 12-15, 1999.
24. Knölker, M., Grossmann-Doerth, U., Schüssler, M. and Weisshaar, E. (1991) *Adv. Space Res.* **11**, 285
25. Kosovichev, A.G. (1996) *Astrophys. J.* **461**, L55
26. Kosovichev, A.G. and Schou, J. (1997) *Astrophys. J.* **482**, L207
27. Kuhn, J.R., Libbrecht, K.G. and Dicke, R.H. (1988) *Science* **242**, 908
28. Kuhn, J.R. and Stein, R.F. (1996) *Astrophys. J.* **463**, L117
29. Lean, J.L., Cook, J., Marquette, W. and Johanneson, A. (1998) *Astrophys. J.* **492**, 390
30. Libbrecht, K.G. and Woodard, M.F. (1990) *Nature* **345**, 779
31. Ludwig, H.-G., Freytag, B. and Steffen, M. (1999) *Astron. Astrophys.* **346**, 111
32. Macris, C.J. and Roesch, J. (1983) *Comptes Rendus, ser II*, **296**, 265
33. Muller, R. (1986) *Solar Phys.*, **119**, 229
34. Muller, R. and Roudier, T. (1984) in *The Hydromagnetics of the Sun*, ESA SP ESA SP 220, p. 51
35. Nesme-Ribes, E. (ed.) (1994) *The solar engine and its influence on terrestrial atmosphere and climate*, NATO ASI **I25**, Kluwer
36. Nordlund, Å. (1982) *Astron. Astrophys.* **107**, 1
37. Nordlund, Å. (1985a) in *Small scale dynamical processes in quiet stellar atmospheres* ed. W. Keil, Sacramento Peak Observatory, Sunspot, NM 88349, USA
38. Nordlund, Å. (1985b) in *Progress in stellar spectral line formation theory*, eds. J.E. Beckman & L. Crivellari (NATO ASI series **152**), Reidel, Dordrecht, p. 215
39. Nordlund, Å. (1986) *Solar Phys.*, **100**, 209
40. Nordlund, Å. (1991) in *Stellar atmosphere: Beyond classical models*, eds. L. Crivellari, I. Hubeny & D.G. Hummer (NATO ASI series **341**), Kluwer, Dordrecht, p. 61
41. Nordlund, Å. and Dravins, D. (1990) *Astron. Astrophys.* **228**, 155
42. Nordlund, Å. and Stein, R.F. (1990) *Comp. Phys. Comm.* **59**, 119
43. Nordlund, Å. and Stein, R.F. (1991) in *Stellar atmospheres: beyond classical models*, eds. L.Crivellari, I. Hubeny & D.G.Hummer (NATO ASI series **341**), Kluwer, Dordrecht, 263
44. Nordlund, Å. and Stein, R.F. (1996) in *Proceedings of the 32nd Liège Int. Astrophys. Colloquium 'Stellar Evolution: What should be done'*, eds. A. Noels et al., 75
45. Richard, O., Vauclair, S., Charbonnel, C. and Dziembowski, W.A. (1996) *Astron. Astrophys.* **312**, 1000

46. Rast, M.P., Fox, P.A., Lin, H., Lites, B.W., Meisner, R.W. and White, O.R., *Nature* **401**, 678
47. Solanki, S.K. and Unruh, Y.C. (1998) *Astron. Astrophys.* **329**, 747
48. Spruit, H.C. (1976) *Solar Phys.* **50**, 269
49. Spruit, H.C. (1977) *Solar Phys.* **55**, 3
50. Spruit, H.C. (1982a) *Astron. Astrophys.* **108**, 348
51. Spruit, H.C. (1982b) *Astron. Astrophys.* **108**, 356
52. Spruit, H.C. (1991) in *The Sun in Time*, eds. C. Sonett, M. Giampapa & M.S. Matthews, University of Arizona Press, Tucson, 118
53. Spruit, H.C. (1992) in *Sunspots: Theory and Observations*, eds. J.H. Thomas and N.O. Weiss, Cambridge: CUP, 163
54. Spruit, H.C. (1997) *Mem. Soc. Astron. It.*, **68**, No. 2, 397
55. Spruit, H.C. and Weiss, A. (1986) *Astron. Astrophys.* **166**, 167
56. Steffen, M., Ludwig, H.-G. and Krüss, A. (1989) *Astron. Astrophys.* **213**, 317
57. Steffen, M. (1993) in *Inside the stars* (IAU Coll 137), eds. W. Weiss & A. Baglin, Astron. Soc. Pac. Conference series **40**, 300
58. Stein, R.F. and Nordlund, Å (1989) *Astrophys. J.* **342**, L95
59. Stein, R.F. and Nordlund, Å. (1991) in *Challenges to Theories of the Structure of Moderate-Mass Stars*, eds. D.O. Gough & J. Toomre, Lecture Notes in Physics **388**, Springer, Berlin, 195
60. Stein, R. F. and Nordlund, Å. (1998) *Astrophys. J.* **499**, 914
61. Steiner, O., Grossmann-Doerth, U., Knoelker, M. and Schüssler, M. (1998) *Astrophys. J.* **495**, 468
62. Topka, K.P., Tarbell, T.D. and Title, A.M. (1997) *Astrophys. J.* **484**, 479
63. Title, A.M., Topka, K.P., Tarbell, T.D., Schmidt, W., Balke, C. and Scharmer, G. (1992) *Astrophys. J.* **393**, 782
64. Ulrich, R.K., in *New eyes to see inside the Sun*, F. Deubner et al., eds. (1997) IAU Symposium **185**, 59
65. Unruh, Y.C., Solanki, S.K. and Fligge, M. (1999) *Astron. Astrophys.* **345**, 635
66. Woodard, M.F., Libbrecht, K.G., Kuhn, J.R. and Murray, N. (1991) *Astrophys. J.* **373**, L81

INDEX

MIX
Papier aus verantwortungsvollen Quellen
Paper from responsible sources
FSC® C105338

If you have any concerns about our products,
you can contact us on
ProductSafety@springernature.com

In case Publisher is established outside the EU,
the EU authorized representative is:
Springer Nature Customer Service Center GmbH
Europaplatz 3, 69115 Heidelberg, Germany

Printed by Libri Plureos GmbH
in Hamburg, Germany